Multiphoton Processes

Multiphoton Processes

Proceedings of an International Conference
at the University of Rochester
Rochester, N.Y., June 6–9, 1977

JOSEPH H. EBERLY
University of Rochester

PETER LAMBROPOULOS
University of Southern California

John Wiley & Sons, New York / Chichester / Brisbane / Toronto

Library of Congress Cataloging in Publication Data:

Main entry under title:
  Multiphoton processes.

  (Wiley series in pure and applied optics)

  Includes index.
  1. Multiphoton processes—Congresses. I. Eberly,
J. H., 1935-   II. Lambropoulos, Peter.

QD461.5.M84   541′.35   77-13021
ISBN 0-471-03790-7

Printed in the United States of America

10 9 8 7 6 5 4 3 2 1

Contributors

J. R. ACKERHALT
Theoretical Division
Los Alamos Scientific Laboratory
Los Alamos, New Mexico

J. A. ARMSTRONG
IBM
Thomas J. Watson Research Center
Yorktown Heights, New York

LLOYD ARMSTRONG
Department of Physics
The Johns Hopkins University
Baltimore, Maryland

JAMES E. BAYFIELD
Department of Physics and
   Astronomy
University of Pittsburgh
Pittsburgh, Pennsylvania

C. D. CANTRELL
Theoretical Division
Los Alamos Scientific Laboratory
Los Alamos, New Mexico

SHIH-I CHU
JILA, University of Colorado and
National Bureau of Standards
Boulder, Colorado

M. J. COGGIOLA
Materials and Molecular Research
   Division
Lawrence Berkeley Laboratory
Berkeley, California

C. COHEN-TANNOUDJI
Ecole Normal Supérieure et
   Collège de France
Paris, France

K. N. DRABOVICH
Moscow State University
Moscow, U.S.S.R.

P. ESHERICK
IBM
Thomas J. Watson Research Center
Yorktown Heights, New York

S. EZEKIEL
Research Laboratory of
   Electronics
Massachusetts Institute of
   Technology
Cambridge, Massachusetts

GY. FARKAS
Central Research Institute for
   Physics
Hungarian Academy of Sciences
Budapest, Hungary

SERGE FENEUILLE
Laboratoire Aime Cotton
Orsay, France

H. W. GALBRAITH
Theoretical Division
Los Alamos Scientific Laboratory
Los Alamos, New Mexico

Y. GONTIER
Service de Physique Atomique
Centre d'Etudes Nucléaires de
  Saclay
Gif-sur-Yvette, France

E. H. A. GRANNEMAN
F.O.M.-Institute for Atomic
  and Molecular Physics
Amsterdam, The Netherlands

E. R. GRANT
Materials and Molecular Research
  Division
Lawrence Berkeley Laboratory
Berkeley, California

YU. G. GRIN
Moscow State University
Moscow, U.S.S.R.

S. E. HARRIS
W. W. Hansen Laboratories of
  Physics
Stanford University
Stanford, California

N. R. ISENOR
Department of Physics
University of Waterloo
Waterloo, Ontario, Canada

YU. N. KARAMZIN
Moscow State University
Moscow, U.S.S.R.

N. V. KARLOV
P. N. Lebedev Physical Institute
U.S.S.R. Academy of Sciences
Moscow, U.S.S.R.

E. KARULE
Physics Institute
Latvian S.S.R. Academy of
  Sciences
Riga, Salaspils, U.S.S.R.

H. J. KIMBLE
Department of Physics and
  Astronomy
University of Rochester
Rochester, New York

J. KRASINSKI
Institute of Experimental
  Physics
University of Warsaw
Warsaw, Poland

Y. T. LEE
Materials and Molecular Research
  Division
Lawrence Berkeley Laboratory
Berkeley, California

V. S. LETOKHOV
The Institute of Spectroscopy
Academy of Sciences of the
  U.S.S.R.
Moscow, U.S.S.R.

G. MAINFRAY
Service de Physique Atomique
Centre d'Etudes Nucléaires de
  Saclay
Gif-sur-Yvette, France

L. MANDEL
Department of Physics and
  Astronomy
University of Rochester
Rochester, New York

H. MITTER
Institut für Theoretische
  Physik der Universität
Graz, Austria

J. A. PAISNER
Lawrence Livermore Laboratory
Livermore, California

EDWIN A. POWER
University College London
London, England

A. M. PROKHOROV
P. N. Lebedev Physical Institute
U.S.S.R. Academy of Sciences
Moscow, U.S.S.R.

WILLIAM P. REINHARDT
JILA, University of Colorado and
National Bureau of Standards
Boulder, Colorado

S. REYNAUD
Ecole Normal Supérieure et
   Collège de France
Paris, France

V. I. RITUS
P. N. Lebedev Physical Institute
U.S.S.R. Academy of Sciences
Moscow, U.S.S.R.

P. A. SCHULZ
Materials and Molecular Research
   Division
Lawrence Berkeley Laboratory
Berkeley, California

Y. R. SHEN
Materials and Molecular Research
   Division
Lawrence Berkeley Laboratory
Berkeley, California

T. P. SHLEGEL
Moscow State University
Moscow, U.S.S.R.

R. W. SOLARZ
Lawrence Livermore Laboratory
Livermore, California

Aa. S. SUDBO
Materials and Molecular Research
   Division
Lawrence Berkeley Laboratory
Berkeley, California

A. P. SUKHORUKOV
Moscow State University
Moscow, U.S.S.R.

M. TRAHIN
Service de Physique Atomique
Centre d'Etudes Nucléaires de
   Saclay
Gif-sur-Yvette, France

I. I. TUGOV
P. N. Lebedev Physical Institute
U.S.S.R. Academy of Sciences
Moscow, U.S.S.R.

M. J. VAN DER WIEL
F.O.M.-Institute for Atomic and
   Molecular Physics
Amsterdam, The Netherlands

An. V. VINOGRADOV
P. N. Lebedev Physical Institute
U.S.S.R. Academy of Sciences
Moscow, U.S.S.R.

H. WALTHER
Sektion Physik
Universität München
Projektgruppe für Laserforschung
   der Max-Planck-Gesellschaft
Garching, West Germany

KENNETH M. WATSON
Lawrence Berkeley Laboratory
Berkeley, California

E. F. WORDEN
Lawrence Livermore Laboratory
Livermore, California

F. Y. WU
Research Laboratory of
  Electronics
Massachusetts Institute of
  Technology
Cambridge, Massachusetts

J. J. WYNNE
IBM
Thomas J. Watson Research Center
Yorktown Heights, New York

J. F. YOUNG
W. W. Hansen Laboratories of
  Physics
Stanford University
Stanford, California

ICOMP Organizing Committee

J. E. BAYFIELD
University of Pittsburgh
Pittsburgh, PA

N. BLOEMBERGEN
Harvard University
Cambridge, MA

K. BOYER
Los Alamos Scientific Laboratory
Los Alamos, NM

J. I. DAVIS
Lawrence Livermore Laboratory
Livermore, CA

A. COLD
Rockefeller University
New York, NY

N. KROO
Central Research Institute for
Physics, Budapest, Hungary

P. LAMBROPOULOS
University of Southern
California, Los Angeles, CA

V. S. LETOKHOV
Institute of Spectroscopy
Moscow, USSR

W. C. LINEBERGER
Joint Institute for Laboratory
Astrophysics, Boulder, CO

M. LUBIN
University of Rochester
Rochester, N.Y

C. MANUS
C.E.N. Saclay, Gif-sur-Yvette
France

B. B. SNAVELY
Kodak Research Laboratories
Rochester, N.Y

H. WALTHER
University of Munich
Garching, Germany

J. H. EBERLY
Conference Chairman
University of Rochester
Rochester, NY

ICOMP Program Committee

J. S. BAKOS
Central Research Institute for
Physics, Budapest, Hungary

J. E. BAYFIELD
University of Pittsburgh
Pittsburgh, PA

C. D. CANTRELL
Los Alamos Scientific Laboratory
Los Alamos, NM

N. B. DELONE
P. N. Lebedev Physical Institute
Moscow, USSR

S. GELTMAN
Joint Institute for Laboratory
Astrophysics, Boulder, CO

N. R. ISENOR
University of Waterloo
Waterloo, Canada

C. ITZYKSON
C.E.N. Saclay, Gif-sur-Yvette
France

G. MAINFRAY
C.E.N. Saclay, Gif-sur-Yvette
France

C. B. MOORE
University of California
Berkeley, CA

C. R. STROUD, JR.
University of Rochester
Rochester, NY

I. I. TUGOV
P. N. Lebedev Physical Institute
Moscow, USSR

P. LAMBROPOULOS
Program Chairman
University of Southern
California, Los Angeles, CA

Preface

This volume is the record of 29 invited papers presented at
the International Conference on Multiphoton Processes (ICOMP),
held at the University of Rochester, Rochester, New York, during
the week of 6-9 June 1977.

It is customary to say that the advent of the laser in 1960
created entirely new possibilities and challenges in atomic and
molecular physics. However, the history of multiphoton phenomena
began long before the laser appeared on the scene. To begin with,
processes such as Rayleigh or Raman scattering are bona fide two-
photon processes. Their theoretical description contains most of
the ingredients necessary for the description of higher-order processes. A much-quoted article by Göppert-Mayer from 1931 dealt
with the theory of two-photon spontaneous and stimulated processes
based on second-order perturbation theory. A less well-known article by Podolsky on a Green's function method for the calculation of two-photon transitions appeared even earlier, in 1928.
The two-photon spontaneous decay of metastable (2S) hydrogen was
first detected in the early 1940s in radiation from stellar nebu-
lae. As for higher-order processes, early observations were carried out successfully in the microwave region of the spectrum
well before 1960.

The observation of a richer variety of optical multiphoton
processes became possible with the appearance of the laser. Stimulated Raman scattering, higher harmonic generation, multiphoton
ionization and gas breakdown, high-resolution multiphoton spectroscopy, and selective excitation of atoms and molecules are a
few representative examples. The large intensity, spectral purity,
and exceptional coherence are some of the features that render the
laser, in its many varieties, a powerful tool in the exploration
of the interaction of radiation with matter. Much of the work in
the field known today as multiphoton processes has been under way
since 1961 or so. But in the last three or four years the field
has undergone an explosive growth, due in part to improvements in
experimental possibilities and in part to interest in potential
applications. It is no longer possible to hold a conference
covering all aspects of multiphoton processes. It may not be

possible two years from now to hold a conference even as broad as ICOMP was, with a manageable number of participants.

Yet there is a body of knowledge and an area of problems pertaining to the more basic aspects of the field, what one might loosely call the physics of multiphoton processes, that will remain at the center of interest. An attempt was made to cover as much of this basic area as possible through the judicious selection of the ICOMP invited papers. To the extent that the attempt was successful, this volume should provide an up-to-date review. Thus even though the field is bound to continue expanding rapidly, this review should serve as an introduction, guide and reference for some time to come.

Multiphoton processes correspond to higher-order terms in the perturbation theory of radiation-matter interactions. As a consequence, their description requires knowledge of more than the single matrix element required in single-photon processes. Moreover, perturbation theory itself is of dubious validity when the radiation field strength becomes comparable to the binding fields of the electron in the atom. These are two features that set multiphoton processes apart from the more traditional single-photon transitions. Intense fields and multiphoton effects are interwoven whether we are dealing with bound or free electrons. Inevitably, the theoretical description and experimental study of these basic effects constitute the necessary ingredients of a review.

We chose to begin the volume, following the Introduction by Professor Prokhorov, with four articles dealing with Hamiltonians and quantum-electrodynamic aspects of intense fields. The second part contains articles concerned with the theory and observations of strong-field resonance fluorescence. This topic provides an example of the behavior of a relatively simple bound system--a two-level atom--in the presence of an intense field. Proceeding with increasingly complex systems, in Part III we have included theoretical and experimental articles on multiphoton processes of arbitrary order in atoms. Part IV addresses problems in which the presence of resonances with intermediate atomic states and/or coherence effects play an important role. (It is known that resonant intermediate states render a multiphoton process more easily observable at the expense of added complexity in the theoretical description.) Included in this part are also articles dealing with the effect of field correlations on two-photon processes. This is another of the novel features of such processes that are not found in single-photon transitions. Part V deals with aspects of multiphoton effects in systems more complicated than atoms: laser interactions with molecules for isotope separation and laser-induced chemistry are areas of high interest for applications. Finally, the last part is concerned with the relatively new subject of collisions in the presence of intense fields.

The above arrangement does not correspond to the order of presentation of the papers at ICOMP. Instead, it is based on the idea of proceeding from fundamental systems and simpler problems to more complex systems and various applications. This ordering is not unique and does have its inconsistencies. But, not surprisingly, it happens to follow by and large the historical development of the field. As it stands, it should serve fairly well as a guided tour for the novice in the field (the expert would probably not notice the order in any case).

The success of ICOMP and the appearance of this volume must be attributed to the hard work of a wide circle of people. The organization of the Conference and its program were the responsibility of the Organizing and Program Committees, whose members are listed elsewhere in the book. We extend our thanks for their essential contributions. In addition we must mention the assistance of David Ham, Gilbert Leppelmeier, Cindy Jay, Andra Cooper, and James Newman in coordinating pre-Conference details. Donald Parry and his staff handled all the arrangements at the University of Rochester with efficiency and patience. We are pleased to acknowledge important financial assistance from the Energy Research and Development Administration, the National Science Foundation, and the Office of Naval Research, as well as the sponsorship of the American Physical Society through its Division of Electron and Atomic Physics, the Optical Society of America, and the International Union of Pure and Applied Physics.

Finally, we reserve especial thanks for the skill and care of Edna Hughes, who prepared the typescript of all of the papers in the volume. We are grateful to Jerome Luine who constructed the combined subject-author index on very short notice.

J. H. Eberly

Rochester, New York

P. Lambropoulos

Los Angeles, California

Contents

xviii   Contents

Multiphoton Processes

Introduction

Multiquantum Processes of Atom Photoionization, Molecule Photodissociation, and Surface Photoeffect in Metals

A. M. PROKHOROV
P. N. Lebedev Physical Institute
Moscow, USSR

My brief talk, which is far from a rigorous scientific report, will be concerned with a rather general, not pretending to be comprehensive, review of the state-of-the art of the problem of multiquantum processes in atoms, molecules, and metals. I will speak about atom photoionization, molecule photodissociation,and about the nonlinear photoelectric effect in metals. The topic of my talk is somewhat broad, but - difficult as it may be - it should reflect the main trends in multiphoton process investigations that have been and are being conducted in the Soviet Union, in the first place at the P.N. Lebedev Physical Institute and particularly at our laboratory.

It is well known that the earlies theoretical predictions of the possible experimental observation of atom photoionization and molecule photodissociation over the optical range, using laser light sources, were made at the P.N. Lebedev Physical Institute in 1964 (see, for example, the works carried out in our laboratory on ionization and dissociation [1,2], and the work of Keldysh on ionization [3]). They give an impetus to thorough experimental and theoretical studies along these lines. In the same year,1964, the group of Delone at our Institute started experiments on multiphoton ionization of atoms. Currently, many research centers (for example, Manus and his associates in France, Isenor and co-workers in Canada, Lambropoulos in the United States, Rapoport at the Voronezh University in the USSR, and the P.N. Lebedev Physical Institute) are engaged in experimental and theoretical investigations of multiphoton ionization over the optical range.

As regards multiphoton molecule dissociation, its extensive study began later due to the extreme complexity of such processes. The first advances in this area were associated, primarily, with accumulation of theoretical results. The principal works of that period belong to Professor Bunkin and Dr. Tugov of our laboratory, who calculated cross-sections for multiquantum dissociation of diatomic molecules. The significance of these works lies not only in deriving general expressions for dissociation cross-sections and establishing their values for certain diatomic molecules; they are of value also because their simplified model representations, which are indispensable in such complicated calculations, have stimulated extensive scientific discussions concerning the optimal theoretical descriptions of multiphoton processes in molecules (in particular, the applicability of the Born-Oppenheimer approximation to such problems). This discussion has given rise to new approaches in the theoretical treatment of some problems of the interaction of light with molecules. But extensive theoretical and experimental investigation of multiphoton and multistep processes occurring in molecules have started only of late, with the development of laser isotope separation and laser photochemistry. The observed processes of the so-called collisionless dissociation of molecules exposed to infrared photons can be assigned only to multiquantum and multistep transitions. The problem of collisionless molecule dissociation is to be discussed at this Conference by Prof. Karlov of our laboratory and Prof. Letokhov of the Spectroscopy Institute [see later papers in this volume].

While multiphoton processes in molecules are at the early stage of their study, the investigation of multiphoton ionization of atoms, thanks to the efforts of the above mentioned groups, is following several trends of development, each of them characterized by its own achievements and prospects.

One of the trends is associated with measurements of the dependence of multiphoton cross-sections on light frequency and polarization, as well as with elucidation, through the use of experimental results, of optimal methods for calculating cross-sections in terms of perturbation theory. The progress made in this area is well known [4]. For instance, nowadays a sufficiently accurate description can be given, say, of nonlinear light absorption in a rarefied atomic medium, arising from its multiphoton ionization. However, one very attractive, theoretically predicted phenomenon has not yet been verified and investigated experimentally. I mean windows of nonlinear transparency which are expected at circular polarization of light. Earlier it has been thought that rather strong light fields always involve nonlinear absorption. But it has turned out that if the initial atomic state is the S-state and the light has circular polarization, an electron can undergo virtual transitions only via states with increasing orbital momentum. There is a frequency in every intraresonance interval,corresponding to zero cross-section for a multiphoton transition (in particular, ionization). This has been theoretically predicted by Rapoport and co-workers of the Voronezh University [5].

Another significant research activity was directed toward
treatment of the ionization process in case of intermediate resonance. Though the phenomenon of resonant ionization was first
observed by Delone et al, at our Institute as long as ten years
ago [4], it is still arresting experimental and theoretical attention for various reasons. Thus, a general theory of this
phenomenon has been developed in the last two years at our Lab
by Fedorov and associates [6]. In the case of resonance, of
greatest importance are perturbations of the resonance state
associated with induced bound-bound and bound-free transitions.
As for ionization, resonance is interesting because of a sudden
increase in the cross-section. The upsurge of interest devoted
to this phenomenon is attributed, however, to its applicability
in diagnostics of the atomic spectrum. One of the possibilities
here is to use rather weak fields when the perturbation of the
resonance state is less than its natural width. Under such conditions, by varying the light frequency and by observing the
resonant increase of ion emission, one can perform detailed
spectroscopic studies of highly-excited atomic states. The possibilities opened up by this method - the so-called resonance
ionization spectroscopy - are evident from a series of works by
Collins, Popescu and their collaborators [7]. This method enjoys
wide usage in spectroscopy of complicated molecules as well [8].
An alternative is to apply fields so strong that would cause perturbations in the resonance state. The same procedure permits
observation and measurement of bound electronic states. Valuable
information about the dynamic Stark effect has been obtained with
the help of this method by Delone and coworkers at our Institute
and by Morellec et al. in France [9].

The results of the investigations on the resonance ionization process make us revise our attitude towards such a well-
known problem as the description of atom interactions with light
in terms of the dipole approximation. I mean the works of
Lambropoulos who has shown that in case of resonance, one should
not neglect quadrupole transitions occurring with a noticeable
probability [10]. That the quadrupole matrix elements are small
compared with the dipole matrix elements is fairly well compensated by the resonance, as compared to the nonresonance case of
dipole transitions.

The resonant process of multiphoton ionization of atoms has
bright prospects as the method for producing completely polarized
electrons. This method has two obvious advantages over the Fano
effect. First, the frequency of ionizing radiation remains
within the light rather than within the ultraviolet range. Second,
there are no other requirements for frequency, but for resonance
establishment. In the main,this method can be represented as

follows [11]. The frequency of the first external field is so
chosen as to excite, in a resonant way, an atom from the ground
state to a certain excited state with a fixed magnetic quantum
number. Thus, a kind of target consisting of polarized excited
atoms is formed. The second field is applied to ionize these
atoms. The first experiments [12] have shown this method to be
relatively simple and effective though it is far from being optimal. In particular, it is necessary to study in further detail
the phenomena that may cause relaxation of resonance states,
thereby reducing degree of polarization of the electrons formed.
Closer investigation of this method is also of considerable practical concern, since polarized electron beams provide an excep-
tionally useful instrument for studying the details of electron-
atom interactions.

    Now I will characterize, also very briefly, the state-of-the
art in the problem of nonlinear photoelectric effect in metals
whose experimental study was initiated in the same "critical"
year of 1964 by Teich, Wolga et al. who observed a two-quantum
photoeffect on the sodium surface [13]. In the years that followed other research labs set themselves to these works. Among
them I should like to mention, in the first place, the research
group headed by Farkas of the Central Institute for Physical
Research in Hungary, that has been successfully experimenting in
this area since 1967 maintaining close contact with the P. N.
Lebedev Physical Institute and the Landau Institute of Theoretical
Physics. In the intervening years such important characteristics
of the phenomenon as angular and energy distribution of emitted
ions, temporal, spectral, thredhold and polarization dependences
of photoelectric current, have received detailed study [14]. The
results on multiphoton surface photoeffect, even to a greater
degree than those on ordinary one-quantum effect, provide valuable
information on structure, electronic spectrum, and properties of
surface layers in solids. In the multiphoton case, however,
similar experiments prove to be rather complicated. One of the
obstacles involved in qualitative experiments is low photoelectric
current and, hence the necessity of using high-power laser pulses.
Photoelectric cathode heating, which is practically unavoidable
under such conditions, would give rise to thermionic emission,
thereby adding further difficulties to studying the fine details
of the photoeffect. The way out of this situation has been suggested in our lab [15] and successfully developed by the Farkas
group with theoretical assistance of Anisimov and his associates
of the Landau Institute of Theoretical Institute [16]. The fact
is that at shorter laser pulses and concurrent higher light intensities thermionic emission should be less competitive. Thus,
it has become evident that more favorable conditions for register-
ing multiphoton emission can be created in the picosecond range.

Thorough calculations have revealed that at pulses under 10^{-11} sec., the observation range for photoeffect on the background of thermionic emission can be extended to cover intensities of an order of 10–100 gigawatt/m^2 (note that for the nanosecond case this is about 2–5 Mwatt/m^2). Along these lines, Farkas and co-workers have obtained, in the observation of four- and five-quantum photoemission processes, a number of interesting new results. The latest results are reported at this Conference [see a later paper in this volume by Farkas].

In conclusion, I would like to say that investigations of multiphoton transitions open up new extensive possibilities for studying properties of microscopic systems and for controlling their states. Our future advances in this area will depend on how we shall utilize these possibilities. I hope that this Conference will contribute to this end.

REFERENCES

1.   F. V. Bunkin, A. M. Prokhorov, ZhETF, 46, 1090 (1964).

2.   F. V. Bunkin, R. V. Karapetyan, A. M. Prokhorov, ZhETF, 47, 216 (1964).

3.   L. V. Keldysh, ZhETF, 47, 1945 (1964).

4.   N. B. Delone, UFN, 115, 361 (1975).

5.   B. A. Zon, N. L. Manakov, L. P. Rapoport, ZhETF, 61, 968 (1971).

6.   A. E. Kazakov, V. P. Makarov, M. V. Fedorov, ZhETF, 70, 38 (1976).

7.   D. Popescu, C. Collins, B. Johnson, I. Popescu, Phys. Rev. 9A, 1182 (1974).

8.   P. Johnson et al., J. Chem. Phys., 62, 2500 (1975).

9.   N. B. Delone, B. A. Zon, V. P. Krainov, V. A. Khodovoy, UFN, 120, 3 (1976).

10.   P. Lambropoulos, G. Doolen, S. Rountree, Phys. Rev. Lett. 34, 636 (1975); M. Lambropoulos, S. Moody, S. Smith, W. C. Lineberger, Phys. Rev. Lett., 35, 159 (1975).

11.   P. Farago, D. Walker, J. Phys. 6B, L 280 (1973); P. Lambropoulos, J. Phys. 7B, L 33 (1974).

12.   E. Granneman et al., J. Phys. 9B, L 87 (1976).

13.   M. C. Teich, J. M. Schoer, G. J. Wolga, Phys. Rev. Lett. 13, 611 (1964).

14.   S. I. Anisimov, V. A. Bendersky, G. Farkas, UFN, $\underline{122}$, 185 (1977).

15.   F. V. Bunkin, A. M. Prokhorov, ZhETF, $\underline{52}$, 1610 (1967).

16.   S. I. Anisimov, B. L. Kapeliovitch, T. L. Perelman, ZhETF, $\underline{66}$, 776 (1974).

Part 1

QUANTUM ELECTRODYNAMICS AND ELECTRONS IN INTENSE FIELDS

A Review of Canonical Transformations as they Affect Multiphoton Processes

EDWIN A. POWER
University College
London, England

I. INTRODUCTION

A straightforward textbook calculation can illustrate the
relevance of canonical transformations in intense field problems.
It is based, rather loosely, on Geltman's recent paper [1] in
which the theory is developed for a one-dimensional 'atom' im-
mersed in an intense, but classically given, radiation field.
The probability

$$P(T) = 1 - \left| <\psi_o | \psi(T)> \right|^2 \tag{1}$$

is calculated that a measurement, made at time T, will find the
system in any of the states orthogonal to a given initial state
ψ_o, an eigenstate of H_o. In principle P(T) is known so finely
that the excitation probability can be followed over times less
than one cycle of the applied field: a probability that can be
appreciable for intense field strengths. If the field strengths
are not too large, the probabilities for this model follow the
golden rule over large time intervals after many cycles have
passed but before saturation. In (1) $\psi(t)$ satisfies

$$i\hbar \frac{\partial \psi}{\partial t} = H\psi , \qquad \psi(o) = \psi_o . \tag{2}$$

What is the infinitesimal generator H? In the electric dipole
approximation the same Heisenberg equation of motion

$$m\ddot{q} = - \frac{dV}{dq} - e\,\varepsilon\cos(\omega t + \phi) \tag{3}$$

applies for the two Hamiltonians

11

$$H_{min} = \frac{p^2}{2m} + V + \frac{epA(t)}{mc} + \frac{e^2A^2(t)}{2mc^2} \tag{4}$$

and

$$H_{mult} = \frac{p^2}{2m} + V + eq\,E(t) \tag{5}$$

where

$$A(t) = -\frac{\varepsilon e}{\omega}\sin(\omega t + \phi), \quad E = -\dot{A}/c . \tag{6}$$

The two Hamiltonians make different predictions for P(T). The textbook case where an exact calculation of P(T) is possible is the forced harmonic oscillator. [Here we depart from Ref. 1 where there is one bound state and a numerical calculation of P(T).] For an oscillator frequency Ω and quoting the result for $\phi = 0$ to make it simple: for H_{min}

$$P(T) = 1 - \exp\left\{ -\frac{e^2\varepsilon^2}{2m\hbar} \frac{\Omega^2}{(\Omega^2 - \omega^2)^2} \left[(1 - 2\cos\omega T\cos\Omega T + \cos^2\omega T) \right.\right.$$
$$\left.\left. - \frac{2\Omega}{\omega}\sin\omega T\sin\Omega T + \frac{\Omega^2}{\omega^2}\sin^2\omega T \right] \right\} , \tag{7}$$

while for H_{mult}

$$P(T) = 1 - \exp\left\{ -\frac{e^2\varepsilon^2}{2m\hbar} \frac{\omega^2}{(\Omega^2 - \omega^2)^2} \left[\frac{\Omega^2}{\omega^2} (1 - 2\cos\omega T\cos\Omega T + \cos^2\omega T) \right.\right.$$
$$\left.\left. - \frac{2\Omega}{\omega}\sin\omega T\sin\Omega T + \sin^2\omega T \right] \right\} . \tag{8}$$

We note the various factors (Ω/ω) in different places. This is symptomatic of the difference between $H_{min}^{(1)}$ and $H_{mult}^{(1)}$.

In non-relativistic quantum electrodynamics the two $H_{int}^{(1)}$ give identical matrix elements for absorption or emission of single photons conserving energy. The conversion of dipole-velocity to dipole-length introduces factors $\Omega = E_{mn}/\hbar$, which on the energy-shell equal $\pm\omega$. Similar considerations make it clear that many other forms of $H_{int}^{(1)}$ would also give the correct transition rates. Increasing the order of the time derivative on q can be compensated by decreasing the order of the time-derivative on the field. In Eq. (9) is displayed a set of energies and matrix elements

$$e \underset{\sim}{q} \cdot \underset{\sim}{E}^{\perp}, \quad \frac{e}{mc} \, \underset{\sim}{p} \cdot \underset{\sim}{A}, \quad \frac{e}{mc^2} \, \nabla V \cdot \underset{\sim}{Z} \; , \quad - \frac{e}{m^2 c^3} \, (\underset{\sim}{p} \cdot \underset{\sim}{\nabla}) \nabla V \cdot \underset{\sim}{Y}, \quad \ldots\ldots$$

$$\pm i\mu^{nm}\sqrt{2\pi\hbar\omega}, \quad \mp i\mu^{nm}\sqrt{2\pi\hbar\omega}\left(\frac{E_{nm}}{\hbar\omega}\right) \; , \quad \pm i\mu^{nm}\sqrt{2\pi\hbar\omega}\left(\frac{E_{nm}}{\hbar\omega}\right)^{2}$$

$$\mp i\mu^{nm}\sqrt{2\pi\hbar\omega}\left(\frac{E_{nm}}{\hbar\omega}\right)^{3} ,\ldots \quad (9)$$

The third entry arises in the so-called space-translation method:
Z is the Hertz vector. Our problem is to consider the alterna-
tives, how they arise and how they differ.

Any theory involving the interaction of radiation with atoms
or molecules must start from some assumptions concerning the dynamics of the complete system, i.e., from a version of electrodynamics and particle dynamics with coupling. In the domain of
energies of interest to multiphoton processes the particle dynamics is that of first-quantized Schrödinger mechanics. There
is less agreement as to what is suitable for the electrodynamics.
The possibilities range from quantum electrodynamics, where the
transverse electromagnetic fields are unbounded non-commuting
operators with equal-time C.R.

$$\left[E_i^{\perp}(\underset{\sim}{r}), \; B_j(\underset{\sim}{r}')\right] = i\hbar c \; 4\pi\varepsilon_{ijk} \frac{\partial}{\partial x_k'} \, \delta(\underset{\sim}{r}-\underset{\sim}{r}') , \qquad (10)$$

to ignoring the dynamics of the electromagnetic field altogether
and discussing dynamically only the quantum mechanics of the
charges in the atoms but including their interaction with given
impressed electromagnetic fields $E(r,t)$, $B(r,t)$, now c-numbers.
Intermediate between these would be to keep the fields as c-numbers but assume they obey Maxwell's equations with c-number currents constructed from charge densities and transition moments in
some way. In this paper we restrict our considerations to
(i) quantum electrodynamics, quanta of radiation in interaction
with non-relativistic charges, (ii) quantum mechanics of nonrelativistic charges with given c-number fields, (this I shall
call semi-classical without prejudice to other theories using the
same adjective) and (iii) radiation in interaction with prescribed
c-number currents. These three theories can be formulated from
Lagrangians. The Lagrangians they share with their entirely
classical progenitors: (i) classical electrodynamics in interaction with classical charges, (ii) classical charges in given external fields, and (iii) classical radiation interacting with
prescribed currents. The conventional Lagrangians are

14   E. A. Power

$$L_{(elec.dyn.)} = \sum_\alpha \tfrac{1}{2} m_\alpha \dot{\underset{\sim}{q}}_\alpha^2 - V(q) + \frac{1}{8\pi} \int \{ \dot{\underset{\sim}{A}}^2/c^2 - (\text{curl } \underset{\sim}{A})^2 \} dV$$

$$+ \sum_\alpha \int \frac{e_\alpha}{c} \dot{q}_{\alpha i} A_j(\underset{\sim}{r}) \delta_{ij}^{\perp} (\underset{\sim}{r} - \underset{\sim}{q}_\alpha) dV \tag{11}$$

$$L_{(pre.fields)} = \tfrac{1}{2} m_\alpha \dot{\underset{\sim}{q}}_\alpha^2 - V(q) + \sum_\alpha \frac{e_\alpha}{c} \dot{\underset{\sim}{q}}_\alpha \underset{\sim}{A}(\underset{\sim}{q}_\alpha, t) \tag{12}$$

$$L_{(pre.currents)} = \frac{1}{8\pi} \int \{ \dot{\underset{\sim}{A}}^2/c^2 - (\text{curl } \underset{\sim}{A})^2 \} dV$$

$$+ \sum_\alpha \int \frac{e_\alpha}{c} \dot{q}_{\alpha i}(t) A_j(\underset{\sim}{r}) \delta_{ij}^{\perp} (\underset{\sim}{r} - \underset{\sim}{q}_\alpha(t)) dV \tag{13}$$

In electrodynamics both the positions of the charges q_α, and the field coordinate A, are generalized coordinates of the problem. L is a function/functional of $q_\alpha, \dot{q}_\alpha$, A and $\dot{A}$ and variation of the q's implies the equations of motion for particles, while variation of A gives Maxwell's equations.

As Newton came before Lagrange (and Hamilton) so, perhaps the equations of motion have precedence over the variation principles (and vector flows on symplectic manifolds). An obvious question arises: given the equations of motion, what is the Lagrangian? This is the inverse problem of the calculus of variations [2]. Likewise we may ask: given the equations of motion, is there a unique Hamiltonian that can be quantized to provide the starting point for (i) quantum electrodynamics, (ii) semi-classical theory, or (iii) photons with classical current sources?

II. NON-UNIQUENESS OF LAGRANGIAN

It is clear that the Lagrangian is not unique. We are all familiar with several possibilities. The most trivial is a change of scale

$$L^{NEW} = \lambda L^{OLD} , \quad (\lambda \neq 0) \tag{14}$$

with the same q's. The equations of motion are unchanged. This is not true for Hamiltonians, for example,

$$H^{NEW}(p,q) = \lambda \left(\frac{p^2}{2} + \frac{\omega^2 q^2}{2} \right) \tag{15}$$

has equation of motion

$$\ddot{q} = -\lambda^2 \omega^2 q. \tag{16}$$

We can easily remedy this situation by changing the variables as well as the form of the Hamiltonian. Let

$$p^{NEW} = \lambda p \tag{17}$$

then

$$H^{NEW}(p(p^{NEW}),q) = \frac{p^{NEW^2}}{2\lambda} + \frac{\lambda \omega^2 q^2}{2} \tag{18}$$

with equation of motion (16). In fact $H^{NEW}(p^{NEW},q)$ is that Hamiltonian constructed out of L^{NEW}.

Another form of non-uniqueness of Lagrangian is well-known. If

$$L^{NEW} = L^{OLD} + \frac{d\,F(q,t)}{dt} \tag{19}$$

then the equations of motion are unaltered. One has

$$p^{NEW} = \frac{\partial L^{NEW}}{\partial \dot{q}} = p^{OLD} + \frac{\partial F}{\partial q} \tag{20}$$

but

$$\frac{dp^{NEW}}{dt} - \frac{\partial L^{NEW}}{\partial q} - \frac{dp^{OLD}}{dt} + \frac{\partial^2 F}{\partial q^2}\,\dot{q} + \frac{\partial^2 F}{\partial t \partial q} - \frac{\partial L^{OLD}}{\partial q} - \frac{d}{dt}\frac{\partial F}{\partial q}$$

$$= \frac{dp^{OLD}}{dt} - \frac{\partial L^{OLD}}{\partial q} = 0. \tag{21}$$

This is a typical canonical transformation since although the equation of motion is unchanged the phase-space coordinates are changed (Eq. 20). Our previous scale change example is of the same character, as we see from Eq. (17). An important special case of (19), with explicit time-dependence in L, is when

$$L^{OLD} = \frac{1}{2}\,m\dot{q}^2 - V(q) - \frac{e\dot{q}}{c}\,A(t) \tag{22}$$

and

$$F = + \frac{eq}{c}\,A(t)\ . \tag{23}$$

Then

$$L^{NEW} = \frac{1}{2}\,m\dot{q}^2 - V(q) + \frac{eq}{c}\,\dot{A}(t)\ . \tag{24}$$

16   E. A. Power

We have

$$p^{OLD} = \frac{\partial L^{OLD}}{\partial \dot{q}} = m\dot{q} - \frac{e\,A(t)}{c} \tag{25}$$

and

$$p^{NEW} = \frac{\partial L^{NEW}}{\partial \dot{q}} = m\dot{q}\ . \tag{26}$$

The equation of motion is

$$m\ddot{q} = -\frac{dV}{dq} + \frac{e\dot{A}}{c} \tag{27}$$

for both old and new Lagrangians, but the Hamiltonians differ:

$$H^{OLD} = \frac{\left(p + e\,\dfrac{A(t)}{c}\right)^2}{2m} + V(q) \tag{28}$$

$$H^{NEW} = \frac{p^2}{2m} + V(q) - \frac{eq\,\dot{A}(t)}{c}\ . \tag{29}$$

For the old Lagrangian the canonical momentum differs from the
kinetic momentum. They are equal for the new Lagrangian. There
is seldom a requirement that the canonical and kinetic momenta
should be the same. Consider the scale change with $\lambda = -1$, then
$p = -\dot{q}$.

In both previous examples the conjugate momenta are changed
but not the generalized coordinates. On the other hand we are
all used to coordinate changes in Lagrangian mechanics: in fact
the art of solving many problems in classical mechanics is to
choose the coordinates so that the equations of motion are
simple (e.g., polar, relative and centre of mass, areal, bipolar,
etc.). Of course then the Lagrangian must change because if it
did not the explicit form of the equations would be the same and
no simplification has occurred. What one solves with the new
Lagrangian is the same problem in different coordinates.*

If

$$q^{NEW} = q^{NEW}(q) \qquad \text{and} \qquad q = q(q^{NEW}), \tag{30}$$

then, (by definition)

---

*In passing we note that if the actual form of L remains unchanged
after a change of coordinates we have a symmetry in the problem.
We are used to this, in rotational invariance for example. This
is not the issue here.

$$L^{NEW}(q^{NEW}, \dot{q}^{NEW}) = L(q(q^{NEW}), \dot{q}(q^{NEW}, \dot{q}^{NEW})) \tag{31}$$

and

$$p^{NEW} = \frac{\partial L^{NEW}}{\partial \dot{q}^{NEW}} . \tag{32}$$

Then

$$(\dot{p} - \frac{\partial L}{\partial q}) = 0$$

implies

$$\dot{p}^{NEW} - \frac{\partial L^{NEW}}{\partial q^{NEW}} = 0 . \tag{33}$$

In fact, of course, the variation principle will imply the invariance of form $\dot{p} - \partial L/\partial q = 0$ for any coordinates that are chosen.

In this type of transformation both the q's and the p's change — unless the transformation of q is the simple space-translation

$$q = q^{NEW} - \alpha$$

$$p = p^{NEW} . \tag{34}$$

We have, vice-versa, if $F = \dagger qA$

$$p^{NEW} = p^{OLD} + A$$

$$q^{NEW} = q^{OLD} , \tag{35}$$

which is a momentum-translation. It is important to be clear at the vital difference in their constructions — although they look completely analogous in phase space.

A. Semi-Classical Lagrangian

Here we apply a dF/dt transormation, Eq. (19), to the semi-classical Lagrangian (12). For the old Lagrangian*

$$L = \frac{1}{2} m\dot{\underset{\sim}{q}}^2 - V(\underset{\sim}{q}) - \frac{e}{c} \dot{\underset{\sim}{q}} \cdot \underset{\sim}{A}(\underset{\sim}{q}, t), \tag{36}$$

---

*We leave out the summation over α the particle label from now on. The charge e_α is taken to be $-e$.

the conjugate momentum is

$$p = m\dot{q} - \frac{e}{c} A(q,t) \tag{37}$$

and the minimal coupling substitution (take the Hamiltonian without the field and replace p by $(p + e/c\ A)$) follows. A choice of $F(q,t)$ that simplifies

$$L^{NEW} = L + \frac{dF}{dt}$$

$$= \frac{1}{2} m\dot{q}^2 - V(q) + \dot{q}\cdot(- \frac{e}{c} A(q,t) + \nabla F) + \frac{\partial F}{\partial t} \tag{38}$$

would be one that minimizes the term in $\dot{q}$. The coefficient of $\dot{q}$ is

$$- \frac{e}{c} A + \nabla F \ . \tag{39}$$

If A is essentially independent of q, i.e., we are in the long-wavelength limit and the electric dipole approximation is valid, then [3]

$$F = - \frac{e}{c} q\cdot A(t) \tag{40}$$

and we are back to Eqs. (22), (23) and (24). The Hamiltonians are those of Eqs. (28) and (29). If the variation of A with q is important - as it will be for shorter wavelengths - we cannot totally eliminate the $\dot{q}$ term in Eq. (38). Higher multipoles can be considered but we leave the detailed discussions until later.

B. Lagrangian for Field with Given Currents

Most of the general remarks made above for particle mechanics and its Lagrangian hold for field mechanics and its Lagrangian density. The Lagrangian density for the transverse electromagnetic field in interaction with time-dependent currents is

$$L(A,\dot{A},t) = \frac{1}{8\pi} \left\{ \dot{A}^2/c^2 - (\text{curl } A)^2 \right\} - \frac{e}{c} \dot{q}_i(t)A_j(r)\delta^{\perp}_{ij}(r-q(t)) \tag{41}$$

and the conjugate field $\Pi(r)$ to $A(r)$ is

$$\Pi = \frac{\partial L}{\partial \dot{A}} = \dot{A}/(4\pi c^2) = - \frac{1}{4\pi c} E^{\perp} \tag{42}$$

and the conjugate momentum field Π is (except for factors $4\pi c$)

the kinetic momentum field $\dot{\underset{\sim}{A}} = -c\underset{\sim}{E}^{\perp}$. However there is no essential reason for this to be so and adding a time derivative to L,

$$\frac{d}{dt}[\frac{e}{c} q_i A_j(\underset{\sim}{r})\delta^{\perp}_{ij}(\underset{\sim}{r} - \underset{\sim}{q}(t))] \tag{43}$$

gives

$$L^{NEW}(\underset{\sim}{A},\dot{\underset{\sim}{A}},t) = \frac{1}{8\pi}\left\{\dot{\underset{\sim}{A}}^2/c^2 - (\text{curl }\underset{\sim}{A})^2\right\} + \frac{e}{c} q_i(t)\dot{A}_j(\underset{\sim}{r})\delta^{\perp}_{ij}(\underset{\sim}{r}-\underset{\sim}{q}(t))$$

$$- \frac{e}{c} q_i\dot{q}_k A_j(\underset{\sim}{r})\nabla_k\delta^{\perp}_{ij}(\underset{\sim}{r}-\underset{\sim}{q}(t)) \tag{44}$$

for which

$$\underset{\sim}{\Pi}^{NEW} = \frac{\partial L^{NEW}}{\partial \dot{A}_i} = \frac{\dot{A}_i}{4\pi c^2} + \frac{e}{c} q_j\delta^{\perp}_{ij}(\underset{\sim}{r}-\underset{\sim}{q}(t)) \tag{45}$$

and the new conjugate momentum field $\underset{\sim}{\Pi}^{NEW}$ is no longer proportional to $\underset{\sim}{E}^{\perp}$. The variations, both for L and L^{NEW}, give the same equation of motion although by differing routes. It is

$$\text{curl }\underset{\sim}{B} = \frac{1}{c}\frac{\partial \underset{\sim}{E}^{\perp}}{\partial t} - \frac{4\pi e}{c}\dot{\underset{\sim}{q}}(t)\cdot\underset{\approx}{\delta}^{\perp}(\underset{\sim}{r} - \underset{\sim}{q}(t)) \ . \tag{46}$$

The multipolar form of interaction is physically appealing here too. Localizing the currents at $\underset{\sim}{R}$, the approximations to L and L^{NEW} are

$$L \simeq \frac{1}{8\pi}\left\{\dot{\underset{\sim}{A}}^2/c^2 - (\text{curl }\underset{\sim}{A})^2\right\} - \frac{e}{c} q_i(t)A_j(\underset{\sim}{r})\delta^{\perp}_{ij}(\underset{\sim}{r} - \underset{\sim}{R}) \tag{47}$$

and

$$L^{NEW} \simeq \frac{1}{8\pi}\left\{\dot{\underset{\sim}{A}}^2/c^2 - (\text{curl }\underset{\sim}{A})^2\right\} + \frac{e}{c} (\underset{\sim}{q}(t) - \underset{\sim}{R})_i\dot{A}_j\delta^{\perp}_{ij}(\underset{\sim}{r} - \underset{\sim}{R}) \ . \tag{48}$$

We have

$$\underset{\sim}{\Pi}^{NEW} = \dot{\underset{\sim}{A}}/(4\pi c^2) + \frac{e}{c} (\underset{\sim}{q}(t) - \underset{\sim}{R})\underset{\approx}{\delta}^{\perp}(\underset{\sim}{r} - \underset{\sim}{R})$$

$$= -\frac{1}{4\pi c} [\underset{\sim}{E}^{\perp} + 4\pi \underset{\sim}{P}^{\perp}] = -\frac{1}{4\pi c}\underset{\sim}{D}^{\perp} \tag{49}$$

where $\underset{\sim}{P}$ is the dipole moment density relative to $\underset{\sim}{R}$

20   E. A. Power

$$\underset{\sim}{P}(\underset{\sim}{r}) = \sum_\alpha e_\alpha (\underset{\sim}{q}_\alpha - \underset{\sim}{R}) \delta(\underset{\sim}{r} - \underset{\sim}{R}) \ . \tag{50}$$

In this case the equation of motion is

$$\text{curl } \underset{\sim}{B} = \frac{1}{c} \frac{\partial \underset{\sim}{D}^{\perp}}{\partial t} \ . \tag{51}$$

The best generalization to higher moments uses a different summand from that of Eq. (43). We use [4]

$$\frac{d}{dt} \left[- \underset{\sim}{P}^{\perp}(\underset{\sim}{r}) \cdot \underset{\sim}{A}(\underset{\sim}{r}) \right] \tag{52}$$

where

$$\underset{\sim}{P}(\underset{\sim}{r}) = \sum_\alpha e_\alpha (\underset{\sim}{q}_\alpha - \underset{\sim}{R}) \int_0^1 \delta(\underset{\sim}{r} - \underset{\sim}{R} - \lambda(\underset{\sim}{q}_\alpha - \underset{\sim}{R})) d\lambda \ . \tag{53}$$

The bonus one gets in using this is that not only are the higher multipoles contributing correctly to $D^{\perp}$ *but also* the $\dot{q}$-terms that remain have the character curl M, where

$$\underset{\sim}{M}(\underset{\sim}{r}) = \sum_\alpha \frac{e_\alpha}{c} (\underset{\sim}{q}_\alpha - \underset{\sim}{R}) \times \dot{\underset{\sim}{q}}_\alpha \int_0^1 \lambda \, \delta(\underset{\sim}{r} - \underset{\sim}{R} - \lambda(\underset{\sim}{q}_\alpha - \underset{\sim}{R})) d\lambda \ . \tag{54}$$

After some tedious algebra the equation of motion can be shown to be

$$\text{curl } \underset{\sim}{H} = \frac{1}{c} \frac{\partial \underset{\sim}{D}^{\perp}}{\partial t} \tag{55}$$

with

$$\underset{\sim}{H} = \underset{\sim}{B} - 4\pi \underset{\sim}{M} \ . \tag{56}$$

C. Quantumelectrodynamical Lagrangian

The application of the transformation $+ dF/dt$ to the full electrodynamical Lagrangian $L_{(elec.dyn)}$ of Eq. (11) is a straightforward generalization of the two previously considered cases. For the total system both q's and A are generalized coordinates and both the particles and the field are dynamical quantities. However, no more is required than the addition of expression (53), to transform

$$L^{OLD} = L_{Atom} + L_{Rad} + \sum_\alpha \frac{e_\alpha}{c} \dot{q}_{\alpha_j} \int A_j(\underset{\sim}{r}) \delta^\perp_{ij}(\underset{\sim}{r} - \underset{\sim}{q}_\alpha) dV \tag{57}$$

to

$$L^{NEW} = L_{Atom} + L_{Rad} - \frac{1}{c} \int \dot{\underset{\sim}{A}}(\underset{\sim}{r}) \cdot \underset{\sim}{P}^\perp(\underset{\sim}{r}) dV + \int \underset{\sim}{A}(\underset{\sim}{r}) \cdot \text{curl } \underset{\sim}{M}(\underset{\sim}{r}) dV. \tag{58}$$

The detailed calculations are worked out in the literature [5]. It is to be emphasized again that the canonical momenta and kinetic momenta are not necessarily identical. For the old Lagrangian they are the same for the field momentum density but not for the particle momentum

$$\left. \begin{aligned} \underset{\sim}{P}_\alpha &= m\dot{\underset{\sim}{q}}_\alpha + \frac{e_\alpha \underset{\sim}{A}(\underset{\sim}{q}_\alpha)}{c} \\ \underset{\sim}{\Pi}(\underset{\sim}{r}) &= + \frac{\dot{\underset{\sim}{A}}(\underset{\sim}{r})}{4\pi c^2} \end{aligned} \right\} \quad \text{OLD} \tag{59}$$

On the other hand for the new Lagrangian, in electric dipole approximation, they are the same for the particles but not the field

$$\left. \begin{aligned} \underset{\sim}{P}_\alpha &= m\dot{\underset{\sim}{q}}_\alpha \\ \underset{\sim}{\Pi}(\underset{\sim}{r}) &= + \frac{\dot{\underset{\sim}{A}}(\underset{\sim}{r})}{4\pi c^2} - \sum_\alpha' e_\alpha (\underset{\sim}{q}_\alpha - \underset{\sim}{R}) \cdot \underset{\sim}{\delta}^\perp(\underset{\sim}{r} - \underset{\sim}{R}) \ . \end{aligned} \right\} \quad \text{NEW} \tag{60}$$

III. CANONICAL TRANSFORMATIONS

A. Classical

In classical analytical mechanics the step from Lagrangian mechanics to Hamiltonian mechanics is from the tangent bundle of configuration space TX to the cotangent bundle T*X which is the phase space of the physicist. X represents the configuration space (q's) and T represents the space of tangent vectors ($\dot{q}$'s). The Lagrangian is defined on TX and the canonical momenta are given by the fibre derivatives of L. The fibre derivative of L maps $(q,\dot{q}) \rightarrow (q,p)$ where $p = \partial L/\partial \dot{q}$ is an element of the dual space T* of T; $(p\dot{q})$ is the scalar defining the dual, and the Legendre-type transformation

$$H(p,q) + L(q,\dot{q}) = p\dot{q} \tag{61}$$

defines the Hamiltonian. The increment equation leads to Hamilton's equations with symplectic structure

$$\frac{d}{dt}\begin{bmatrix} p \\ q \end{bmatrix} = \begin{bmatrix} 0 & 1 \\ -1 & 0 \end{bmatrix}\begin{bmatrix} \nabla_p H \\ \nabla_q H \end{bmatrix} . \tag{62}$$

In classical mechanics a *canonical transformation* is a transformation of phase space to new coordinates p^{NEW}, q^{NEW} (where in general the transformation can be time-dependent) in which *for any* H there exists H^{NEW} (p^{NEW}, q^{NEW}) such that the truth of (62) implies

$$\frac{d}{dt}\begin{bmatrix} p^{NEW} \\ q^{NEW} \end{bmatrix} = \begin{bmatrix} 0 & 1 \\ -1 & 0 \end{bmatrix}\begin{bmatrix} \nabla_{p^{NEW}} H^{NEW} \\ \nabla_{q^{NEW}} H^{NEW} \end{bmatrix} . \tag{63}$$

As is proved in textbooks on classical mechanics this requires

$$\Gamma\begin{bmatrix} 0 & 1 \\ -1 & 0 \end{bmatrix}\Gamma^T = \lambda\begin{bmatrix} 0 & 1 \\ -1 & 0 \end{bmatrix} \tag{64}$$

for some constant* scalar where

$$\Gamma = \begin{bmatrix} \dfrac{\partial p^{NEW}}{\partial p} & \dfrac{\partial p^{NEW}}{\partial q} \\[2ex] \dfrac{\partial q^{NEW}}{\partial p} & \dfrac{\partial q^{NEW}}{\partial q} \end{bmatrix} \tag{65}$$

Then, for the particular Hamiltonian $H^{OLD}(p,q)$,

$$H^{NEW} = \lambda H^{OLD}(p(p^{NEW},q^{NEW}),q(p^{NEW},q^{NEW})) + R(p^{NEW},q^{NEW}) \tag{66}$$

when R satisfies the equation

---

*Thus the amplitude-phase transformation for the simple harmonic oscillator $\frac{1}{2}(p^2 + \omega^2 q^2)$: $p = R\cos\omega\phi$, $q = R/\omega \sin\phi$, is not canonical since $\lambda = R$ (not constant), but square root amplitude-phase is canonical $p = \sqrt{p^{NEW}}\cos\omega q^{NEW}$
$$q = \sqrt{p^{NEW}/\omega}\ \sin\omega q^{NEW}$$
with $\lambda = \frac{1}{2}$.

$$\frac{\partial}{\partial t}\begin{bmatrix} p^{NEW} \\ \\ q^{NEW} \end{bmatrix} = -\begin{bmatrix} 0 & 1 \\ \\ -1 & 0 \end{bmatrix}\begin{bmatrix} \dfrac{\partial R}{\partial p^{NEW}} \\ \\ \dfrac{\partial R}{\partial q^{NEW}} \end{bmatrix}. \tag{67}$$

When the transformation is independent of t one may have $R \equiv 0$.
If, in addition to $R = 0$ one has $\lambda = 1$, the transformation is called
[6] *completely canonical*. The example, Eq. (17), for $\lambda \neq 1$, defines a non-completely canonical transformation. In that example
the explicit relation is

$$H^{NEW} = \lambda H^{OLD}(p(p^{NEW}),q) . \tag{68}$$

In fact the Lagrangian scale change $(\lambda \neq 1)$ is never completely
canonical. The change of Lagrangian by $+ \, dF/dt$ of Section B has

$$\Gamma = \begin{bmatrix} 1 & \dfrac{\partial^2 F}{\partial q^2} \\ \\ 0 & 1 \end{bmatrix} \tag{69}$$

and has $\lambda = 1$ for any F. However it is not completely canonical
unless F is independent of t. In the specific example worked
through in Eqs. (22) to (29)

$$\text{and}\qquad \left.\begin{array}{l} p^{NEW} = p^{OLD} + \dfrac{eA(t)}{c} \\ \\ R = - \dfrac{eq\dot{A}}{c} \end{array}\right\}, \tag{70}$$

satisfy Eq. (68). Then Eq. (66) gives the new Hamiltonian,
Eq. (29).

One difference between the full electrodynamical Lagrangian
(11) and the prescribed field Lagrangian (12) is that for the
first it is possible to work with *completely* canonical transformations but that for the second these are not sufficient.

The type of change generated through the Lagrangian by a
change of generalized coordinates, when independent of time, is
always a completely canonical transformation. One has

24   E. A. Power

$$q_i^{NEW} = q_i^{NEW}(q_k^{OLD})$$

$$p_i^{NEW} = p_j^{OLD} \frac{\partial q_j^{OLD}}{\partial q_i^{NEW}} \tag{71}$$

and

$$
\begin{aligned}
H^{NEW} &= p_i^{NEW}\,\dot{q}_i^{NEW} - L^{NEW} \\
&= p_j^{OLD}\,\frac{q_j^{OLD}}{q_i^{NEW}}\,\frac{q_i^{NEW}}{q_k^{OLD}}\,\dot{q}_k^{OLD} - L\left(q_k^{OLD}\,(q^{NEW}),\ \frac{q_k^{OLD}}{q_j^{NEW}}\,\dot{q}_j^{NEW}\right) \\
&= p_j^{OLD}\,\dot{q}_j^{OLD} - L^{OLD} \\
&= H^{OLD}\ .
\end{aligned}
\tag{72}
$$

So $\lambda = 1$ and $R = 0$.

If the time occurs explicitly in the transformation so that

$$q_i^{NEW} = q_i^{NEW}(q_k^{OLD},\ t) \tag{73}$$

then R is no longer zero. An example is the time-dependent space translation discussed much in the literature of multiphoton processes under the name Henneberger transformation [7]. Here, for the one-dimensional case,

$$q^{NEW} = q^{OLD} + \alpha(t)$$

$$p^{NEW} = p^{OLD}$$

$$R = \dot{\alpha}(t)p\ . \tag{74}$$

Then

$$H^{NEW} = H^{OLD}\ \text{(expressed in new coordinates)} + \dot{\alpha}p, \tag{75}$$

and this transformation is not completely canonical. In the prescribed field situation where

$$H_{min} = \frac{\left(\underset{\sim}{p} + \frac{e}{c}\underset{\sim}{A}(t)\right)^2}{2m} + V(\underset{\sim}{q}) \tag{76}$$

one may take

$$\underset{\sim}{\alpha} = - \frac{e \, \underset{\sim}{Z}(t)}{mc^2} \quad , \tag{77}$$

$\underset{\sim}{Z}$ is a Hertz potential, and one has

$$\underset{\sim}{q}^{NEW} = \underset{\sim}{q} - \frac{e \, \underset{\sim}{Z}(t)}{mc^2}$$

$$\underset{\sim}{p}^{NEW} = \underset{\sim}{p}$$

$$R = - \frac{e \underset{\sim}{p} \cdot \underset{\sim}{A}(t)}{mc} \tag{78}$$

which makes

$$H^{NEW} = \frac{\underset{\sim}{p}^2}{2m} + \frac{e^2}{2mc^2} \, \underset{\sim}{A}^2(t) + V\left(\underset{\sim}{q}^{NEW} + \frac{e\underset{\sim}{Z}(t)}{mc^2} \right) \quad . \tag{79}$$

The $\dot{\underset{\sim}{\alpha}} \cdot p$ term cancels the $e p \cdot A(t)/mc$ by the above choice of α. It is to be noted that this transformation is not completely canonical and, unlike the momentum translation method (Eq. (29)), the quadratic term $e^2 A^2(t)/2mc^2$ remains in the new Hamiltonian.

B. Quantum

The physicist's usual prescription to go from the classical Hamiltonian to the quantum operator is to "promote" the p's and q's to operators obeying the commutation relations

$$[p_i, q_j] = - i\hbar \delta_{ij} \quad , \tag{80}$$

and the transverse fields $\underset{\sim}{\Pi}(\underset{\sim}{r})$, $\underset{\sim}{A}(\underset{\sim}{r})$ to operators obeying the commutation relations

$$[\Pi_i(\underset{\sim}{r}), \, A_j(\underset{\sim}{r}')] = - i\hbar \delta_{ij}^{\perp}(\underset{\sim}{r} - \underset{\sim}{r}') \quad . \tag{81}$$

Equation (81), which is equivalent to Eq. (10), may be made to depend on (80) by normal mode expansions. However, one does not expect every Hamiltonian that gives the classical equations of motion to be suitable for quantization. Well-known counter examples arise from using angular coordinates as generalized coordinates: one such is the use of $p^{NEW}/2$ of footnote on page 22

for the harmonic oscillator. A straightforward and practical
point of view is to demand quantization at the stage involving
the cartesian coordinates and their conjugate momenta and the
equivalent pairs for the oscillators making up the radiation
field. This means that the general classes of canonical transformations considered in the classical case above are too wide
and are not, a priori, of immediate relevance to the quantum
situation [8]. Canonical transformations should be reconsidered
after quantization.

Transformations of scale ($\lambda \neq 1$) cannot be quantum canonical
transformations because with $p^{NEW} = \lambda p$ and q unchanged the commutation relation (80) is not invariant. However the type of
transformation where in classical mechanics an addition
$dF(q,A(r))/dt$ is made to the Lagrangian (density) is always equivalent to a completely canonical transformation.

$$q^{NEW} = q^{OLD}$$

$$p^{NEW} = p^{OLD} + \frac{\partial F}{\partial q}$$

$$A^{NEW}(r) = A^{OLD}(r) \tag{82}$$

$$\Pi^{NEW}(r) = \Pi^{OLD}(r) + \frac{\partial F}{\partial A(r)}$$

This in turn can be generaged, in quantum theory, by

$$p^{NEW} = e^{iS} p e^{-iS}$$

$$\Pi^{NEW}(r) = e^{iS} \Pi(r) e^{-iS} \tag{83}$$

with S a function of q and $A(r)$. If

$$F = -\hbar S, \tag{84}$$

then

$$p^{NEW} = p - \hbar \frac{\partial S}{\partial q}$$

$$\Pi^{NEW}(r) = \Pi(r) - \hbar \frac{\partial S}{\partial A(r)} \tag{85}$$

The commutative diagram holds:-

$$
\begin{array}{ccc}
L^{OLD} & \xrightarrow{\hspace{2cm}} & L^{NEW} \\
\downarrow & & \downarrow \\
H^{OLD} & \xrightarrow{\; e^{-iS}He^{iS}\;} & H^{NEW}
\end{array}
\tag{86}
$$

We shall say that the transformation is a *quantum completely canonical transformation* when

$$
H^{NEW}\{\underset{\sim}{p}^{NEW},\; \underset{\sim}{q}^{NEW};\; \underset{\sim}{\Pi}^{NEW},\; \underset{\sim}{A}^{NEW}\}
$$

$$
= H\{\underset{\sim}{p}(\underset{\sim}{p}^{NEW},\; \underset{\sim}{q}^{NEW},\; \underset{\sim}{\Pi}^{NEW},\; \underset{\sim}{A}^{NEW}),\; \underset{\sim}{q}(\;);\underset{\sim}{\Pi}(\;),\; \underset{\sim}{A}(\;)\}.
\tag{87}
$$

So the functional form of H^{NEW} is, from (83),

$$
H^{NEW} = H(e^{-iS}\underset{\sim}{p}e^{iS},\; e^{-iS}\underset{\sim}{q}e^{iS};\; e^{-iS}\underset{\sim}{\Pi}e^{iS},\; e^{-iS}\underset{\sim}{A}e^{iS})
$$

$$
= e^{-iS}\,H\,e^{iS}.
\tag{88}
$$

A conventional way to develop quantum canonical transformations is through this operator transformation (88). It should be noted that the transformation (88) is contravariant to the transformation (83) and also that neither of these refers to any bases on the Fock-Hilbert space nor transformations of such bases.

It is obvious that the commutation relations (80) and (81) hold for the new canonical pairs. This suggests a further generalization to cases where S is not solely a function of $\underset{\sim}{q}$ and $\underset{\sim}{A}(\underset{\sim}{r})$ but also $\underset{\sim}{p}$ and $\underset{\sim}{\Pi}(\underset{\sim}{r})$; for in this case too the commutation relations will be invariant. Now the new variables are defined by

$$
\underset{\sim}{q}^{NEW} = e^{iS}\,\underset{\sim}{q}\,e^{-iS}
$$

$$
\underset{\sim}{A}^{NEW}(\underset{\sim}{r}) = e^{iS}\,\underset{\sim}{A}(\underset{\sim}{r})\,e^{-iS},
\tag{89}
$$

in addition to Eqs. (83). A simple subclass of this generalization is when S is a function only of the $\underset{\sim}{p}$'s and $\underset{\sim}{\Pi}(\underset{\sim}{r})$. Then

$$
\underset{\sim}{q}^{NEW} = \underset{\sim}{q} + \hbar\,\frac{\partial S}{\partial \underset{\sim}{p}} \qquad ; \qquad \underset{\sim}{p}^{NEW} = \underset{\sim}{p}
\tag{90}
$$

$$
\underset{\sim}{A}^{NEW}(\underset{\sim}{r}) = \underset{\sim}{A}(\underset{\sim}{r}) + \hbar\,\frac{\partial S}{\partial \underset{\sim}{\Pi}(\underset{\sim}{r})} \qquad ; \qquad \underset{\sim}{\Pi}^{NEW}(\underset{\sim}{r}) = \underset{\sim}{\Pi}(\underset{\sim}{r}),
$$

and we have a space translation. In classical dynamics these must arise from a transformation of variables. With the

restriction to completely canonical transformations the relation
(88) defines the new Hamiltonian operator. There is, of course,
no guarantee that the new Hamiltonian can be got from a new
Lagrangian in the manner of diagram (86). Conversely too there
is no guarantee that an arbitrary new Lagrangian obtained through
using transformation (3) gives rise to a new Hamiltonian that is
related to the old by a quantum canonical transformation. If the
L-H diagram is closed and commutative, the transformation is
called a *Lagrangian-induced* quantum completely-canonical transformation.

C. Quantum With Externally-Driven Fields

The theoretical situation is different for a system of
charges in the presence of given fields from quantum electrodynamics. With given driving fields the Hamiltonians are explicitly
time-dependent. One of the simplest cases follows the discussions
in Section II-A, an atom in interaction with an electromagnetic
field and in electric dipole approximation. The transformation
(4) to (5) is not completely canonical because the generating
function F is an explicit function of time. H^{NEW} is not unitarily
related to H^{OLD} and there is no guarantee that calculations even
of energy shifts will give identical answers. The relationship
between H_{OLD} and H_{NEW} is

$$H^{NEW} = e^{-iS} H^{OLD} e^{iS} + \hbar \frac{\partial S}{\partial t} , \tag{91}$$

which is the quantum analogue of the classical Eq. (66), and the
diagram

$$
\begin{array}{ccc}
L^{OLD} & \xrightarrow{\ +\frac{dF}{dt}\ } & L^{NEW} \\
\downarrow & & \downarrow \\
H^{OLD} & \xrightarrow{\ e^{-iS}He^{iS}+\hbar\frac{\partial S}{\partial t}\ } & H^{NEW}
\end{array}
\tag{92}
$$

is commutative.

Note that the difference between Eqs. (4) and (5) can be
interpreted as a *gauge* difference rather than as a canonical
transformation. An electric field with no space variation can
be obtained either from the transverse vector potential $\underset{\sim}{A}$ or from
the longitudinal potential $-\nabla\phi$ where $\phi = -\underset{\sim}{q}\cdot\underset{\sim}{E}(t)$. So the gauge
transformation

$$
\begin{aligned}
\underset{\sim}{A} &\rightarrow \underset{\sim}{A} - \nabla\chi \\
\phi &\rightarrow \phi + \frac{1}{c}\frac{\partial\chi}{\partial t}
\end{aligned}
$$

where $\chi = q \cdot A(t)$ turns A to 0 and $-e\phi$ to $-e\phi + eq \cdot E(t)$. This is
the basis for several published [9] discussions on the $-\mu \cdot E$,
$-ep \cdot A/mc$ problem. A recent paper by Forney, Quattropani and
Bassani [10] has extended the discussion to gauge freedom that extends beyond the Coulomb gauge (or that of Lorentz). In the presence of an electromagnetic wave the Hamiltonian for a one-charge
system is extended to include the presence of a fixed charge of
opposite sign at the origin. The vector potentials for both
charges are added so that

$$m \dot{q} = p - \frac{e}{c} \left(A(q,t) - A(0,t) \right).$$

On the other hand a compensating term for this addition is provided by subtracting the energy of the dipole formed by the pair
of charges in the electromagnetic field in the potential term
$V - eEz \cos \omega t$. Any quantum canonical transformation of extended
momentum translation form can be reinterpreted as a gauge transformation. This is because, since

$$p^{NEW} = p^{OLD} + \frac{\partial F}{\partial q}$$

we have

$$\dot{q} = \frac{p^{NEW} + \frac{e}{c} A + \nabla F}{m}$$

and

$$H^{NEW} = \frac{\left(p^{NEW} + \frac{eA}{c} + \nabla F \right)^2}{2m} + V - \frac{\partial F}{\partial t}.$$

In the electric dipole case we can remain in the Coulomb gauge
because $\nabla^2 F = 0$.

For the electric dipole approximation

$$S = -F/\hbar = \frac{-e}{\hbar c} q \cdot A(t) \tag{93}$$

It is very easy to extend this to higher multipoles. For the
Lagrangian

$$\frac{1}{2} m \dot{q}^2 - V(q) - \frac{eq_i}{c} \left\{ A_i(R,t) + q_j \nabla_j A_i(R,t) \right\} \tag{94}$$

with

$$F = \frac{e}{c} \left\{ q_i A_i + \frac{1}{2} q_i (q_j \nabla_j) A_i \right\} \tag{95}$$

we obtain both the electric quadrupole and the magnetic dipole.

30   E. A. Power

We cannot have one without the other. The only step where care
is needed is to subtract *half* of the term $- e/c \; \dot{q}_i q_j \nabla_i A_i$ from the
Lagrangian. The result that takes the place of Eq. (5) is

$$H^{NEW} = H_o + e q \cdot E(t) + e Q : \nabla E(t) + m \cdot B(t) + \frac{e^2}{8mc^2} \, [q \times B(t)]^2 \quad (96)$$

with $Q = \tfrac{1}{2} q \, q$, $m = -(e/2mc) \; q \times p$. The general case can be done in
the complete quantum electrodynamical formalism: this is discussed
in the next section and the semi-classical results are readily
generalized too - if required.

The analogous problem for the space-translated Hamiltonian
in the semi-classical and electric-dipole approximation (the
Henneberger transformation [7]), is to transform to

$$H^{NEW} = \frac{p^2}{2m} + V\left(q + \frac{e \, Z(t)}{mc^2}\right) + \frac{e^2}{2mc^2} \, A^2(t) \; . \quad (97)$$

This can be accomplished by using Eq. (92) with

$$S = - \frac{e}{mc^2} \, p \cdot Z(t) \; . \quad (98)$$

Then

$$\hbar \, \frac{\partial S}{\partial t} = - \frac{e}{mc^2} \, p \cdot Z = \frac{-e \; p \cdot A(t)}{mc} \quad (99)$$

which cancels the $e \; p \cdot A(t)/mc$ in Eq. (4). The diagram

$$
\begin{array}{ccc}
L^{OLD} & \xrightarrow{\;\; q(q^{NEW})\;\;} & L^{NEW} \\
\Big\downarrow & & \Big\downarrow \\
H^{OLD} & \xrightarrow{\;\; e^{-iS} H e^{iS} + \hbar \frac{\partial S}{\partial t}\;\;} & H^{NEW}
\end{array}
\quad (100)
$$

is commutative.

If H^{OLD} and H^{NEW} are related as expressed in Eq. (92) we
shall call the transformation a *quantum canonical transformation*.
If in addition the L-H diagram is commutative we shall call it a
Lagrangian-induced canonical transformation. In quantum electrodynamics a quantum completely-canonical transformation is a quantum canonical transformation with a time-independent S. That not
all quantum canonical transformations are Lagrangian-induced is
easily seen by examining the so-called "picture" changes as
canonical transformations. For example, the Heisenberg picture
results from the quantum canonical transformation (91), with

$$S = -\frac{1}{\hbar} Ht$$

when

$$H^{NEW} = e^{iHt/\hbar} H e^{-iHt/\hbar} - H = 0 \tag{101}$$

and in this picture the state vector is constant. Another example is the interaction picture with $S = - (1/\hbar) H_o t$. Neither of these is Lagrangian-induced.

Some authors [11] refer to the transformation Eq. (98), as a change to the Henneberger picture. It is a Lagrangian-induced quantum canonical transformation with L^{NEW} in diagram (100)

$$L^{NEW} = \frac{1}{2} m\dot{\underset{\sim}{q}}^2 - V\left(\underset{\sim}{q} + \frac{e\underset{\sim}{Z}(t)}{mc^2}\right) - \frac{e^2\underset{\sim}{A}^2(t)}{2mc^2} \tag{102}$$

obtained by the coordinate change

$$\underset{\sim}{q} = \underset{\sim}{q}^{NEW} + \frac{e\underset{\sim}{Z}(t)}{mc^2} \; . \tag{103}$$

If the canonical transformation is Lagrangian-induced, the transformation between L^{OLD} and L^{NEW} can be of either type. If of the form of diagram (92) we shall call it an *extended momentum translation* while, if it follows diagram (100) it is an *extended space translation* or *point transformation*. The momentum translation, Eq. (93), and the space translation, Eq. (98), are both Lagrangian-induced and fit the nomenclature. Neither is a quantum completely-canonical transformation.

IV. TRANSFORMATIONS IN QUANTUM ELECTRODYNAMICS AND CALCULATIONS OF PHOTON PROCESSES

In this section we consider two specific transformations remaining entirely within the domain of quantum electrodynamics, so the fields are themselves dynamical variables. First we analyze the Power-Zienau [12] canonical transformation to the multipolar Hamiltonian. Second, the generalized Henneberger [13] transformation is presented. These are generalizations to the field theoretical case of the semi-classical situation discussed in Section III-C. Several physical processes, including some involving multiphoton absorptions and others important to quantum optics are discussed.

The starting Lagrangian is Eq. (11) with

$$L = L_{Atoms} + L_{Rad} + L_{Int}$$

where

$$L_{Int} = \sum_\alpha \frac{e_\alpha}{c} \int \dot{q}_{\alpha_i} A_j(\underset{\sim}{r}) \delta^\perp_{ij}(\underset{\sim}{r} - \underset{\sim}{q}_\alpha) dV$$

(104)

and the Hamiltonian is

$$H = H_{Atoms} + H_{Rad} + H_{Int}$$

where

$$H_{Atoms} = \sum_\alpha \frac{\underset{\sim}{p}_\alpha^2}{2m_\alpha} + V(\underset{\sim}{q})$$

$$H_{Rad} = \frac{1}{8\pi} \int \left[(-4\pi c\ \underset{\sim}{\Pi}(\underset{\sim}{r}))^2 + (\text{curl}\ \underset{\sim}{A}(\underset{\sim}{r}))^2 \right] dV$$

(105)

$$H_{Int} = H^{(1)}_{int} + H^{(2)}_{int} = - \sum_\alpha \frac{e_\alpha \underset{\sim}{p}_\alpha \cdot \underset{\sim}{A}(\underset{\sim}{q}_\alpha)}{m_\alpha c} + \sum_\alpha \frac{e_\alpha^2\ \underset{\sim}{A}^2(\underset{\sim}{q}_\alpha)}{2m_\alpha c^2}$$

The conjugate momentum field is the transverse electric field

$$-4\pi c\ \underset{\sim}{\Pi}(\underset{\sim}{r}) = - \dot{\underset{\sim}{A}}(\underset{\sim}{r})/c = \underset{\sim}{E}^\perp(\underset{\sim}{r})$$

(106)

and the particle conjugate momenta are those necessary for minimal coupling.

A. Power-Zienau

This transformation is a Lagrangian-induced, quantum completely-canonical transformation. The generator is

$$F = -\hbar S = - \frac{1}{c} \int \underset{\sim}{P}(\underset{\sim}{r}) \cdot \underset{\sim}{A}(\underset{\sim}{r}) dV$$

(107)

where $\underset{\sim}{P}(\underset{\sim}{r})$ is a polarization field depending only on the position coordinates of the charged particles. A typical multipole form, for a single system centered at $\underset{\sim}{R}$, would be

$$\underset{\sim}{P}(\underset{\sim}{r}) = \sum_\alpha e_\alpha(\underset{\sim}{q}_\alpha - \underset{\sim}{R})\delta(\underset{\sim}{r} - \underset{\sim}{R}) + \sum_\alpha \frac{e_\alpha}{2}(\underset{\sim}{q}_\alpha - \underset{\sim}{R})(\underset{\sim}{q}_\alpha - \underset{\sim}{R}) \cdot \nabla\delta(\underset{\sim}{r} - \underset{\sim}{R}) + \ldots$$

(108)

Then

$$L^{NEW}_{int} = \int (\text{curl}\ \underset{\sim}{M}(\underset{\sim}{r})) \cdot \underset{\sim}{A}(\underset{\sim}{r}) dV - \frac{1}{c} \int \underset{\sim}{P}^\perp(\underset{\sim}{r}) \cdot \dot{\underset{\sim}{A}}(\underset{\sim}{r}) dV$$

(109)

where $M(r)$ is a magnetization field. The multipolar form, for a single system, is

$$M(\underset{\sim}{r}) = \sum_{\alpha} \frac{e_{\alpha}}{2c} (\underset{\sim}{q}_{\alpha} - \underset{\sim}{R}) \times \underset{\sim}{\dot{q}}_{\alpha} \; \delta(\underset{\sim}{r} - \underset{\sim}{R}) + \ldots \tag{110}$$

and the Hamiltonian is

$$H^{NEW} = H'_{Atoms} + H_{Rad} + H_{Pol} + H_{Mag} + H_{Dia}$$

where

$$H_{Pol} = - \int \underset{\sim}{P}(\underset{\sim}{r}) \cdot (-4\pi c \; \underset{\sim}{\Pi}(\underset{\sim}{r})) dV \simeq - \sum_{\alpha} e_{\alpha} (\underset{\sim}{q}_{\alpha} - \underset{\sim}{R}) \cdot \underset{\sim}{D}^{\perp}(\underset{\sim}{R}) + \ldots$$

$$H_{Mag} = - \int \underset{\sim}{M}(\underset{\sim}{r}) \cdot \text{curl } \underset{\sim}{A}(\underset{\sim}{r}) dV \simeq - \frac{e}{2mc} \sum_{\alpha} (\underset{\sim}{q}_{\alpha} - \underset{\sim}{R}) \times \underset{\sim}{P}_{\alpha} \cdot \underset{\sim}{B}(\underset{\sim}{R}) + \ldots$$

$$H_{Dia} = \frac{1}{2} \int O_{ij}(\underset{\sim}{r},\underset{\sim}{r}') (\text{curl } \underset{\sim}{A}(\underset{\sim}{r}))_i (\text{curl } \underset{\sim}{A}(\underset{\sim}{r}'))_j dVdV'$$

$$\simeq \frac{e^2}{8mc^2} \sum_{\alpha} [(\underset{\sim}{q}_{\alpha} - \underset{\sim}{R}) \times \underset{\sim}{B}(\underset{\sim}{R})]^2 + \ldots \tag{111}$$

In the new system we have, in contradistinction to Eq. (106)

$$-4\pi c \; \underset{\sim}{\Pi}(\underset{\sim}{r}) = \underset{\sim}{E}^{\perp}(\underset{\sim}{r}) + 4\pi \; \underset{\sim}{P}^{\perp}(\underset{\sim}{r}). \tag{112}$$

The conjugate momentum field is not the transverse electric field but the transverse displacement vector $\underset{\sim}{D}^{\perp}(r)$. In this transformation there is a cancellation of both $\tilde{H}^{(1)}_{int}$ and $H^{(2)}_{int}$, a complete cancellation which is not confined to the electric dipole approximation. The corrections to H_{Atoms}, arising in the last column and implied by the notation H'_{Atoms} in Eq. (111), are not particularly important in multiphoton calculations. They are essential in any analysis of interatomic or intermolecular forces, and, for a single atom, in calculations to predict the Lamb shift starting from H^{NEW}.

Since the transformation is completely-canonical it is necessarily unitary and the spectra of H^{NEW} and H^{OLD} are identical. Clearly using either in calculations of energy shifts, correct to any given order, for the complete system field + particles will give the same answers. The relevant question is then the pragmatic one: which is the most convenient? Many examples involving different kinds of energy shifts, including

such esoteric ones as specific discriminating forces between
molecular enantiomorphs – specific to the chirality of the enantiomorphs – where both electric and magnetic transitions are
allowed between the same levels, all point to the advantage in
using the multipolar Hamiltonian H^{NEW}.

It is easy to convince oneself that the transition rates, if
they are calculated by means of the Fermi rule, are also unchanged
as between either Hamiltonian (105) or (111). The rate $\Gamma_i = 2\pi/\hbar$
$|M_{fi}|^2 \rho_f$ from state $|i>$ to $|f>$ (density of states per unit energy
ρ_f) is calculated from Fig. 1, together with resonant contribu-
tions to the energy shift of state $|i>$.

$$\Delta_{\text{non-resonant terms}} \simeq \sum_f M_{if} \frac{1}{E_i - H_o + i\epsilon} M_{fi}$$

$$= \sum_f M_{if} (-2\pi i \delta^{(+)}(H_o - E_i)) M_{fi}$$

$$= -i\pi \sum_f |M_{if}|^2 \delta(E_i - E_f) + \sum_f M_{if} \frac{P}{E_i - E_f} M_{fi}$$

$$= -\frac{i\hbar}{2} \Gamma_i + \Delta E_i \, . \tag{113}$$

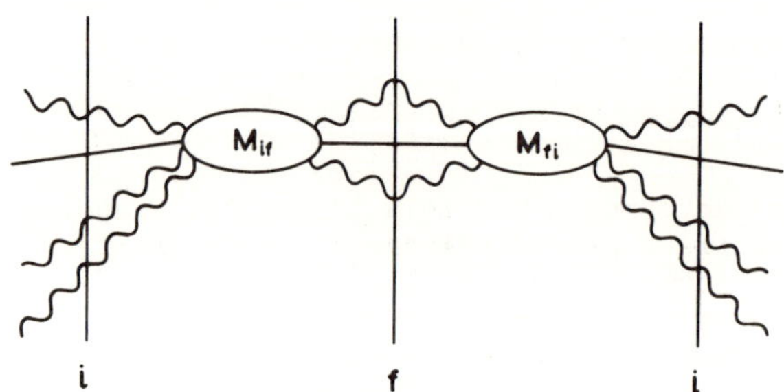

Figure 1

Calculations of transition rates evaluated through on-the-energy-
shell matrix elements for multiphoton absorption (whether or not
there are resonances within M_{fi} itself) always give the same result using either Hamiltonian. Certain sum rules, judiciously
chosen, may be necessary to demonstrate this in an explicit manner. Geltman [14] did this for two-photon electric dipole absorption some time ago. Recently we, in London, have checked
this for three photons and we have also checked the two-photon
case for higher multipoles involved in the virtual transitions.

When all the photons are from the same mode the calculations are easier than with different photons. An indication of the three-photon single-mode calculation follows (Fig. 2). For the $-\mu\cdot E^\perp$ interaction (suppressing the polarization indices and the intensity-correlation factors that lead to $g^{(3)}I^3$):

$$M_{fi} = \sum_{1,2} \mu^{f2}\mu^{21}\mu^{10}\left(\frac{2\pi\hbar\omega}{V}\right)^{3/2} \frac{1}{\hbar\omega - E_{10}} \cdot \frac{1}{2\hbar\omega - E_{20}} \quad . \quad (114)$$

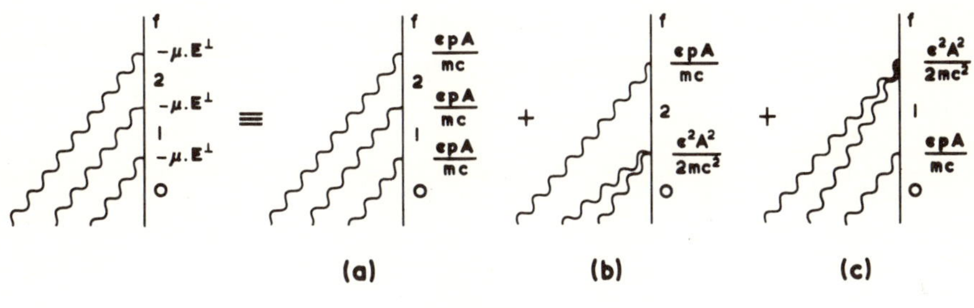

FIGURE 2

Note the $(\hbar\omega)^{1/2}$ factor from each $E^\perp$ term. On the other hand using the vector potential the single-vertices graph gives, after converting dipole velocities to dipole lengths,

$$\sum_{1,2} \mu^{f2}\mu^{21}\mu^{10}\left(\frac{2\pi}{V\hbar\omega}\right)^{3/2} E_{f2}E_{21}E_{10}\frac{1}{\hbar\omega - E_{10}} \cdot \frac{1}{2\hbar\omega - E_{20}} \quad . \quad (115)$$

The other two graphs give

$$\sum_{1,2} \mu^{f2}\mu^{21}\mu^{10}\left(\frac{2\pi}{V\hbar\omega}\right)^{3/2} \frac{E_{f2}(E_2 - 2E_1 + E_0)}{2(2\hbar\omega - E_{20})}$$

$$+\sum_{1,2} \mu^{f2}\mu^{21}\mu^{10}\left(\frac{2\pi}{V\hbar\omega}\right)^{3/2} \frac{(E_f - 2E_2 + E_1)E_{10}}{2(\hbar\omega - E_{10})} \quad . \quad (116)$$

For each A we have one $(\hbar\omega)^{-1/2}$ factor and for the $e^2A^2/2mc^2$ term we use the closure relation

$$\sum_1 \{(E_2 - E_1)q^{21}q^{10} - (E_1 - E_0)q^{21}q^{10}\} = (-i\hbar)^2 <2|0> \quad .$$

If we take out a common factor from (115) and (116) we have

$$\sum_{1,2} \mu^{f2}\mu^{21}\mu^{10}\left(\frac{2\pi\hbar\omega}{V}\right)^{3/2}\frac{1}{(\hbar\omega)^3}\frac{1}{\hbar\omega-E_{10}}\cdot\frac{1}{2\hbar\omega-E_{20}}$$

$$\left\{(3\hbar\omega-E_{20})(E_{20}-E_{10})E_{10}+\frac{1}{2}(3\hbar\omega-E_{20})(E_{20}-2E_{10})(\hbar\omega-E_{10})\right.$$

$$\left.+\frac{1}{2}E_{10}(3\hbar\omega-2E_{20}+E_{10})(2\hbar\omega-E_{20})\right\} \tag{117}$$

Now let $D_1 = \hbar\omega - E_{10}$, $D_2 = 2\hbar\omega - E_{20}$, the matrix element has the factor

$$\frac{1}{D_1D_2}\left\{(\hbar\omega+D_2)(\hbar\omega+D_1-D_2)(\hbar\omega-D_1)+\frac{1}{2}(\hbar\omega+D_2)(-D_2+2D_1)D_1\right.$$

$$\left.+\frac{1}{2}(\hbar\omega-D_1)(2D_2-D_1)D_2\right\}=\frac{(\hbar\omega)^3}{D_1D_2}+\frac{D_1-D_2}{2} \tag{118}$$

The second term in Eq. (118) has no denominators and the closure relation

$$\sum_{1,2}\mu^{f2}\mu^{21}\mu^{10}(E_{21}-E_{10}-E_{f2}+E_{21})$$

$$=\sum_{1,2}\mu^{f2}\mu^{21}\mu^{10}\,3(D_1-D_2)=0 \tag{119}$$

shows that it makes no contribution to (117). Finally when the first term in (118) is inserted into Eq. (117) the $(\hbar\omega)^3$ term cancels, leaving the matrix element (114) that arises from $-\underset{\sim}{\mu}\cdot\underset{\sim}{E}^{\perp}$.

B. Henneberger

This transformation is usually considered within quantum optics in the semi-classical situation. However the generalization to the case where the radiation field is part of the dynamical system has been discussed [13] by Henneberger and by Power and Thirunamachandran. In quantum electrodynamics it was used by Kramers and Schwinger in the context of mass renormalization. The unitary transformation of the Hamiltonian to

$$H^{NEW} = e^{-iS} \, H \, e^{iS} \, , \qquad S = - \frac{e}{\hbar mc^2} \, \underset{\sim}{p} \cdot \underset{\sim}{Z}(\underset{\sim}{q}) \tag{120}$$

ensures that the transformation is completely canonical. Hence, using the arguments outlined in the previous subsection, the calculations of energy shifts and of transition rates using matrix elements on the energy shell must give results identical with those obtained with conventional Hamiltonians. These have been verified for two-photon absorption [15] and, recently, for atomic level shifts in laser fields [16]. The semi-classical equivalent of two-photon absorption was discussed in Henneberger's original paper [7]. The generalized transformation, unlike its semi-classical progenitor, is not, even in the electric dipole approximation, Lagrangian-induced. One way of confirming this is to check that the new field coordinate is

$$e^{-i(e/\hbar mc^2)\underset{\sim}{p}\cdot\underset{\sim}{Z}(\underset{\sim}{R})} \; A_{\underset{\sim}{i}}(\underset{\sim}{r}) \; e^{(ie/\hbar mc^2)\underset{\sim}{p}\cdot\underset{\sim}{Z}(\underset{\sim}{R})}$$

$$= A_{\underset{\sim}{i}}(\underset{\sim}{r}) + \frac{e p_j}{mc} \int \frac{\delta\frac{1}{ij}(\underset{\sim}{r}' - \underset{\sim}{R})}{|\underset{\sim}{r}' - \underset{\sim}{R}|} \, dV' \tag{121}$$

and involves the particle momentum. This property is essential for the mass-renormalization programme for which it was first developed. In fact the field-independent term arising from the transformation of $H^{(1)}_{int} = e\underset{\sim}{p}\cdot\underset{\sim}{A}/mc$ is

$$\frac{ie}{8\pi\hbar m^2 c^3} \, p_i p_j \int \frac{1}{|\underset{\sim}{r}' - \underset{\sim}{q}|} \, [E_i(\underset{\sim}{r}'), \, A_j(\underset{\sim}{q})] dV'$$

$$\simeq \frac{p^2}{2m} \left(- \frac{4}{3\pi} \frac{e^2}{\hbar c} \frac{\hbar}{mc} \int dK \right) = \frac{p^2}{2m} \cdot R \quad .$$

The formally divergent factor is the mass renormalization term of order e^2. We note the following points:

1)   The cancellation of the $H^{(1)}_{int} = e\underset{\sim}{p}\cdot\underset{\sim}{A}/mc$ term arises from the transformation of H_{Rad} . This is in contradistinction to the multipolar case where the cancellation arises from the kinetic energy.

2)   The $H^{(2)}_{int} = e^2 A^2/2mc^2$ term remains uncancelled and hence must be used in multiphoton process calculations involving the Henneberger Hamiltonian.

3)   The Baker-Hausdorff type series does not terminate. Only
in the electric dipole approximation does it simplify. For higher
multipoles the disentangling is cumbersome. It would be hopeless
to deal with multiphoton absorption, other than electric dipole
transitions, for more than two or three photons.

4)   Away from the electric dipole approximation the transformed
kinetic energy produces an H_{int} linear in e (in addition to the
quadratic term needed for mass-renormalization). The calculation
of photon absorption to include electric quadrupole or magnetic
dipole transitions requires this term in addition to the conventional momentum translation interaction energy to give correct
results.

The new Hamiltonian (showing explicitly the terms quadratic
in e for electric dipole and linear in e for electric quadrupole
and magnetic dipole) is

$$H^{NEW} = \frac{\underset{\sim}{p}^2}{2m_R} + V(\underset{\sim}{q}) + H_{Rad} + \frac{e}{mc^2}\, \underset{\sim}{Z} \cdot \underset{\sim}{\nabla}\, \underset{\sim}{V}$$

$$- \frac{e}{m^2 c^2} [p_i p_j - m q_j \nabla_i V] \nabla_j Z_i + \frac{\underset{\sim}{e}^2 \underset{\sim}{A}^2}{2mc^2}$$

$$+ \frac{e^2}{2mc^2} (\underset{\sim}{Z} \cdot \underset{\sim}{\nabla})^2 V + \ldots \tag{122}$$

where $m_R = m(1-R)$. As mentioned in 3) above, if we remain in the
electric dipole approximation there is the finite form

$$H^{NEW} = \frac{\underset{\sim}{p}^2}{2m_R} + V\left(\underset{\sim}{q} + \frac{e\underset{\sim}{Z}(\underset{\sim}{R})}{mc^2}\right) + \frac{e^2}{2mc^2}\, \underset{\sim}{A}^2 \ . \tag{123}$$

We illustrate the use of the new Hamiltonian for calculating some
elementary processes. First the calculation of single-photon absorption for an electric quadrupole transition $a \to b$. The matrix
element using $-\underset{\approx}{Q} : \underset{\sim}{\nabla}\, \underset{\sim}{E}^\perp(\underset{\sim}{R})$ of Power and Zienau [12] gives immediately

$$- \langle b| \frac{e}{2} q_i q_j |a\rangle \sqrt{\frac{2\pi\hbar\omega n}{V}}\, e_i^{(\lambda)}\, k_j \tag{124}$$

with rate (for random oriented quadrupole)

$$\Gamma = B^{quad.}\, I, \qquad B^{quad.} = \frac{4\pi}{5\hbar^2} |Q|^2 k^2 \tag{125}$$

We note the extra k in Eq. (124) compared with the electric di-
pole, leading to k^2 in (125), c.f. $(2\pi/3\hbar^2)|\underset{\sim}{\mu}|^2$. With the
Hamiltonian (122) the matrix element is

$$<b\left|\frac{e}{m^2 c^2}\, p_i p_j\right|a> \qquad \frac{2\pi\hbar cn}{Vk^3}\ e_i^{(\lambda)}\ k_j \tag{126}$$

where we have the $k^{-3/2}$ to be compensated with having quadrupole
(velocity)^2 form rather than quadrupole (length)^2 form. We have
(the ½ is interesting!),

$$<b\left|\frac{p_i p_j}{m^2 c^2}\right|a> = -\ \frac{E_{ba}E_{ba}}{\hbar^2 c^2}\ <b\left|\tfrac{1}{2}\, q_i q_j\right|a> \tag{127}$$

in analogy with

$$<1\left|\frac{p_i}{mc}\right|0> = \frac{iE_{10}}{\hbar c}\ <1\left|q_i\right|0>.$$

On the energy shell we have $E_{ba} \simeq \hbar ck$ and Eq. (126) is identical
with Eq. (124). The magnetic dipole term arises from the antisymmetric part of $\nabla_j Z_i$ in Eq. (122). We have evaluated two-
photon absorption up to electric quadrupole and magnetic dipole
using the Henneberger Hamiltonian and verified it gives the same
as the multipolar calculation. Since this is in the literature
[15] for a second calculation here let us consider 3-photon
electric dipole absorption and tie it in with calculations in
the previous section.

Considering point 2) above, since the $e^2A^2/2mc^2$ term remains
in the new Hamiltonian, one cannot expect the matrix element
(114) from the terms in Z alone. One can show that the $\underset{\sim}{Z}$ terms
(Figs. 3(a), (d), (e), ($\tilde{f}$)) give (115). The mixed graphs (Figs.
3(b), (c)), those with two-photon $\underset{\sim}{A}^2$ and single-photon Z vertices, add as before to give (116). This breakdown of the total
is rather surprising at first sight because the graphs 3(b), (c)
differ from graphs 2(b), (c) in the single-photon vertex just by
the factor (E/$\hbar\omega$) and there is no energy conservation at what
are now virtual transitions. However, the difference between
the contributions 3(b), (c) and 2(b), (c) is

$$\sum_{1,2} \mu^{f2}\mu^{21}\mu^{10}\left(\frac{2\pi}{V\hbar\omega}\right)^{3/2}\frac{1}{2\hbar\omega}\left[-\frac{(E_{f2}-\hbar\omega)E_{f2}(E_{20}-2E_{10})}{(2\hbar\omega-E_{20})}\right.$$

$$\left.+\frac{(E_{f2}-E_{20})E_{10}(E_{10}-\hbar\omega)}{(\hbar\omega-E_{10})}\right]$$

$$=\sum_{1,2}\mu^{f2}\mu^{21}\mu^{10}\left(\frac{2\pi}{V\hbar\omega}\right)^{3/2}\frac{1}{2\hbar\omega}\left[3\hbar\omega(D_1-D_2)-(D_1^2+D_2^2-4D_1D_2)\right]$$

$$=0$$

because of the sum rule (119) and

$$\sum_{1,2}\mu^{f2}\mu^{21}\mu^{10}\{E_{f2}E_{21}-2E_{f2}E_{20}+E_{21}E_{10}\}=0$$

a consequence of $p[p,q]-[p,q]p=0$. To complete the calculation
we must show that Fig. 3(a), (d), (e), (f) contribute (115).
They are

$$\sum_{1,2}\mu^{f2}\mu^{21}\mu^{10}\left(\frac{2\pi}{V\hbar\omega}\right)^{3/2}\frac{1}{(\hbar\omega)^2}E_{f2}E_{21}E_{10}\left\{\frac{E_{f2}E_{21}E_{10}}{(\hbar\omega-E_{10})(2\hbar\omega-E_{20})}\right.$$

$$\left.+\frac{E_{f2}(E_2-2E_1+E_0)}{2(2\hbar\omega-E_{20})}+\frac{(E_f-2E_2+E_1)E_{10}}{2(\hbar\omega-E_{10})}+\frac{E_f-3E_2+3E_1-E_0}{6}\right\}$$

which can be simplified using (118) to

$$\sum_{1,2}\mu^{f2}\mu^{21}\mu^{10}\left(\frac{2\pi}{V\hbar\omega}\right)^{3/2}\frac{1}{(\hbar\omega)^3}E_{f2}E_{21}E_{10}\left\{\frac{(\hbar\omega)^3}{D_1D_2}\right\} \tag{115}$$

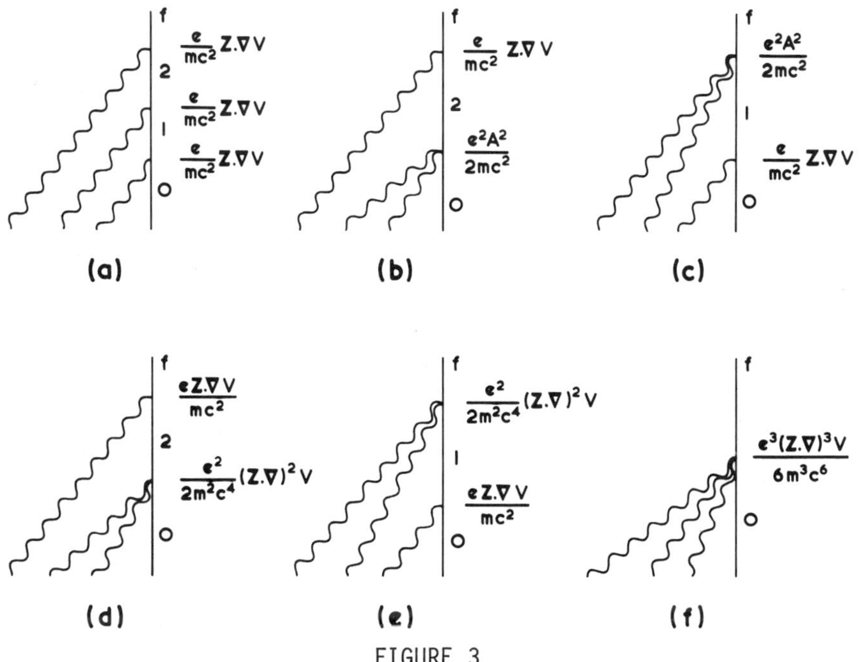

FIGURE 3

As a third, and last, calculation using the generalized
Henneberger transformation I will mention the minor controversy
over the energy-level shifts of an atom in the presence of an
electromagnetic field as calculated with the Henneberger
Hamiltonian. There were several papers [16] claiming that the
shifts predicted by this method were only valid at high frequency.
Lambropoulos, Power and Thirunamachandran have shown [17] that
this is not so. The dressing due to emission and absorption of
photons must be considered in addition to the D.C. term (leading
term is quadratic in Z) in the new Hamiltonian. When the contributions of both uncrossed (Fig. 4(a)) and crossed (Fig. 4(b))
graphs are added to the direct shift (remember $H_{int}^{(2)}$ NEW includes
A^2 and Z^2) the total, up to terms linear in intensity, is that
proportional to the dynamic polarizability for the relevant state
$|n>$. That is to say, the same as that predicted using $-\mu \cdot E^{\perp}$ interaction. It is clear that if higher intensity contributions
are considered too – leading to light shifts, etc. – the result
of careful use of the space-translated Hamiltonian will give the
same answer as using the conventional one.

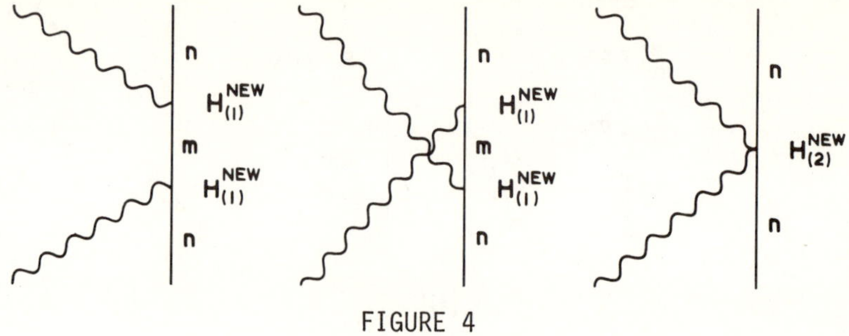

FIGURE 4

C. Remark

Certain approximations used in model calculations do not always commute with the canonical transformation considered.

$$H \longrightarrow H^{NEW}$$
$$\downarrow \qquad\qquad\qquad\qquad \downarrow$$
$$H^{Approx} \longrightarrow (H^{Approx})^{NEW} / (H^{NEW})^{Approx}$$

An example is the two-level atom model with the minimal-multipolar transformation.

(i) External field case

$$H^{Approx} = \begin{bmatrix} E_1 & -\mu E(t) \\ -\mu E(t) & E_2 \end{bmatrix} = \frac{E_1+E_2}{2} + \frac{E_1-E_2}{2}\,\sigma_z - \mu E(t)\sigma_x \tag{128}$$

Now

$$(H^{Approx})_{min} = e^{(i\mu A/\hbar c)\sigma_x}\,H^{Approx}\,e^{-(i\mu A/\hbar c)\sigma_x} - \frac{\mu \dot{A}}{c}\,\sigma_x \tag{129}$$

and

$$e^{(i\mu A/\hbar c)\sigma_x}(-\mu E\sigma_x)e^{-(i\mu A/\hbar c)\sigma_x} = -\mu E\sigma_x \tag{130}$$

which cancels $-\hbar\,(\partial S/\partial t)$. But the remainder is

$$\frac{E_1 - E_2}{2}\cos\left(\frac{2\mu A}{\hbar c}\right)\sigma_z + \frac{E_2 - E_1}{2}\sin\left(\frac{2\mu A}{\hbar c}\right)\sigma_y \tag{131}$$

and only in *lowest* power of μ is it the transformed Hamiltonian

$$
\begin{bmatrix}
E_1 & -i\mu\,\dfrac{(E_2 - E_1)A(t)}{\hbar c} \\[2ex]
i\mu\,\dfrac{(E_2 - E_1)A(t)}{\hbar c} & E_2
\end{bmatrix}
\tag{132}
$$

which one expects from $(H_{min})^{Approx}$.

(ii) The same sort of difference occurs with the quantized electromagnetic field interaction with 2-level system. Essentially we add H_{Rad} to H^{Approx} and interpret E and A as quantized fields according to previous prescription. Two examples where difference is seen:

      (α) Energy shift in external field case
      (β) 3-photon absorption in quantized field case.

(α)   Since the Hamiltonian is explicitly dependent on time we must use a definition of energy shifts other than the spectrum of H. This has been shown by Langhoff, Epstein and Karplus [*18*] to be given by

$$
\delta E_n \; = \; \frac{\overline{(\Phi_n(t), (H(t) - H_o)\Psi_n(t))}}{(\Phi_n(t), \Psi_n(t))}
$$

$$
\Psi_n(t) \; = \; P\; e^{-i\int^t H(t')dt'}\;\psi_n
\tag{133}
$$

$$
\Phi_n(t) \; = \; e^{-i\int^t_o H_o dt'}\;\psi_n \quad .
$$

To the lowest order, for $\varepsilon\cos(\omega t + \phi)$, with $\Delta E = (E_2 - E_1)$

$$
\delta E_1 \; = \frac{\mu^2\varepsilon^2\,\Delta E}{2((\hbar\omega)^2 - \Delta E^2)} \quad .
\tag{134}
$$

Using $(H_{min})^{Approx}$

$$
\delta E_1 \; = \frac{\mu^2\varepsilon^2\,\Delta E}{2((\hbar\omega)^2 - \Delta E^2)}\left(\frac{\Delta E}{\hbar\omega}\right)^2
\tag{135}
$$

and we must add the correction term from Eq. (129), namely $\mu^2 \varepsilon^2 \Delta E / 2(\hbar\omega)^2$ to get the same as for $-\mu E$.

(β)   The matrix element for 3-photon absorption using $-\mu \cdot E$ proceeds as for Eq. (114) but without the sums.

$$M_{if} = i\mu^3 \left(\frac{2\pi\hbar\omega}{V}\right)^{3/2} \frac{1}{\hbar\omega - \Delta E} \cdot \frac{1}{2\hbar\omega} \times (i.c.f.). \tag{136}$$

On the other hand with $e\underset{\sim}{p} \cdot \underset{\sim}{A}/mc$,

$$-i\mu^3 \left(\frac{2\pi\hbar\omega}{V}\right)^{3/2} \left(\frac{\Delta E}{\hbar\omega}\right)^3 \frac{1}{\hbar\omega - \Delta E} \cdot \frac{1}{2\hbar\omega} \times (i.c.f.) \tag{137}$$

which is (-27) times M_{if}. There are however, in Eq. (129), both quadratic and cubic terms in A (see Fig. 5). These contribute $28\, M_{if}$ and the final rate is that for $-\mu \cdot E(t)$.

(iii) More subtle is the approximation to the given field Hamiltonian from the quantum electrodynamical transformations:

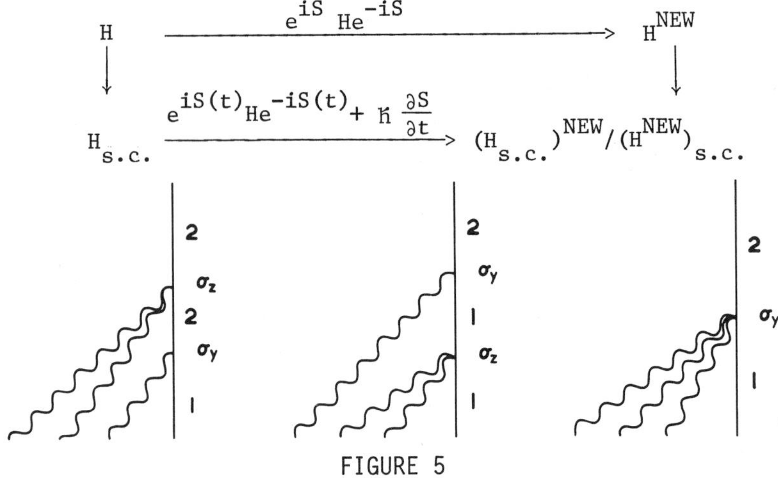

FIGURE 5

This difference becomes important with several atoms [19]. An amusing exercise is to examine the three possible $H_{s.c}^{NEW}$ for identity and difference (2 are the same, one different!) in the following scheme

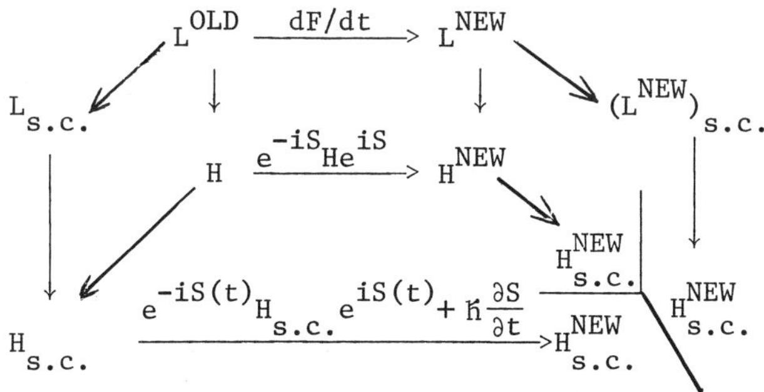

The assymetry position of the ambiguity in the diagram is because the prescribed motion is that of the field, and $L_{s.c.}$ is Eq. (12), not Eq. (13).

ACKNOWLEDGMENTS

It is with great pleasure that I acknowledge the collaboration with T. Thirunamachandran and thank him for many discussions. I also wish to thank many friends and colleagues for helpful comments and conversations at various times. These include M. Babiker, G. Barton, D. P. Craig, S. Geltman, W. P. Healy, W. C. Henneberger, P. Lambropoulos, M. O. Scully, R. G. Woolley and, my first co-worker on this topic, the late Sigurd Zienau. The financial support of the S.R.C. is also welcome.

REFERENCES

1.   S. Geltman, J. Phys. B10, 831 (1977). See also; M. D. Burrows and W. R. Salzman, Phys. Rev. 15, 1636 (1977).

2.   R. M. Santilli, Annals Phys. 103, 354 (1977).

3.   M. Göppert-Mayer, Ann. Phys. Lpz., 9, 273 (1931); P. I. Richards, Phys. Rev. 73, 254 (1948).

4.   J. Fiutak, Canad. J. Phys. 41, 12 (1963).

5.   M. Babiker, E. A. Power, and T. Thirunamachandran, Proc. Roy. Soc., London, A338, 235 (1974).

6.   A. Wintner, "The Analytical Foundations of Celestial Mechanics", (Princeton University Press, 1947) p. 26.

7.   W. C. Henneberger, Phys. Rev. Letts. $\underline{21}$, 838 (1968).

8.   E. N. Glass, J. J. G. Scanio, Am. J. Phys. $\underline{45}$, 344 (1977).

9.   C. Cohen-Tannoudji, et al., Phys. Rev. $\underline{48}$, 2747 (1973);
     M. Sargent, M. O. Scully, and W. E. Lamb, "Laser Physics"
     (Addison-Wesley, MA. 1974) p. 15.

10. J. J. Forney, A. Quattropani, and F. Bassani, Nuovo Cimento
     $\underline{37B}$, 78 (1977).

11. S. K. Vermani and B. L. Beers, Phys. Rev. $\underline{A12}$, 715 (1975).

12. E. A. Power and S. Zienau, Phil. Trans. Roy. Soc. $\underline{A251}$,
     427 (1959).

13. W. C. Henneberger, Nucl. Phys. $\underline{B23}$, 365 (1970); E. A. Power
     and T. Thirunamachandran, J. Phys. $\underline{B8}$, L167 (1975).

14. S. Geltman, Phys. Lett. $\underline{4}$, 168 (1963).

15. E. A. Power and T. Thirunamachandran, J. Phys. $\underline{B8}$, L170
     (1975).

16. R. F. O'Connell, Phys. Rev. $\underline{A12}$, 1132 (1975); G. W. Ford,
     and R. F. O'Connell, Phys. Rev. $\underline{A13}$, 1281 (1976).

17. P. Lambropoulos, E. A. Power and T. Thirunamachandran, Phys.
     Rev. $\underline{A14}$, 1910 (1976); P. L. Knight, Phys. Lett. $\underline{60A}$, 182
     (1977).

18. P. W. Langhoff, S. T. Epstein and M. Karplus, Rev. Mod.
     Phys. $\underline{44}$, 602 (1972).

19. See papers by Eberly and Wodkiewicz and by Yamanoi and
     Takatsuji in Proceedings of 4th Rochester Conference on
     Coherence and Quantum Optics (Plenum, New York, 1978).

Intense Fields and Quantum Electrodynamics

H. MITTER
Institut für Theoretische Physik der Universität
Graz, Austria

I.   INTRODUCTION

Quantum electrodynamics (QED) deals with electromagnetic interaction processes of particles. If these processes take place
in the presence of an intense electromagnetic wave field (such
as produced by a laser), the particles will be influenced by the
laser wave: Massive particles interact with the laser via their
charge and magnetic moment, non-laser-photons are coupled via
virtual electron-positron pairs to the laser wave. As a consequence, absorption and emission of laser quanta can take place
both in the initial and final state as well as in virtual intermediate states of any interaction process described by QED.
Thus the probability for these processes will be changed in the
presence of the laser field: there may be intensity-dependent
corrections, which can be viewed as a stimulation; eventually a
process will even show a resonant behaviour. The theory of these
intensity-dependent corrections has been developed in the past
decade in some detail and it is fair to say, that the basic facts
are now well understood. There has been little (if any) experimental effort so far to verify the theory by measurements, in
spite of the fact that the situation is not quite hopeless with
strong lasers. The present paper is intended as a short review
of theoretical results without many details. Formal developments
are contained in another review [1]. For details on individual
processes the reader is referred to the literature. The list
given at the end of this paper is not exhaustive, especially
with respect to older papers. A rather complete list of these
can be found in Ref. [2].

*
Supported in part by Fonds zur Förderung der wissenschaftlichen
in Österreich, Proj. No. 3225.

II. PARTICLES IN LASER FIELDS

Quantum electrodynamics in intense fields (IFQED) is based on a semiclassical approximation scheme. All particles, whose interaction processes are studied (e.g. electrons, non-laser-photons) are described according to QED, i.e. by quantized fields. All interactions except those with the laser wave are studied in the framework of perturbation theory (expansion in powers of the fine structure constant α). The interaction with the laser is treated exactly (i.e. an arbitrary number of laser quanta may be involved). The laser field is, however, described as an external (prescribed, classical) plane wave field.

It is this latter approximation, on which some comments must be made. On the basis of physical arguments the approximation seems quite well founded. Even a short laser pulse contains a large number of photons, if the intensity is high enough. Thus a classical description seems appropriate. The validity of the external field description can, however, be justified even in the framework of the fully quantized theory [3]. The use of plane waves may be less convincing, since one would like to focus the laser beam in order to obtain maximum intensity. The only argument at hand is, that the only electromagnetic scale of a particle is its Compton wavelength, in terms of which even the field in the focus of a laser beam looks rather "undeformed". With similar arguments one may justify the use of infinitely extended wave trains instead of pulses of finite duration.

A plane wave is characterized by a propagation vector[*]

$$k^\mu = \omega(1,\vec{n}), \qquad \omega = \frac{2\pi}{\lambda} \quad (\lambda \text{ is the wavelength}),$$

$$k^2 = 0 ,$$

and two mutually orthogonal polarization vectors

$$e_i^\mu, \quad i = 1,2; \qquad e_i \cdot e_j = -\delta_{ij}; \qquad k.e_i = 0.$$

The most general ansatz for the corresponding vector potential reads

$$A^\mu(x) = a \sum_{i=1}^{2} e_i^\mu \, a_i(\xi), \qquad \xi = k \cdot x. \tag{1}$$

---

[*]We use four-vector notation with

$$a.b = a_o b_o - (\vec{a}.\vec{b}), \qquad a^2 = a_o^2 - \vec{a}^2$$

corresponding to arbitrary (elliptic) polarization.   a is an
amplitude factor and the arbitrary functions a_i characterize the
frequency decomposition. For a monochromatic wave train of infinite extent and circular polarization we have, e.g.,

$$a_1 = \cos \xi, \qquad\qquad a_2 = -\sin \xi . \tag{2}$$

For a finite pulse the a_i's are zero outside a finite domain in
ξ and for polychromatic waves one would have to consider spectral
superpositions in ω. Most of the calculations have been done,
however, with infinitely extended trains and only one frequency.
Circular polarization is particularly attractive, since all formulae are considerably simplified due to additional symmetries.
     In order to describe the interaction of particles with the
laser field exactly, one must solve the corresponding relativistic
wave equation (Dirac- or Klein-Gordon equation) in the presence
of the vector potential (1). This has been done a long time
ago [4]. We shall not write down the solution in detail here,
but shall rather discuss its main features for the special case
of an infinite, monochromatic wave train, e.g. of the type (2).
The solution may , in this case, be written as an infinite sum
of plane waves with wave vectors

$$P_{(n)} = \tilde{p} - nk , \qquad n = 0, \pm1, \pm2, \ldots \tag{3}$$

where $\tilde{p}$ is an "effective" momentum of the particle in the field

$$\tilde{p} = p + \frac{\nu^2}{\rho} k , \tag{4}$$

p is the four-momentum free of a free particle

$$p^2 = \kappa^2, \; \kappa = \frac{mc}{\hbar} \;\text{ (inverse Compton wavelength) } \kappa = \frac{2\pi}{\lambda_c} \tag{5}$$

and the dimensionless parameters ν and ρ are characteristic of
the modification of the particle's properties due to the interaction with the wave field. They are given by

$$\nu = \frac{e\,a}{mc^2} \;\text{ (e is the charge of the particle)} \tag{6}$$

$$\rho = 2\,\frac{(pk)}{\kappa^2} \tag{7}$$

Thus the effective momentum depends on the intensity of the field
via the (classical) parameter ν. Thus the laser influences the
particle in two ways:

(a)   the momentum p is replaced by $\tilde{p}$, where we have

$$\tilde{p}^2 = \tilde{\kappa}^2 = \kappa^2 \cdot (1 + \nu^2) \qquad \text{(intensity-dependent} \atop \text{mass shift)} \qquad (8)$$

(the shift is in general polarization-dependent; for linear polarization we would obtain, e.g. $\nu^2/2$ instead of ν^2).

(b)   the particle may absorb or emit an arbitrary number n of laser momenta (quanta) k.

This connection is formulated usually in terms of the concept of quasi-levels [5]: an electron with momentum (3) occupies a quasi-level characterized by an integer n, which has the dispersion law

$$(\tilde{p}_n + nk)^2 = \tilde{\kappa}^2$$

and the wave function gives a coherent superposition of these quasi-levels (which are, however, not eigenstates of the Hamiltonian).

Since this concept is quite useful for understanding the physics of IFQED, it seems appropriate to comment on its limitations. If the wave train is of infinite extent, the particle is never separated from the laser beam, which is not a very realistic situation. One may, however, study pulses of finite extent in ξ and work with Volkov wave packets, at least in simple cases [6]. The conclusions reached with ordinary Volkov solutions remain essentially true and refer to the maximum of the momentum distribution of the packet. The limitations due to finite spatial extension of the laser wave have not been studied thoroughly. Some consequences of limitations due to non-monochromacy will be mentioned later in connection with resonances. If self-energy corrections are taken into account [7,8], the effective mass in Eq. (8) becomes complex. The real part of the correction corresponds to a shift, the imaginary part to a finite width. Both quantities are proportional to α and therefore very small.

The parameters ν and ρ are the only gauge- and Lorentz-invariant combinations, which can be constructed from the vector potential (1) (for an infinitely extended wave) and the particle momentum. Together with α, they control the order of magnitude of all effects in IFQED, as long as one sticks to (1). In terms of the laser wavelength λ (cm) and the illumination density $I(W/cm^2)$, we have (for electrons)

$$\nu^2 \simeq 7.5 \times 10^{-11} \lambda^2 I$$

Thus one should be able to reach values near or above one with
very strong lasers. The other parameter is very small in general.
The maximum value is obtained, if the particle momentum and the
laser propagation vector are antiparallel

$$\rho \sim 4 \; \frac{\lambda_\tau}{\lambda} \; \frac{|\vec{p}|}{\kappa}$$

For a neodymium-glass laser and electrons we need a particle
energy of 50 GeV in order to reach $\rho \sim 1$. Here intense X-ray-
or gamma lasers would improve the situation considerably.

Let us now look briefly to what extent the propagation of
non-laser photons is modified by the intense external field. QED
teaches us, that the vacuum is not empty: there are always virtual
particle-antiparticle pairs. Any external field will polarize
these pairs, and if a photon propagates through the medium
"vacuum + external field", it will experience a dielectric medium.
In order to learn what actually happens, one has to compute the
vacuum polarization current induced by the external field and to
solve Maxwell's equations with this current as a source term.
This has been done approximately with the lowest order ($\sim \alpha$)
polarization current [9] induced by a laser wave. For circular
laser polarization the result can be described approximately by
two complex indices of refraction. The real parts are extremely
small for conventional lasers, so that the vacuum is indeed very
transparent. The imaginary parts correspond to pair creation and
are appreciable only for extremely high energies of the non-laser
photon. These features are due to the parameter ρ, which enters
vitally. In order to observe pair production from the vacuum
in collisions between a γ-ray and an intense laser ray one has to
go to γ-energies of 20 GeV and more. Nice dispersive effects
would show up at energies an order of magnitude higher. Details
can be found in Refs. [9] and [10]. For lower energies one may
almost always neglect these phenomena and treat non-laser photons
as particles with $p^2 = 0$.

III. GENERAL FEATURES OF CROSS-SECTIONS

From the concept of quasi-levels it seems reasonable to extend the calculational tools of QED to IFQED and this is indeed
possible, i.e., one may adapt the concept of Feynman graphs to
this situation. This can even be done in two different ways.
Either the graphs are considered in configuration space: then we
have to replace the free wave functions for the ingoing and out-
going particles by Volkov functions and the propagators by corresponding ones in the laser field (which can be calculated; the

results are complicated). Or we read the graphs in momentum
space: then we can use the conventional substitutions for the
particle lines, but have to modify the vertices appropriately
(which is not simple either). Details on each of these formulations are contained in Ref. [1]. The cross-sections calculated
in this way have several aspects in common.

The first one is connected with kinematics. The kinematic
of any process is expressed by (modified) energy-momentum conservation laws, which read

$$\Sigma \ (\tilde{p}^{(i)} - \tilde{p}^{(f)}) - rk = 0, \qquad r = 0, \pm 1, \pm 2, \ldots \tag{9}$$

Here the sum extends on all quasimomenta of the particles in the
initial (i) and final (f) state (for non-laser photons one has to
take the ordinary momenta in lowest order). These momenta have
to be taken on the mass-shell: this means we have to fulfill
Eq. (8) for massive particles and $p^2 = 0$ for non-laser photons.
The integer r denotes the net number of laser momenta taken from
the external field. Thus the external field redistributes energies and momenta of the particles in a characteristic way, which
depends on the intensity, frequency and polarization of the laser
wave. It has to be observed, that the kinematical relations for
the energies and momenta can become quite complicated, since
these quantities occur also in the denominator via ρ.

Another common feature is, that the cross-sections always
take the form of an infinite sum of incoherent contributions

$$d\sigma = \sum_{r=-\infty}^{+\infty} d\sigma_r \tag{10}$$

Every term corresponds to one of the conservation equations (9).
The terms with $r \neq 0$ can be envisaged as higher harmonic analogies
to the lowest one. Since r appears also in the conservation equations, there are (for given energies and momenta) kinematical re-
strictions on the possible harmonic terms. The individual contributions $d\sigma_r$ are in general very complicated and involve infinite sums of products of Bessel functions, which must be estimated or computed numerically. The consistency of the external
field description can be checked to some extent from (10). To
describe the laser by an external field means to neglect the reaction of the particles on that field. Thus the laser beam should
not be depleted considerably during the interaction process
(clearly one should not use this theory for laser-plasma interactions!). This means, that the sum in (10) must converge rapidly
enough, since very high values of r (order of magnitude; number

of laser quanta in the beam) would contradict our original external field description. In practical cases this r-convergence
is usually found; from certain values of r on the terms become
extremely small. It may, however, happen that several hundred
terms contribute significantly.

As a last general feature we have to mention the appearance
of resonances. Whenever the transition from the initial to the
final state can proceed via an intermediate state, which corresponds to a real particle, the cross-section becomes infinite for
the corresponding values of the quasi-momenta. That is, we
should observe a resonance, if the square of

$$q = \Sigma\ \tilde{p}^{(i)} - nk = \Sigma\ \tilde{p}^{(f)} + (r{-}n)k, \quad n = 0, \pm 1, \pm 2, \ldots$$

can assume "on-shell" values (e.g. κ^2 or zero). The infinities
disappear, if self-energy- and/or vacuum polarization effects
are taken into account; the resonances acquire then a finite
width $\sim \alpha$ and are also slightly shifted. At first this resonance
pattern seems to be quite attractive for a possible experimental
verification. The resonance region is, however, not always easy
to reach and it can happen, that the resonances lie close to each
other. In addition, there can be an additional radiative width
due to non-monochromaticity. The time for switching the field
on and off may also play an essential role in this connection.
Calculations [11] have shown that the radiative width may be much
larger than the corresponding one due to QED-type radiative corrections. The resonance features will therefore probably suffer
from the limitations of the present formalism of IFQED.

IV. SOME INDIVIDUAL PROCESSES

The simplest nontrivial process is the emission or absorption of a non-laser quantum by an electron travelling in the
laser field.

$$e + L \leftrightarrow e + L + \gamma$$

(L stands for the laser beam), usually referred to as high-intensity Compton scattering. The kinematics of the process is rather
simple; the ordinary Compton formula (which results, if we use
Eq. (9) with $r = 1$ and put all particles on their free mass shell)
is replaced by a modified one displaying an intensity-dependent
frequency shift. The calculation of $d\sigma_r$ is also not too complicated in this case, since the sums over the Bessel functions can
be performed exactly. Since the details of the Compton process

54   H. Mitter

as well as the related one of pair production or annihilation in
the vacuum

$$L + \gamma \leftrightarrow e + \bar{e} + L$$

are well-known for a long time, they will not be repeated here.
Some recent results are, however, available on Møller scattering

$$2e + L \rightarrow 2e + L \; .$$

The process has already been studied long ago [12], the resonance
structure has been described and renormalization has been performed. It has to be noted, that the kinematics is much more
complicated than for ordinary Møller scattering. In the latter
case there are quadratic relations between the energies of the
particles before and after the interaction and one may use the
center-of-mass system, in which these energies are equal. In the
presence of the laser the relation involves fourth powers of the
energies and depends also on r; the cm energies of the particles
before and after the collision are not equal. In addition, the
formulae for the partial cross-sections $d\sigma_r$ are quite formidable.
In Ref. [12] an auxiliary reference system has been chosen, in
which the cross sections can be evaluated as far as possible.
The disadvantage is that the transformation to a physical reference system can be done only approximately and it is hardly possible to make statements about the kinematics. Another policy is
chosen in a recent approach [13]; the calculations are carried
out in the cm-system of the incoming particles (this is the system which would correspond to the actual situation met in storage
rings). The cross-sections are calculated numerically, which is
a difficult task. The kinematical situation can be characterized
by the angles

ψ laser beam relative to incoming particle beams

ϕ = " =   = " =   one of scattered particles

δ scattering angle

and the energies before and after the collision. An interesting
parameter in this respect is the relative loss in kinetic energy
of one particle

$$Z = \frac{E'_{1K} - E_K}{E_K}$$

For nonrelativistic particle energies and r = 0 this parameter
varies between

0 $(\phi = 0)$ and ν^2 $(\phi = \pi/2)$ for $\psi = 0$ (practically: $\psi < 5^{\circ}$)

$-\dfrac{\nu^2}{2}$ $(\phi = 0)$ and 0 $(\phi = \pi/2)$ for $\psi = \pi/2$.

For higher energies the maximum is shifted towards smaller angles ϕ. Thus there may be large effects for $\nu^2 \simeq 1$, which should be found by observing scattered electrons in different directions. In the low energy region the change is entirely due to the mass shift, which could thus be measured indirectly. For $r \neq 0$ the energy change is expected to be still higher, since additional laser momenta are involved. Since the laser energy is small compared to the rest energy of the particle, this effect can play a role only in the nonrelativistic region; for higher electron energies one would need very high r's in order to change the kinetic energy appreciably. If now $d\sigma_r$ is of the same order of magnitude as $d\sigma_0$ for some r-values, one will measure for the same angles a mixture of different electron energies $E'(r)$ with probabilities $d\sigma_r/d\Omega$ and will thus observe a "smearing" of the energy. If $Z(r)$ is approximately the same for a whole range of r, the differential cross-section for given parameters will be larger than $d\sigma_0$ alone and this will facilitate the measurement. Numerical calculations have shown that the effective maximum of Z may increase considerably in the nonrelativistic region, especially for larger values of ν, if all r values are included, which contribute appreciably to the cross section.

Unfortunately, such a study is not (yet) available for the stimulated Compton effect

$$\gamma + e + L \rightarrow \gamma + e + L$$

for which results are only known [7] in the auxiliary reference frame and for the special case, in which the primary γ is parallel to the laser [13a].

V. OTHER PARTICLES

The modification of the mass shell by the laser wave has also consequences for weak decay processes such as

$$\mu \rightarrow e + \nu + \bar{\nu}, \; e \rightarrow e + \nu + \bar{\nu}, \; \pi \rightarrow \mu + \nu, \; \pi \rightarrow \pi^{o} + e + \nu \; .$$

The probabilities for these processes are substantially changed in the presence of the laser for high intensity ($\nu \sim 0.25 - 2$) and very high particle energies ($\rho \gtrsim 1$). For instance, the probability for μ decay is about four times larger for $\nu \sim 0.5$, $\rho \sim 1$.

Details are found in Refs. [14]. Another modification is introduced, if we consider particles with anomalous magnetic moments
(e.g. nucleons). In this case the interaction of the anomalous
moment μ with the laser causes a splitting of the quasilevels
into triplets with very small separation $\sim(\nu\mu)^2$. This has consequences for high-intensity Compton scattering of these particles:
in addition to the higher harmonics appearing for normal particles
there should be a "zeroth" harmonic of very low (radio) frequency
and there is also some characteristic pattern in the angular distribution of the other harmonics. Since the magnetic interaction is very weak in general, it will be very hard to observe
these effects. A careful study of the experimental situation
seems desirable. Details on the theory are contained in Refs.
[15].

VI.   STANDING WAVE FIELDS

The generalization of the theory for several different laser
waves is quite complicated. There is one situation which has
found interest; if a laser wave is reflected from a mirror, we
may obtain a standing wave. A particle should thus see a lattice
structure and we should obtain a Bragg-type diffraction pattern
[16]. Since this review is devoted to relativistic quantum
mechanics, we shall neither comment on the description of the
effect by classical [17] or nonrelativistic quantum mechanics [18].
Attempts to solve the Dirac equation for this case [19] have not
provided explicit solutions. It is, however, possible to solve
the Klein-Gordon equation [20] for a spinless particle in a standing wave field made up from two beams propagating in opposite
directions and carrying (a) opposite or (b) equal polarization,
if the particle propagates only in the same direction as the
optical beams. This is due to additional symmetries, which result in the conservation of the particle's (a) energy or
(b) momentum component along the beam direction, respectively.
In case (a) the solutions show stability properties known
from parametrized oscillations in mechanics or Bloch waves in
solids; the energy has a band-type structure; between the allowed
bands the solution is not stable (i.e. not normalizable). Calculation of the transmission coefficient [21] for the whole
arrangement shows, that only particles with "correct" energies
can penetrate, i.e. we may in principle construct an "energy
selector". It is relatively easy to obtain very narrow bands:
the condition is $\nu \cdot \lambda/\lambda_c \gg 1$. The energy has then almost a
level structure

$$E_n \simeq mc^2 + (1 + n)\Delta E \qquad n = 0,1,2,\ldots$$

$$\Delta E = \nu.\hbar\omega \ .$$

This phenomenon is (in contrast to the effective mass for a particle in a propagating wave) essentially quantum mechanical in nature.

In case (b) we have an analogous situation for the momentum component along the beam direction. Since the solution contains positive and negative energies, one would expect pair production. It is possible to show that the unstable solutions correspond to charge density zero.

If the particle has spin, the solution is not easily obtained. One can, however, prove that unstable solutions are completely ruled out in case (b). On the contrary there are stability zones for opposite polarization (a), which are calculated numerically at present.

An experimental investigation of this phenomenon is perhaps easier as far as the laser is concerned. The difficulties are connected with the electron beam. Theoretically, it remains to be seen to what extent the limitations of IFQED affect these results.

VII. MORE COMPLICATED FIELDS

It is interesting to consider a situation in which another external field is present. If this field is the Coulomb field of a nucleus, one has to apply a perturbative treatment, since exact solutions for the Dirac equation for Coulomb-plus-laser field are not available. Already in the earlier times of IFQED a number of processes such as stimulated Coulomb-scattering, pair creation, Bremsstrahlung have been investigated [22]. For an experimental investigation the dependence of the modified Coulomb scattering cross section on the polarization of the laser beam seems most promising (see Ref. [22] , last paper). In a recent paper [23], the Weizsäcker-Williams method has been applied to stimulated Bremsstrahlung, whereby a resonance pattern has been found.

There is a case in which solutions in closed form may be obtained [24], namely, if we have a combination of a laser wave and a magnetic field parallel to it. The combination of the quasi-levels due to the laser wave and the Landau levels due to the magnetic field gives a structure which depends on the polarization of the laser and is band-like [25]. This has consequences for the corresponding Compton [26] and pair production process [25,27].

VIII. CONCLUSION

The effects of intense wave fields on particle processes, which involve in fact genuine multi-photon transitions, may – according to theoretical investigations – become quite large at high intensities $\nu^2 \simeq 1$. For all processes involving particles at low energies it is essentially the intensity-dependent mass shift, which causes the deviations from the corresponding formulae for low intensities and becomes manifest in the kinematical aspects of scattering processes. Since the quasi-level picture underlies almost all investigations made so far, it seems worthwhile to look both theoretically and experimentally, to what extent this picture can account for the physics happening in realistic laser beams.

The more specific nonlinear properties of the vacuum in QED (which involve possibility of pair production from the vacuum and dispersive properties) will, even at high intensities, show up only at very high energies of the probing (non-laser) photon. The measurement of certain resonance patterns would exhibit the self-energy structure of the particle, another typical aspect of QED. The corresponding experiments could in principle be done even at low energies, where the effective mass can be seen. It is, however, just here, where the limits of the concepts used at present put the theoretical predictions on unsafe grounds.

Thus, in spite of all the efforts, a lot remains to be done, until IFQED can compete with ordinary QED. both with respect to the safety of its basis and the degree of precision of its predictions.

REFERENCES

1.   H. Mitter, Acta Phys. Austr. Suppl. XIV, 397 (1975).

2.   J. H. Eberly, Progress in Optics, Vol. 7, p. 361 (1969).

3.   I. and Z. Białynicki-Birula, Phys. Rev. A8, 3147 (1973).

4.   D. V. Volkov, Zs. f. Phys. 94, 250 (1935).

5.   V. I. Ritus, JETP 51, 1544 (1966).

6.   R. Neville, F. Rohrlich, Phys. Rev. D3, 1692 (1970).

7.   V. P. Oleinik, JETP 53, 1997 (1967).

8.   V. N. Baier, V. M. Katkov, A. I. Mil'shtein, V. M. Strakhovenko, JETP 69, 783 (1975); W. Becker, H. Mitter, Journ. Phys. A9, 2171 (1976).

9. W. Becker, H. Mitter, Journ. Phys. A8, 1638 (1975); V. N. Baier, A. I. Mil'shtein, V. M. Strakhovenko, JETP 69, 1893 (1975).

10. D. A. Morozow, N. B. Naroshnyi, JETP 72, 44 (1977).

11. M. V. Fedorow, JETP 68, 1209 (1975).

12. V. P. Oleinik, JETP 52, 1049 (1967).

13. J. Bös, W. Brock, to be published.

13a. R. Gush, H. P. Gush, Phys. Rev. D12, 404 (1975).

14. A. I. Nikishow, V. I. Ritus, JETP 46, 1768 (1964); N. B. Narozhnyi, A. I. Nikishow, V. I. Ritus, JETP 47, 930 (1964); V. I. Ritus, JETP 56, 986 (1969); V. A. Lyul'ka, JETP 67, 1638 (1974); V. A. Lyul'ka, JETP 69, 800 (1975).

15. W. Becker, H. Mitter, Journ. Phys. A7, 1266 (1974); W. Becker, ibid. A8, 160 (1975); W. Becker, V. Koch, H. Mitter, ibid. A8, 1480 (1975).

16. P. L. Kapitza, P.A.M. Dirac, Proc. Camb. Phil. Soc. 29, 297 (1933).

17. F. Ehlotzky, C. Leubner, Journ. Phys. A8, (1975) and other papers quoted therein.

18. R. Gush, H. P. Gush, Phys. Rev. D3, 1712 (1971).

19. N. D. Sen-Gupta, Z. Phys. 200, 13 (1967).

20. I. Berson, J. Valdmanis, J. Math. Phys. 14, 1481 (1973).

21. The material presented in this section is based on calculations by R. Meckbach, W. Becker and H. Mitter, which will be published in due course.

22. V. P. Yakovlev, JETP 49, 318 (1965); F. V. Bunkin, M. V. Fedorow, ibid. 49, 1215 (1965); V. P. Yakovlev, ibid. 51, 411 (1966); M. M. Denisov, M. V. Dedorov, ibid. 53, 779 (1967); I. V. Lebedew, Opt. Spectrosc. 32, 59 (1970); F. Ehlotzky, Nuovo Com. 69B, 73 (1970).

23. A. V. Borisow, V. Ch. Zhukovskij, JETP 70, 477 (1976).

24. P. J. Redmond, Journ. Math. Phys. 6, 1163 (1965).

25. V. P. Oleinik, JETP 61, 27 (1971).

26. V. P. Oleinik, V. A. Sinyak, JETP 69, 94 (1975); I.M. Ternov, V. G. Bagrov, V. R. Khalilov, V. N. Rodionov, Yad. Fiz. 22, 1040 (1975).

27. V. N. Rodionov, I.M.Ternov,V.R.Khalilov,JETP 69, 1148 (1975).

On the Connection of Quantum Electrodynamics of Intensive Field with Quantum Electrodynamics at Small Distances

V. I. RITUS
P. N. Lebedev Physical Institute
Moscow, USSR

I. INTRODUCTION

As is well known the behaviour of quantum electrodynamics at small distances and the problem of its self-consistency are usually studied with the aid of the exact photon propagator (see classical works by Landau, Abrikosov and Khalatnikov [1] and by Gell-Mann and Low [2]). Proceeding from renormalizability of quantum electrodynamics Gell-Mann and Low have shown that in the region of a large momentum squared the photon propagator becomes a function of one variable: if the exact propagator $D = K^{-2}d$, the renorminvariant quantity $\alpha d(K^2/m^2,\alpha) \rightarrow \Phi_1(\phi(\alpha)K^2/m^2)$ - the contribution of all radiative corrections is reduced to the scale factor $\phi(\alpha)$ at the dynamical variable K^2/m^2. This means that the form of charge distribution of small distances from the centre of charge is α-independent.

For the self-consistency of quantum electrodynamics it is essential whether or not the limit of the function $\Phi_1(z)$ at $z \rightarrow \infty$ is finite. Indeed, the fact that $\Phi_1(z)$ tends to the constant α_* implies that for very high values of K^2 the exact photon propagator becomes free up to the constant factor α_*/α, which changes the fine structure constant α by the limiting interaction constant α_*. In this case the charge density at small distances is described by the function $\sqrt{4\pi\alpha_*}\,\delta(x)$, i.e. it corresponds to a finite point bare charge $e_* = \sqrt{4\pi\alpha_*}$. If at $z \rightarrow \infty, \Phi_1(z)$ tends to infinity, the charge density at small distances has a stronger singularity at zero than the δ-function, and this singularity has no definite physical meaning. The situation when $\Phi_1(z)$ has a pole at a finite z is still more contradictory. In this case the theory can "exist" at a zero physical charge only (see the papers by Landau and Pomeranchuk [3] and Fradkin [4]).

The most economical description of the situation is reached with the aid of the Gell-Mann-Low function $\psi_1(y)$, which is an invariant characteristic of the shape of the function $\Phi_1(z)$, i.e., does not depend on the multiplicative change $z \to tz$ of the last function argument. The finite limit α_* of the function $\Phi_1(z)$ at $z \to \infty$ corresponds to a rather strong zero $y = \alpha_*$ of the Gell-Mann-Low function $\psi_1(y)$. In a series of papers [5-8] Johnson, Baker and Willey have shown that the zero of the Gell-Mann-Low function $\psi_1(y)$ is also the zero of a simpler function $F_1(\alpha)$, which is the single-loop contribution into the first coefficient of the expansion of the inverse propagator d^{-1} in powers of $\ln(K^2/m^2)$ at $K^2/m^2 \to \infty$. Only three first expansion terms of these functions obtained in perturbation theory are known at present.

Along with the photon propagator the Lagrange function of an electromagnetic field plays an essential role in quantum electrodynamics. As is well known in classical (Maxwell) electrodynamics the Lagrange function is quadratic in field. The quantum (radiation) effects of interaction between electromagnetic field and charged particle field of vacuum lead to the appearance of essentially nonlinear terms in the Lagrange function.

The first radiative correction to the Lagrange function of a constant electromagnetic field was found by Heisenberg and Euler [9]. An essential contribution to the investigation of this correction was made by Weisskopf [10] and Schwinger [11]. Schwinger has formulated an elegant method of considering the problem of vacuum polarization by an external electromagnetic field.

The author has found [12] a radiative correction to the Lagrange function of a constant electromagnetic field following the Heisenberg-Euler correction and has shown that due to renormalization invariance the exact Lagrange function of an electromagnetic field that takes into account all the radiative corrections becomes a function of one variable in the asymptotical region of a strong field: if $\ell = L/L^{(0)}$ is the ratio of the exact Lagrange function to the Maxwell one, then $\alpha\ell^{-1} \to \Phi_2(\phi(\alpha)eF/m^2)$ with the same scale factor $\phi(\alpha)$ at the dynamical variable eF/m^2 as in the case of an exact photon propagator. This makes the Lagrange function of a strong field a convenient object of the investigation of questions of principle of quantum electrodynamics. In particular, the Lagrange function may serve as a source of information about the universal function of quantum electrodynamics $\phi(\alpha)$ that reflects the role of radiative corrections in extreme conditions when the going beyond the framework of perturbation theory is essential.

In the present paper it is shown that the functions $\Phi_1(z)$ and $\Phi_2(z)$ are different and cannot be made similar by multiplicative change of their arguments: $\Phi_1(z) \neq \Phi_2(tz)$, $t = $ constant. This results from the fact that the corresponding Gell-Mann-Low functions $\psi_a(y)$, $a = 1,2$, differ in the third terms of their expansions. If, however, at $z \to \infty$, $\Phi_1(z)$ tends to the constant α_*, $\Phi_2(z)$ tends to α_* too. Indeed, if for very large K^2 the photon propagator becomes free up to the factor α_*/α , then both the field equations and the Lagrange function generating them must differ from Maxwellian ones by the factor α/α_* only. The Maxwellianness must be here just in the region of a very strong field where, as is seen from the calculations, the distances of the order of $(eF)^{-1/2} \ll m^{-1}$, small in comparison with the Compton ones, are essential for the formation of the Lagrange function. To such quantities tending to α_* pertains also the bare constant α_o, which is considered as a function of α and of a dimensionless regularization parameter $x \to \infty$ (see Sec. III). The question of whether or not the constant α_* is finite can be investigated by one of the functions $\Phi_a(z)$ or by one of the Gell-Mann-Low functions, $\psi_a(y)$, which is more convenient for this purpose. An advantage of the Lagrange function is first of all the simplicity of its calculation, the fact that it is gauge-invariant at each stage and also the fact that the corresponding Gell-Mann-Low function $\psi_2(y)$ has a negative third-order term, which promotes the existence of zero of this function (as distinct from the Gell-Mann-Low function for photon propagator).

The structure of the paper and the main results are as follows. In Sec. II the method of calculating the Lagrange function of an electromagnetic field is developed on the example of scalar charge particle electrodynamics. A more detailed description of this method is given in the paper of the present author [12] devoted to the Lagrange function of an electromagnetic field in spinor electrodynamics. This time, however, we proceed from the closed functional connection between an exact action for an electromagnetic field that takes into account all the radiative corrections and the first approximation to it. Besides, scalar particle electrodynamics has some specific features as compared with spinor electrodynamics because of a complicated structure of the current (which is proportional to the momentum operator π_α and not to the c-number γ_α-matrix) and higher (linear in K^2 or in inverse proper time) divergences. In spite of this our method turns out to be rather flexible and compact for an adequate handling with these pecularities. The second-order correction in α to the Lagrange function, Z_3-factor in the corresponding approximation and the first two expansion terms of the Callan-Symanzik β-function are obtained.

In Sec. III the behaviour of the exact Lagrange function of an electromagnetic field is found for an extremely strong field with the aid of a renormalization group and different renorminvariant charges are compared in the asymptotical region of their dynamical variables. Along with the usual renorminvariant charge determined by the photon propagator the invariant charge determined

by the exact Lagrange function of a constant electromagnetic
field and the bare charge determined by the Z_3-factor are considered. In spinor electrodynamics for the two last charges the
Gell-Mann-Low functions $\psi_2(y)$ and ψ_3 are found in α^3-approxima-
tion and turn out to be different from one another and from the
Gell-Mann-Low function $\psi_1(y)$ for the first charge, i.e.,

$$\psi_a(y) = y \left\{ \frac{1}{3} \frac{y}{\pi} + \frac{1}{4}\left(\frac{y}{\pi}\right)^2 + \begin{bmatrix} 0.049991... \\ -0,079785... \\ -0,238631... \end{bmatrix} \left(\frac{y}{\pi}\right)^3 + ... \right\}, \quad a=1,2,3.$$

Their negative third terms promote the existence of zeros in
these functions which is necessary for self-consistency of quan-
tume electrodynamics.

Finally, in Sec. IV some integral transformations of renorminvariant charges are carried out which do not change either
boundary α or asymptotical α_* values of these charges but transform their Gell-Mann-Low functions. Integral transformations are
shown to be able to make negative the third term of the expansion
of the Gell-Mann-Low function.

II.   LAGRANGE FUNCTION OF AN ELECTROMAGNETIC FIELD

The Lagrange function of an electromagnetic field can be
calculated in two ways. One of them is to integrate the expression for action variation at the variation of the potential
$A_\mu(x)$ of the external field on the amount $\delta A_\mu(x)$:

$$\delta W = \int d^4x \; \delta A_\mu(x) <j_\mu(x)> , \tag{1}$$

in which $<j_\mu(x)>$ is the mean value of the current density induced in vacuum by external field. In scalar electrodynamics

$$<j_\mu(x)> = e(x|\pi_\mu G + G\pi_\mu|x), \qquad \pi_\mu = p_\mu - eA_\mu. \tag{2}$$

Using this expression for the current and the relation $e\delta A_\mu =
-\delta\pi_\mu$, one can write (1) in the form

$$\delta W = iTr \, \delta G_o^{-1} G , \tag{3}$$

where $G_o = -i(\pi^2 + m^2)^{-1}$ is the Green function of a particle in
an external field without radiative corrections, and Tr is a

symbol of diagonal summing over space-time coordinates. Thus, in the first approximation, when $G \approx G_o$, we have $W^{(1)} = -i\mathrm{Tr}\,\ln G_o$. This expression, as well as (3), differ in sign from the corresponding expression in spinor electrodynamics. Further radiative corrections to the action can be obtained with account taken of the following corrections to G at the integration of (1) or (3).

It is more convenient, however, to use a closed functional relation for the exact vacuum conservation amplitude in the presence of the field, see, e.g., ref. [13],

$$e^{iW[eA,\alpha]} = \exp\left[\frac{1}{2} \int d^4\xi d^4\xi' \; \frac{\delta}{\delta A_\alpha(\xi)} \; \mathcal{D}_o(\xi-\xi') \; \frac{\delta}{\delta A_\alpha(\xi')}\right] e^{iW^{(1)}[eA]},$$

$$(4)$$

in which $\mathcal{D}_o$ is the propagation function of photon in vacuum.

In this relation the universality of electromagnetic interaction is embodied: the amplitude $\exp iW^{(1)}[eA]$, taking into account the interaction of electrons with external field only, defines the exact amplitude $\exp iW[eA, \alpha]$, which takes into account also the mutual radiative interaction of electrons with the aid of quantum electromagnetic field. Both the interactions are characterized by one and the same charge e. Specifically for $W^{(2)}$ we obtain

$$W^{(2)} = ie^2 \int d^4x d^4x' \mathcal{D}_o(x-x')\{(x|\pi_\alpha G_o|x')(x'|\pi_\alpha G_o|x) +$$

$$+ (x|\pi_\alpha G_o \pi_\alpha|x')(x'|G_o|x) + 4i\delta(x-x')(x|G_o|x') +$$

$$+ (x|\pi_\alpha G_o|x)(x'|\pi_\alpha G_o|x')\}. \qquad (5)$$

For a constant and homogeneous field $F_{\mu\nu}$ the last term in (5) does not contribute to $W^{(2)}$ since the average current induced in vacuum by such a field is zero.

It is convenient to use hereafter the Fock-Schwinger [14,11] proper-time representation for the Green function of a scalar particle in a constant field

$$G_o(x,x') \equiv (x|G_o|x') = \frac{-ie^{if}}{(4\pi)^2} \int_o^\infty \frac{ds}{s^2} \, e^{-im^2 s - L(s) + iz\beta z/4}, \qquad (6)$$

in which

$$\beta = eF\,\mathrm{cth}\,eFs, \qquad L(s) = \frac{1}{2}\,\mathrm{tr}\,\ln\left(\frac{\mathrm{sh}\,eFs}{eFs}\right)$$

are matrix and scalar functions of the matrix $F_{\mu\nu}$, $z=x-x'$, and f is a non-diagonal phase of the Green function, equal to a line integral of the potential $eA_\mu(y)$ along the straight line joining the points x' and x.

It is easily seen that the matrix elements $(x|\pi_\alpha G_0|x')$, $(x|G_0\pi_\alpha|x')$ differ from (6) by the additional factor $(1/2)(\beta+eF)_{\alpha\beta}z_\beta$, $(1/2)(\beta-eF)_{\alpha\beta}z_\beta$ and the matrix element $(x|\pi_\alpha G_0 \pi_\alpha|x')$ can be represented in the form

$$(x|\pi_\alpha G_0 \pi_\alpha|x')=-i\delta(x-x')-m^2(x|G_0|x')+\tfrac{1}{2}e^2 zFFz(x|G_0|x') \qquad (7)$$

Taking into consideration the above remarks and using for $\mathcal{D}_0$ the proper-time representation

$$\mathcal{D}_0(z) = -i(4\pi)^{-2}\int_0^\infty dt\, t^{-2}\exp(iz^2/4t)$$

we obtain from (5) the following expression for $L^{(2)}$:

$$L^{(2)}=\frac{-e^2}{(4\pi)^6}\iiint_0^\infty \frac{dsds'dt}{(ss't)^2}\, e^{-im^2(s+s')-L-L'}\int d^4z\, e^{\,izAz/4}[-\tfrac{1}{4}z(\beta'\beta$$

$$-3e^2 FF)z-m^2]+\frac{3e^2}{(4\pi)^4}\iint_0^\infty \frac{dsdt}{(st)^2}\, e^{-im^2 s-L}\;. \qquad (8)$$

Here $A = \beta + \beta' + t^{-1}$ is a symmetric 4×4 matrix and β', L' are obtained from β, L by the substitution $s\to s'$.

Integration over z in the first term and further integration over t are carried out in the same way as in ref. [12]. In the second term the integral over t is regularized by introduction of a lower limit t_0 and the integral over s is regularized by subtraction of the field-independent term. As a result we get

$$L^{(2)}=\frac{-i\alpha}{64\pi^3}\iint_0^\infty \frac{dsds'\; e^{-im^2(s+s')}\; e^4\eta^2\varepsilon^2}{\sin e\eta s\;\; \sin e\eta s'\;\; {\rm sh}\,e\varepsilon s\;\; {\rm sh}\,e\varepsilon s'}\,(-m^2 I_0 - \tfrac{i}{2}\,I) -$$

$$-\frac{\alpha^2}{16\pi^2 im^2 t_0}\,L^{(0)}+\frac{3\alpha}{4\pi i t_0}\,\frac{\partial L_R^{(1)}}{\partial m^2}\;, \qquad (9)$$

where $L^{(0)}$ is a Lagrange function of the Maxwell field, $L_R^{(1)}$ is a nonlinear correction to it found by Schwinger [11]:

$$L^{(0)} = \frac{\varepsilon^2 - \eta^2}{2} \, , \quad L_R^{(1)} = \frac{-1}{16\pi^2} \int\limits^{\infty} \frac{ds}{s} \, e^{-im^2 s} \left[\frac{e^2 \eta\varepsilon}{\sin e\eta s \; \sh e\varepsilon s} \right.$$

$$\left. - \frac{1}{s^2} - \frac{e^2(\eta^2 - \varepsilon^2)}{6} \right] \tag{10}$$

and the functions I_o, I are determined by formula (36) of Ref. [12], where the parameters a, b are the same as in the spinor case and the parameters p, q are different:

$$p = 2e^2\eta^2(\ctg e\eta s \; \ctg e\eta s' + 3), \quad q = 2e^2\varepsilon^2(\cth e\varepsilon s \; \cth e\varepsilon s' - 3). \tag{11}$$

Finally, the η and ε are the magnetic and electric fields in a system where they are parallel.

   · The integral term in (9) does not vanish as the field is switched off and must be regularized. Regularization is carried out in the same way as in Ref. [12]. As a result $L^{(2)}$ is represented in the following form

$$L^{(2)} = L^{(0)} \frac{3\alpha^2}{16\pi^2} \left(\ln \frac{1}{i\gamma m^2 s_o} - \frac{11}{9} \ln 2 - \frac{10}{9} - \frac{1}{3im^2 t_o} \right) +$$

$$+ \delta m^2 \frac{\partial L_R^{(1)}}{\partial m^2} + L_R^{(2)} \, , \tag{12}$$

where

$$L_R^{(2)} = \frac{-i\alpha}{32\pi^3} \int\limits_o^{\infty} ds \left\{ \int\limits_o^{\infty} ds' \left[K(s,s') - \frac{K_o(s)}{s'} \right] + K_o(s) \left(\ln i\gamma m^2 s - \frac{7}{6} \right) \right\} , \tag{13}$$

$$K(s,s') = \frac{e^4 \eta^2 \varepsilon^2 e^{-im^2(s+s')}}{\sin e\eta s \; \sin e\eta s' \; \sh e\varepsilon s \; \sh e\varepsilon s'} \left(-m^2 I_o - \frac{i}{2} I \right)$$

$$+ \frac{m^2 e^{-im^2(s+s')}}{ss'(s+s')} \left[1 + \frac{i}{m^2(s+s')} + \frac{e^2(\eta^2 - \varepsilon^2)}{6} \left((s+s')^2 - ss' \right. \right.$$

$$\left. \left. + \frac{11ss'}{m^2(s+s')} \right) \right] , \tag{14}$$

$$K_o(s) = e^{-im^2 s}\left(-m^2 + \frac{i}{2}\frac{\partial}{\partial s}\right)\left(\frac{e^2 \eta\varepsilon}{\sin e\eta s \ \ \mathrm{sh}\ e\varepsilon s} - \frac{1}{s^2} - \frac{e^2(\eta^2-\varepsilon^2)}{6}\right),$$

$$(15)$$

and the radiative correction to the particle mass is equal to

$$\delta m^2 = \frac{3\alpha m^2}{4\pi}\left(\ln\frac{1}{i\gamma m^2 s_o} + \frac{7}{6} + \frac{1}{im^2 t_o}\right) \qquad \ln\gamma = 0.577\ldots .$$

$$(16)$$

Note the divergences in $L^{(2)}$ and δm^2 linear in t_0^{-1}.

The first-order radiative correction in α to the Lagrange function is equal to [11]

$$L^{(1)} = L^{(0)}\frac{\alpha_o}{12\pi}\ln\frac{1}{i\gamma m_o^2 s_o} + L_R^{(1)}(m_o^2)$$

$$(17)$$

where $L_R^{(1)}$ is defined by formula (10) and unrenormalized values of the fine structure constant and the mass are marked by the index zero. Thus, to the accuracy of radiative corrections of the order of α^2 we have

$$L = L^{(0)} + L^{(1)} + L^{(2)} = L_R^{(0)} + L_R^{(1)} + L_R^{(2)},$$

$$(18)$$

the expression in the left-hand side being determined by nonrenormalized and in the right-hand side by renormalized values of the field, charge and mass. The renormalized mass is connected with the bare one by the relation $m^2 = m_0^2 + \delta m^2$. The relation $L_R^{(0)} = L^{(0)}Z_3^{-1}$ between renormalized and nonrenormalized Lagrange functions of the Maxwell field that leads to the renormalization of the field $\eta = \eta_0 Z_3^{-1/2}$, $\varepsilon = \varepsilon_0 Z_3^{-1/2}$ and charge $e = e_0 Z_3^{1/2}$ is realized by the Z_3-factor

$$Z_3^{-1} = 1 + \frac{\alpha_o}{12\pi}\ln\frac{1}{i\gamma m_o^2 s_o} + \frac{\alpha_o^2}{4\pi^2}\left(\ln\frac{1}{i\gamma m_o^2 s_o} - \frac{11}{12}\ln 2 - \frac{13}{24}\right).$$

$$(19)$$

Owing to the use of renormalized mass in Z_3^{-1} there remained only logarithmic divergences. The coefficient of the first logarithm is established reliably [15,16], whereas the available calculations [17,18] of the coefficients of the second logarithm disagree with each other. The coefficient found in our paper coincides with that obtained by Bialynicka-Birula [17].

Using the expression obtained for Z_3 and formula (31) of Sec. III, we find the following expansion

$$\beta(\alpha) = \frac{1}{12}\frac{\alpha}{\pi} + \frac{1}{4}\left(\frac{\alpha}{\pi}\right)^2 + \ldots \tag{20}$$

for the Callan-Symanzik β-function in scalar electrodynamics.

The correction $L_R^{(2)}$ possesses the following asymptotical properties

$$L_R^{(2)} = \frac{\alpha^3}{\pi m^4}\left[\frac{275}{2592}(\eta^2-\varepsilon^2)^2 + \frac{4}{81}(\eta\varepsilon)^2\right] \, , \, \frac{e\eta}{m^2} \, , \, \frac{e\varepsilon}{m^2} \ll 1, \tag{21}$$

$$L_R^{(2)} = \frac{\alpha^2\eta^2}{8\pi^2}\left(\ln\frac{2e\eta}{\gamma\pi m^2} + a_2\right), \, \frac{e\eta}{m^2} \gg 1, \, \frac{e\varepsilon}{m^2} \lesssim 1 \, . \tag{22}$$

The imaginary part, which is exponentially small compared with the real one, is absent from formula (21). In the case of a weak electric field the imaginary part of the Lagrange function (18) is equal to

$$\mathrm{Im}\, L_R = \mathrm{Im}\left(L_R^{(1)}+L_R^{(2)}\right) = \frac{\alpha\varepsilon^2 e^{-\pi m^2/e\varepsilon}}{4\pi^2}(1+\pi\alpha+\ldots), \, \frac{e\varepsilon}{m^2} \ll 1. \tag{23}$$

Note that in spinor electrodynamics the analogous limit is twice as large as that due to the doubled number of states of the pair with a zero projection of the orbital momentum.

The behaviour of the real part of the Lagrange function in a weak field determines the field and charge renormalization, whereas the behaviour of its imaginary part for a weak electric field determines unambiguously the radiative correction δm^2 to the bare mass squared, namely it fixes the constant 7/6 in the expression (16). If in the expression for δm^2 we chose another constant b instead of 7/6, it would appear instead of 7/6 in the last term of (13) too. Then the term $(3\alpha m^2/4e\varepsilon)(b-7/6)$ would be added in the parenthesis of expression (23) for the imaginary part of the Lagrange function of a weak field. Being transferred into the exponent this term would change the parameter m^2 by the quantity $-(3\alpha m^2/4\pi)(b-7/6)$ finite at $\varepsilon \to 0$. This would mean that m is not a real particle mass, since according to the physical renormalization principle all the radiative corrections enter in the observed mass. Thus the parameter m connected with the bare mass m_0 by the relation $m^2 = m_0^2 + \delta m^2$ has the meaning of a real particle mass only if $b = 7/6$. The expression for δm^2 was controlled by a consideration of the mass operator and of the position of the pole of the modified Green function of a particle in vacuum. Thus, renormalization of field, charge and mass is determined unambiguously by the behaviour of the exact Lagrange function in a weak field limit, i.e., by boundary condition - its real part must be a Maxwell one and the imaginary part must be equal to $(e\varepsilon)^2 f(\alpha, \eta/\varepsilon) \exp(-\pi m^2/e\varepsilon)$.

The coefficient f characterizes the spin of a polarizable charged field and is not essential for mass renormalization.

Note that in the calculation of the exact propagator of a photon in vacuum the charge and mass renormalization can also be determined only by the boundary condition on the propagator behaviour: Re $d \to 1$ at $K^2 \to 0$, and Im $d \to -(1+4m^2K^{-2})^{1/2} g_1(\alpha)$ or $-(1+4m^2K^{-2})^{3/2} g_2(\alpha)$ at $K^2 \to -4m^2$ for a spinor or scalar charge field, respectively.

As is seen from formula (18) the Lagrange function is a renorminvariant quantity - it does not change under the substitution of its renormalized parameters for the nonrenormalized ones. This property results from *a priori* invariance of the vacuum conservation amplitude $\exp i (W^{(0)} + W)$ under such a substitution and along with the known fundamental properties of the Lagrange function makes it a convenient object for the investigation of questions of principle of quantum electrodynamics.

III. COMPARISON OF DIFFERENT RENORMINVARIANT CHARGES IN QUANTUM ELECTRODYNAMICS

Consider three quantities invariant under renormalization group in quantum electrodynamics (spinor or scalar).

1) The invariant charge $\alpha d_R(x,\alpha)$ determined by the exact photon propagator $K^{-2} d_R$ and dependent on the fine-structure constant α and on the ratio $x = K^2/m^2$ of the photon momentum squared to the electron mass squared [2].

2) The invariant charge $\alpha \ell_R(x,y,\alpha)^{-1}$ determined by the ratio $\ell_R = L_R/L_R^{(0)}$ of the Lagrange function L_R of a constant intense electromagnetic field to the Lagrange function $L_R^{(0)}$ of a Maxwell field and depending on α and magnetic η and electric ε

fields in a system where they are parallel and made dimension-less in a proper way: $x = e\eta/m^2$, $y = e\varepsilon/m^2$ [12].

3) The bare coupling constant $\alpha_0 = \alpha Z_3(x,\alpha)^{-1}$ dependent on α and on the cutoff parameter $x = (im^2 s_0)^{-1}$ for which was taken the proper time s_0, more exactly the space-like interval is_0 [12].

As a result of renormalizability of quantum electrodynamics, all these quantities in the limit $x \to \infty$ are described by asymptotic (index ∞) functions dependent on the variable $z = x\psi(\alpha)$ only:

$$\alpha d_{R\infty}(x,\alpha) = \Phi_1(z), \quad \alpha\ell_{R\infty}(x,\alpha)^{-1} = \Phi_2(z), \quad \alpha Z_3(x,\alpha)^{-1} = \Phi_3(z), \quad (24)$$

where $\phi(\alpha)$ is a universal function in quantum electrodynamics which appears at the resolving of the equation $\alpha_0 = \alpha Z_3(x,\alpha)^{-1}$ with respect to the cutoff parameter: $x^{-1} \equiv im^2 s_0 = \phi(\alpha)\chi(\alpha_0)$. Factorization of the right-hand side into the product of α and α_0-functions is a result of the same renormalizability.

Indeed, relation (18) can be written in the form

$$\alpha_o^{-1}\ell\left(\frac{eF}{m^2}, \alpha_o, im^2 s_o\right) = \alpha^{-1}\ell_R\left(\frac{eF}{m^2}, \alpha\right), \quad (25)$$

if in the left-hand side the mass renormalization is carried out and the invariance of the product $e_o F_o = eF$ is used. If one of the field parameters, e.g., $x = e\eta/m^2$, tends to infinity, the asymptotic expression for ℓ in the approximation found: $\ell = 1 + \ell^{(1)} + \ell^{(2)}$ depends neither on m^2 nor on the second parameter $e\varepsilon$. We assume that this important property is valid for ℓ in any order in α_0. Then, $\lim \ell = \ell_\infty(ie\eta s_0, \alpha_0)$. Substituting the quantity $is_0 = m^{-2}\phi(\alpha,\alpha_0)$ found from the relation $\alpha = \alpha_0 Z_3(im^2 s_0, \alpha_0)$ into ℓ_∞ , we obtain according to (25)

$$\alpha_o^{-1}\ell_\infty\left(\frac{e\eta}{m^2}\,\phi(\alpha,\alpha_o), \alpha_o\right) = \alpha^{-1}\ell_{R\infty}\left(\frac{e\eta}{m^2}, \alpha\right). \quad (26)$$

Since the right-hand side does not depend on α_0, the left-hand side must not depend on α_0 either. This may happen only if $\phi(\alpha,\alpha_0)$ is the product of $\alpha-$ and α_0-functions:

$$im^2 s_o = \phi(\alpha,\alpha_o) = \phi(\alpha)\chi(\alpha_o). \quad (27)$$

As soon as α_0 vanishes, the left-hand side of (26) becomes the function $(e\eta/m^2)\phi(\alpha)$ only and thus from formulae (26) and (27) we obtain for $\alpha\ell_{R\infty}^{-1}$ and $\alpha_o = \alpha Z_3^{-1}$ expressions (24).

Since, as we assume on the basis of universality of electromagnetic interactions, the boundary conditions on the exact Lagrange function and on the exact propagator define one and the same renormalized interaction constant, the fine structure constant, then with one and the same way of regularization there exists the only possible connection between α and α_0 which is independent of whether it was found in the calculation of the Lagrange function or of the photon propagator. Therefore, in the calculation of $\alpha d_{R\infty}$ we shall obtain expression (24) with the same function $\phi(\alpha)$.

It is clear that the function $\phi(\alpha)$ is determined to within a multiplicative constant on which the invariant charges (24) do not depend while the functions $\Phi_a(z)$ do depend. This constant is a parameter of the renormalization group. Since

$$x\phi(\alpha) = \Phi_1^{-1}[\alpha d_{R\infty}(x,\alpha)] \ , \ \text{etc. for } a = 2,3, \tag{28}$$

the functions $\Phi_a^{-1}(y)$, inverse of the functions $\Phi_a(z)$, are determined to within the same multiplicative constant. By the logarithmic derivatives

$$[\ln \phi(\alpha)]' = 1/\alpha\beta(\alpha) \ , \ [\ln \Phi_a^{-1}(y)]' = 1/\psi_a(y), \tag{29}$$

defining the Callan-Symanzik β-function and the Gell-Mann-Low ψ_a-functions are determined unambiguously and are renorminvariant characteristics of the representation (24).

From (28)-(29) important functional relations follow

$$\alpha\beta(\alpha) \ \frac{\partial}{\partial\alpha}[\alpha d_{R\infty}(x,\alpha)] = x \ \frac{\partial}{\partial x} \ [\alpha d_{R\infty}(x,\alpha)]=\psi_1[\alpha d_{R\infty}(x,\alpha)],$$

$$\text{etc. for } a = 2,3. \tag{30}$$

The first of them is the Callan-Symanzik equation [19,20] which is satisfied by the functions $\alpha d_{R\infty}$, $\alpha\ell_{R\infty}^{-1}$, αZ_3^{-1}. The second relation is a differential form of the Gell-Mann-Low equation [2]. Their consequence is the functional connection between β and ψ_a-functions [20,21].

If in the relation (27) α is considered as a function of $x \equiv (im^2 s_0)^{-1}$ and α_0, then we obtain by differentiation

$$\alpha\beta(\alpha) = -x\left(\frac{\partial\alpha}{\partial x}\right)_{\alpha_o} = -\frac{\alpha}{Z_3} x \left(\frac{\partial Z_3}{\partial x}\right)_{\alpha_o}. \tag{31}$$

In general, from Eq. (28) we obtain in a similar way

$$\alpha\beta(\alpha) = -x\left(\frac{\partial\alpha}{\partial x}\right)_{\alpha_{a\infty}} , \tag{32}$$

where $\alpha_{a\infty}$ is the a-th renorminvariant charge in the asymptotical region of x.

Thus, if the physical meaning of the parameters α, m is fixed by renormalization, then any renorminvariant charge and Z_3-factor obtained by any regularization (cut-off in momentum, in proper time etc.) bear universal information about quantum electrodynamics in the form of one and the same function $\alpha\beta(\alpha)$. As α, m we use the fine structure constant and the physical mass of an electron. This physical meaning of α, m is provided by quite definite boundary conditions for the exact Lagrange function at $F \to 0$ or for the exact photon propagator at $K^2 \to 0$ and $K^2 \to -4m^2$, see Sec. II. Another physical meaning of α, m would correspond to another function $\alpha\beta(\alpha)$. In general, it may be said about the Callan-Symanzik β-function and the Gell-Mann-Low ψ_a-function that their form depends on the physical meaning of their argument.

An explicit form of the functions considered is known only in the framework of perturbation theory according to which the functions $d_{R\infty}^{-1}$, $\ell_{R\infty}$, Z_3 are expanded in power series in α and $\ell n\, x$:

$$\ell_{R\infty}(x,\alpha)=1+\frac{\alpha}{\pi}\,(a_{10}+a_{11}\ell n\, x) + \sum_{n=2}^{\infty}\left(\frac{\alpha}{\pi}\right)^n \sum_{\kappa=0}^{n-1} a_{n\kappa}\,\ell n^\kappa x . \tag{33}$$

According to Callan-Symanzik, Eq. (30), the coefficients $a_{n\kappa}$ are expressed through the a_{no} and the power series coefficients for $\beta(\alpha)$*

$$\beta(\alpha) = \sum_{\kappa=1}^{\infty} \beta_\kappa\left(\frac{\alpha}{\pi}\right)^\kappa = \frac{1}{3}\frac{\alpha}{\pi} + \frac{1}{4}\left(\frac{\alpha}{\pi}\right)^2 - \frac{121}{288}\left(\frac{\alpha}{\pi}\right)^3 + \ldots \tag{34}$$

due to recurrence relations

---

*The values of the first three coefficients in the β-function expansion are known at the present time for spinor electrodynamics and are presented in (34) [22]. For scalar electrodynamics $\beta_1 =$ $= 1/12$, $\beta_2 = 1/4$ as follows from formulae (19), (31) or (22), (36). The coefficient β_1 in spinor electrodynamice is four times as large as in scalar electrodynamics owing to four times larger number of states of spinor pair.

$$\kappa a_{n\kappa} = -\beta_n \delta_{\kappa 1} + \sum_{i=\kappa}^{n-1} (i-1)a_{i\kappa-1}\beta_{n-i}, \qquad \kappa \geqslant 1. \tag{35}$$

From these relations it follows that in terms of the order of α^n, $n \geqslant 2$, this highest exponent of the logarithm is $n-1$. The coefficients of these leading logarithms in (33) are determined by the coefficients β_1, β_2 only

$$a_{11} = -\beta_1, \quad a_{nn-1} = -\beta_2 \beta_1^{n-2}/(n-1), \tag{36}$$

while the coefficients a_{nn-r}, $2 \leqslant r \leqslant n-1$, of the remaining logarithms are determined by the coefficients $\beta_1, \beta_2, \ldots, \beta_{r+1}$ and by the constants $a_{20}, a_{30}, \ldots, a_{r0}$. The constants a_{no} show individual features of the functions $\Phi_a(z)$ and their one-parametric arbitrariness due to the renormalization group.

It is only in the constants a_{no} that the series (33) for the functions $d_{R\infty}^{-1}$, $\ell_{R\infty}$, Z_3 may differ from each other. On the other hand under group transformation $\phi(\alpha) \to t^{-1}\phi(\alpha), \Phi_a(z) \to \Phi_a(tz)$ the constants a_{no} are transformed

$$a_{10} \to a_{10} - \beta_1 \ell n\, t, \quad a_{20} \to a_{20} - \beta_2 \ell n\, t,$$

$$a_{30} \to a_{30} - (\beta_2 - \beta_1 a_{20})\ell n\, t - \frac{1}{2}\beta_1\beta_2\, \ell n^2 t,$$

$$a_{40} \to a_{40} - (\beta_4 - \beta_2 a_{20} - 2\beta_1 a_{30})\,\ell n\, t - (\frac{1}{2}\beta_2^2 + \beta_1\beta_3$$

$$- \beta_1^2 a_{20})\,\ell n^2 t - \frac{1}{3}\beta_1^2\beta_2\,\ell n^3 t, \ldots . \tag{37}$$

Such a transformation does not change the shape of $\Phi_a(z)$ which is described by the invariant function $\psi_a(y)$. Therefore, the differences in shape of the functions $\Phi_a(z)$ that do not reduce to a multiplicative change of their argument are reflected by the difference of their Gell-Mann-Low functions. Expanding the latter in series

$$\psi_a(y) = y \sum_{n=1}^{\infty} \psi_n^{(a)} \left(\frac{y}{\pi}\right)^n \tag{38}$$

and connecting with the aid of (30) the coefficients $\psi_n^{(a)}$ with the β_n and $a_{no}^{(a)}$, we obtain (index a at ψ_n and a_{no} is implied):

$$\psi_1 = \beta_1, \quad \psi_2 = \beta_2, \quad \psi_3 = \beta_3 + \beta_2 a_{10} - \beta_1 a_{20} \, ,$$

$$\psi_4 = \beta_4 + 2\beta_3 a_{10} + \beta_2 a_{10}^2 - 2\beta_1 (a_{30} + a_{10} a_{20}) ,$$

$$\psi_5 = \beta_5 + 3\beta_4 a_{10} + \beta_3 (3a_{10}^2 + a_{20}) + \beta_2 (a_{10}^3 - a_{30})$$
$$- \beta_1 (6a_{10} a_{30} + 3a_{10}^2 a_{20} + 2a_{20}^2 + 3a_{40}) , \ldots \, . \quad (39)$$

The constants a_{no} enter into the right-hand sides of relations
(39) in combinations invariant under transformation (37). It
is just by these combinations that the functions $\psi_a(y)$ differ
from the universal function $y\beta(y)$ and from each other; the difference begins only from the third-order terms $n \geqslant 3$ and in the
n-th order it is determined by combinations of β_κ and $a_{\kappa o}$ of the
order of $\kappa \leqslant n-1$. The table presents the constants a_{10}, a_{20} and
the coefficients ψ_3 found for the functions $d_{R\infty}^{-1}$, $\ell_{R\infty}$, Z_3 from
(39) and (34).

	x	a_{10}	a_{20}	ψ_3
$d_{R\infty}^{-1}$	K^2/m^2	$\dfrac{5}{9}$	$\dfrac{5}{24} - \zeta(3)$	$\dfrac{1}{3} \zeta(3) - \dfrac{101}{288} = 0.049991..$
$\ell_{R\infty}$	$en/\gamma\pi m^2$	$-2\pi^{-2}\zeta'(2)$	-0.878572	$-0.079785 \ldots$
Z_3	$(i\gamma m^2 s_o)^{-1}$	0	$\dfrac{5}{12} \ell n\, 2 - \dfrac{5}{6}$	$-\dfrac{41}{288} - \dfrac{5}{36} \ell n2 = -0.238631..$

Here $\ell n\gamma = 0.577..$ is the Euler's constant, $\zeta(x)$ is the Riemann
zeta-function. The results for the $d_{R\infty}^{-1}$ -function are obtained
in Refs. [23-26, 27], for $\ell_{R\infty}$ and Z_3 - in the present paper based
on Ref. [12]; see also preliminary report [27]. The constant a_{20}
for $\ell_{R\infty}$ is determined by the following expression that contains
a double integral

$$a_{20} = - \frac{5}{12} \ell n\, \pi - \frac{5}{4} + \frac{1}{\pi^2} \zeta'(2) + \frac{1}{2} \int_o^{1/2} d\xi \int_o^{\infty} du \, \ell n\, u \, Q'(u,\xi), \quad (40)$$

in which the function $Q(u,\xi)$ is equal to

$$Q(u,\xi)=\frac{5}{3}+\frac{2\left[\operatorname{ch} u(1-2\xi)-\frac{uW}{\operatorname{sh} u}\right]}{(W-1)\operatorname{sh} n\xi\,\operatorname{sh} u\,(1-\xi)}\left(1-\frac{\ell nW}{W-1}\right)+\frac{1}{\xi(1-\xi)}\left[\frac{2W}{\operatorname{sh}^2 u}-\frac{2}{u^2}\right.$$

$$\left.-\frac{\operatorname{cth}z}{z}-\frac{1}{\operatorname{sh}^2 z}+\frac{2}{z^2}\right]\,,\quad W=\frac{\xi(1-\xi)u\,\operatorname{sh} u}{\operatorname{sh} u\,\xi\,\operatorname{sh} u(1-\xi)}\,,\quad z=u(1-\xi),$$

$$(41)$$

and the derivative is taken with respect to u. This integral was found numerically

$$\int_0^{1/2} d\xi\int_0^\infty du\,\ell n\,u\,Q'(u,\xi)=\int_0^\infty du\left[\frac{5}{6(1+u)}-\frac{1}{u}\int_0^{1/2}d\xi Q(u,\xi)\right]=1.886784\ldots$$

Thus, all the three functions $\psi_a(y)$ are different, $\psi_2(y)$ and $\psi_3(y)$ as distinct from $\psi_1(y)$ having third-order negative terms. The latter circumstance favours a possible vanishing of these functions at y ~ 1. As is known, the existence of a rather strong zero of the Gell-Mann-Low function at $y=\alpha_*$ implies that the corresponding invariant charge is equal to α_* in the limit x → ∞ [2]. The finiteness of the bare charge limit $\alpha_* = \lim\alpha_o\,(x,\alpha)$, x → ∞, in which the initial perturbation theory is developed, is necessary for the self-consistency of quantum electrodynamics. If this fact has a physical meaning, the limit α_* does not depend on the cutoff method, which defines the values of constants a_{no} in the representation (33) for Z_3, i.e., the shape of the $\Phi_3(z)$-function. One may change the cutoff method so that the $\Phi_3(z)$ function will coincide with $\Phi_1(z)$ or $\Phi_2(z)$ and, consequently, the limits of these functions at x → ∞ must be finite and equal to α_* as well. The finiteness of all the limits is possible if $\beta(\alpha)=0$ at $\alpha=1/137$ (see Eq. (30), and Ref. [21]) or at $\alpha=\alpha_*$, which is more plausible, or if there is another general reason for all the $\psi_a(y)$-functions to vanish at one and the same $y=\alpha_* ~ 1$.

IV.   INTEGRAL TRANSFORMATIONS OF RENORMINVARIANT CHARGES

In the introduction and in Sec. III general arguments were presented in favor of the fact that all the renorminvariant charges have one and the same limit if it exists even in one of them. It would be desirable to prove this statement mathematically.* In this section we can only point out some
*However, the equality of the limits of d- and Z_3^{-1}-functions follows from spectral representations.

renorminvariant charges, which differ in the asymptotic region too, i.e., they have different Gell-Mann-Low functions but, obviously have similar asymptotic limits.

Consider the charge $eq(m^2 r^2, \alpha)$, which is inside the sphere of radius r surrounding the point field source with the physical charge e:

$$\frac{eq(m^2 r^2, \alpha)}{4\pi r} = e \int \frac{d^3 K}{(2\pi)^3} e^{iKx} \frac{d(K^2/m^2, \alpha)}{K^2} \tag{42}$$

or

$$q(t, \alpha) = \frac{2}{\pi} \int_0^\infty \frac{dz}{z} \sin z \, d(\frac{z^2}{t}, \alpha), \quad t = m^2 r^2 . \tag{43}$$

From this integral transformation it follows that $\alpha q(t, \alpha)$ is a renorminvariant quantity, which at $t = \infty$ and $t = 0$ has values coinciding with $\alpha d(0, \alpha) = \alpha$ and $\alpha d(\infty, \alpha) = \alpha_*$. If we use the Källen-Lehmann representation [28,29]

$$d(x, \alpha) = 1 + x \int_4^\infty \frac{dz \, \rho(z, \alpha)}{z(z+x-i\varepsilon)} \quad , \quad \rho(z, \alpha) = - \frac{1}{\pi} \text{Im } d(-z, \alpha), \tag{44}$$

we shall obtain the following expression for $q(t, \alpha)$

$$q(t, \alpha) = 1 + \int_4^\infty \frac{dz}{z} \rho(z, \alpha) e^{-\sqrt{zt}} . \tag{45}$$

The first term of the expansion in α of the spectral function $\rho(z, \alpha) = \rho^{(1)}(z) + \rho^{(2)}(z) + \ldots$ was found by Schwinger [30], and the second one by Källen and Sabry [24]:

$$\rho^{(1)}(z) = \frac{\alpha}{3\pi}(1 + \frac{2}{z})\sqrt{1 - \frac{4}{z}} , \tag{46}$$

$$\rho^{(2)}(z) = -(\frac{\alpha}{\pi})^2 \left\{ (\frac{13}{108} + \frac{7}{216\tau} + \frac{1}{9\tau^2})\sqrt{1 - \tau^{-1}} + (-\frac{22}{9} + \frac{1}{3\tau} + \frac{5}{8\tau^2} \right.$$

$$+ \frac{1}{9\tau^3}) \ell n(\sqrt{\tau} + \sqrt{\tau-1}) + (\frac{2}{3} + \frac{1}{3\tau})\sqrt{1 - \tau^{-1}} \ \ell n[8\sqrt{\tau}(\tau-1)]$$

$$+ (-\frac{2}{3} + \frac{1}{6\tau^2}) \int_\tau^\infty dy \left[\frac{3y-1}{y(y-1)} \ell n(\sqrt{y} + \sqrt{y-1}) - \frac{\ell n[8\sqrt{y}(y-1)]}{\sqrt{y}(y-1)} \right] \right\} ,$$

$$z = 4\tau, \tag{47}$$

With the aid of this spectral function we find according to (45) the following expansion for q in the asymptotic region of $t = m^2 r^2 \to 0$ or $x \equiv (\gamma m r)^{-2} \to \infty$:

$$q_\infty(t,\alpha)^{-1} = 1 + \frac{\alpha}{\pi}\left[\frac{5}{9} - \frac{1}{3}\ln x\right] + \left(\frac{\alpha}{\pi}\right)^2\left[\frac{5}{24} - \frac{\pi^2}{27} - \zeta(3) - \frac{1}{4}\ln x\right] + \ldots$$

$$(48)$$

This expansion has the common structure (33) of asymptotic behaviour of invariant charges. The coefficient ψ_3 extracted from this expansion and formula (39) is equal to $\psi_3 = 1/3\,\zeta(3) - 101/288 + \pi^2/81 = 0.171838 \ldots$. Thus, the invariant charges αd and αq have different Gell–Mann–Low functions but similar limits at $x \to \infty$ if they do exist.

It is easily seen that the Källen–Lehman representation (44) can be written also in the form of the integral Laplace transform

$$d(x,\alpha) = 1 + x\int_0^\infty dt\, e^{-xt}\,\tilde{\rho}(t,\alpha) = x\int_0^\infty dt\, e^{-xt}\, q_1(t,\alpha) \quad (49)$$

in which

$$q_1(t,\alpha) = 1 + \tilde{\rho}(t,\alpha) = 1 + \int_4^\infty \frac{dz}{z}\,\rho(z,\alpha)e^{-zt}. \quad (50)$$

It is obvious that $\alpha q_1(t,\alpha)$ is a renorminvariant quantity that has the values α and α_* at $t = \infty$ and $t = 0$. The calculation of $q_1(t,\alpha)$ with the aid of (50) and the above spectral function leads to the following result in the asymptotic region $t \to 0$ or $x \equiv (\gamma t)^{-1} \to \infty$:

$$q_{1\infty}(t,\alpha)^{-1} = 1 + \frac{\alpha}{\pi}\left[\frac{5}{9} - \frac{1}{3}\ln x\right] + \left(\frac{\alpha}{\pi}\right)^2\left[\frac{5}{24} + \frac{\pi^2}{54} - \zeta(3) - \frac{1}{4}\ln x\right] + \ldots$$

$$(51)$$

Thus the invariant charge αq_1 corresponds to the Gell–Mann–Low function with a negative coefficient $\psi_3 = 1/3\,\zeta(3) - 101/288 - \pi^2/162 = -0.010932 \ldots$.

Formulae (45) and (50) admit the generalization

$$q_\nu(t,\alpha) = 1 + \int_4^\infty \frac{dz}{z}\,\rho(z,\alpha)e^{-(zt)^\nu}, \quad (52)$$

where ν is a positive parameter. It is obvious that αq_ν is a renorminvariant quantity, which has the values α and α_* at the points $t = \infty$ and $t = 0$. It coincides with αq at $\nu = 1/2$ and with

αq_1 at $\nu = 1$. In the asymptotic region $t \to 0$ or $x \equiv (\gamma^{1/\nu} t)^{-1} \to \infty$ the function $q_\nu(t,\alpha)$ behaves as

$$q_{\nu\infty}(t,\alpha)^{-1} = 1 + \frac{\alpha}{\pi}\left[\frac{5}{9} - \frac{1}{3}\ln x\right] + \left(\frac{\alpha}{\pi}\right)^2\left[\frac{5}{24} - \zeta(3) + \frac{\pi^2}{27}\left(1 - \frac{1}{2\nu^2}\right) - \frac{1}{4}\ln x\right] + .. \tag{53}$$

Thus, the invariant charge αq_ν possesses the Gell-Mann-Low function with the third expansion coefficient

$$\psi_3 = \frac{1}{3}\zeta(3) - \frac{101}{288} - \frac{\pi^2}{81}\left(1 - \frac{1}{2\nu^2}\right) , \tag{54}$$

which depends on ν. As ν changes from 0 to ∞ the coefficient ψ_3 decreases monotonically from ∞ to $1/3\ \zeta(3) - 101/288 - \pi^2/81 = -0.071876...$. Note that the integral transformation under consideration makes it impossible for ψ_3 to become a large negative number - even if the Gell-Mann-Low function has a positive zero α_*, it cannot be small. Nevertheless a richer family of transformations can still diminish ψ_3 even more.

The author would like to thank Dr. V. O. Papanian for checking the calculations of the last section.

REFERENCES

1.   L. D. Landau, A. A. Abrikosov and I. M. Khalatnikov, Dokl. Akad. Nauk SSSR, 95, 773, 1177 ; 96, 261 (1954).

2.   M. Gell-Mann, F. E. Low, Phys. Rev. 95, 1300 (1954).

3.   L. D. Landau, I. Ya. Pomeranchuk, Dokl. Akad. Nauk SSSR, 102, 489 (1955).

4.   E. S. Fradkin, ZhETF, 28, 750 (1955).

5.   K. Johnson, M. Baker, R. Willey, Phys. Rev. 136, B1111, (1964).

6.   K. Johnson, R. Willey, M. Baker, Phys. Rev. 163, 1699 (1967).

7.   M. Baker, K. Johnson, Phys. Rev. 183, 1292 (1969); D3, 2516, 2541 (1971).

8.   K. Johnson, M. Baker, Phys. Rev. D8, 1110 (1973).

9.   W. Heisenberg, H. Euler, Zs. Phys. 98, 714 (1936).

10.   V. Weisskopf. Kgl. Dan. Vid. Selsk. Mat.-Phys. Medd., 14, No. 6, (1936).

11. J. Schwinger, Phys. Rev. $\underline{82}$, 664 (1951).

12. V. I. Ritus, ZhETF, $\underline{69}$, 1517 (1975).

13. H. M. Fried, "Functional Methods and Models in Quantum Field Theory," M.I.T. Press, Cambridge, (1972). I.Bialynicki-Birula, Z. Bialynicka-Birula, "Quantum Electrodynamics," Pergamon Press, Oxford, (1975).

14. V. Fock. Sow. Phys., $\underline{12}$, 404 (1937).

15. M. Neumann, W. H. Furry, Phys. Rev. $\underline{76}$, 1677 (1949).

16. L. P. Gor'kov, I. M. Khalatnikov, ZhETF, $\underline{31}$, 1062 (1956).

17. Z. Bialynicka-Birula, Bull. Acad. Polon. Sci. $\underline{13}$, 369 (1965).

18. I.-J. Kim, C. R. Hagen, Phys. Rev. $\underline{D2}$, 1511 (1970).

19. C. Callan, Jr., Phys. Rev. $\underline{D2}$, 1541 (1970).

20. K. Symanzik, Comm. Math. Phys. $\underline{18}$, 227 (1970); $\underline{23}$, 49 (1971).

21. S. Adler, Phys. Rev. $\underline{D5}$, 3021 (1972).

22. E. Rafael, J. L. Rosner, Ann. Phys. $\underline{82}$, 369 (1974).

23. E. Jost, J. M. Luttinger, Helv. Phys. Acta, $\underline{23}$, 201 (1950).

24. G. Källen, A. Sabry, Kgl. Dan. Vid. Selsk. Mat.-Fys. Medd., $\underline{29}$, No. 17 (1955).

25. C. R. Hagen, M. A. Samuels, Phys. Rev. Lett. $\underline{20}$, 1405 (1968).

26. B. E. Lautrup, E. Rafael, Phys. Rev. $\underline{174}$, 1835 (1968).

27. V. I. Ritus. Phys. Lett. $\underline{65B}$, 355 (1976).

28. G. Källen, Helv. Phys. Acta. $\underline{25}$, 417 (1952).

29. H. Lehmann, Nuovo Cimento, $\underline{11}$, 342 (1954).

30. J. Schwinger, Phys. Rev. $\underline{75}$, 651 (1949).

Multiphoton Phenomena in Photoelectron Emission Processes of Metals at High Laser Intensities

GY. FARKAS

Central Research Institute for Physics
11 - 1525 Budapest, P.O.B. 49 Hungary

The surface multiphoton photoelectric effect of metals (MPE) has proved to be a useful experimental method for investigations into the general laws governing multiphoton processes. Since several exhaustive review papers [1-5] have already been published in the course of the last few years, here, after providing a brief historical background, we shall concern ourselves primarily with the more recent developments of the problem; namely, the behavior of the intensity dependence of the surface photoeffect at high laser intensities, the multiphoton resonance in the case of metals, and the energy absorption from the laser beam by the already emitted "free" electrons.

BRIEF HISTORICAL INTRODUCTION

The multiphoton surface photoeffect occurs when the work function, A, of a metal characterized by the Fermi energy of the Sommerfeld-type conduction electron gas is higher than the photon energy, $h\nu$, of the irradiating laser light. The electrons move freely in the metal except in the direction normal to the surface, therefore the photoeffect depends only on the laser electric field component $E_\perp$ which is perpendicular to the surface, so the effect is a vertical one. The first theoretical works applying the higher order perturbation approximation [6-12], determined the photocurrent by calculating the transition probability from the fundamental bound state to the free electron final state. The well known result showed that the photocurrent, j, is proportional to the n_o-th power of the square of the perpendicular laser electric field component $E_\perp$ or the corresponding laser intensity I

$$j \propto E_\perp^{2n_o} \propto I^{n_o} \quad , \text{ with } n_o = [A/h\nu + 1] \ . \tag{1}$$

Since the validity of the simple perturbation calculation is limited to the relatively lower light intensities, Keldysh and others [13-17] elaborated a more general method in which the final state is a Volkov state instead of a pure free electron state, by taking into account the presence of the strong laser field outside the metal. The results are summarized in Fig. 1. It can be seen that depending on the parameter $\gamma = 2\pi\nu\sqrt{2mA}/E_\perp e$ (e and m are the electron charge and mass, respectively), in the case $\gamma \gg 1$ the older (1) perturbation n-th power law is reobtained while for $\gamma \ll 1$ the optical tunnel emission is predicted.

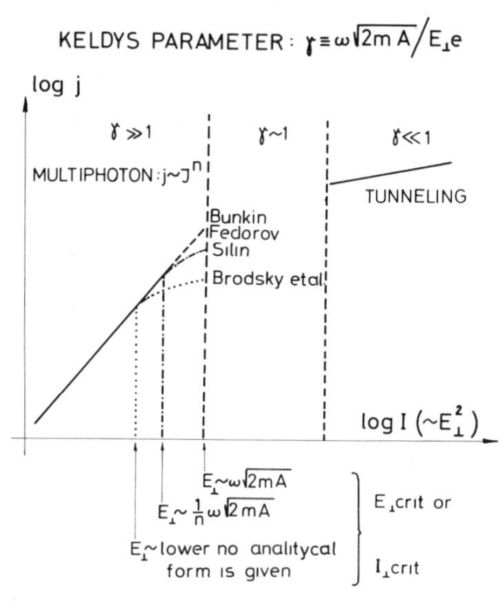

FIGURE 1

For the critical light intensities at which the power law loses its validity and a deviation from it occurs, the different theories give different values depending on their initial assumptions for the process (e.g. taking into account the static long range Coulomb-field of the ionized metal [16,17] etc.).

The overall conclusion drawn from these theories was that the existence of the multiphoton surface photoeffect might be verified by experiment taking into consideration several phenomena predicted by the theories, namely:

- Polarization dependence;
- Intensity dependence;
- Work function dependence;

    - Photoelectron energy distribution dependence;
    - Coherence dependence.
It should be noticed, however, that all these idealized theories
ignored the influence of the temperature effects of the interaction processes which, as we shall see, play a very important role.

    In fact, the first pioneering works [4] of Ready in the USA
investigating the interaction of a laser beam with metals has
shown that only thermionic electron emission took place - the
metal being heated by the irradiating laser beam.

    Later on, in the painstaking experiments performed mostly by
E.M. Logothethis and P.L. Hartman [20,21] and by M.C.Teich et al.
[12,19] in the USA, Budapest [22-25] and Moscow [5,26], and in
Saclay [27], the masking thermionic emission was reduced and the
earlier mentioned experimental evidence for the existence of the
multiphoton effect (MPEO) with nanosec laser pulses was verified
for Au, Ag, Ni, Hg, etc. Figures 2 and 3 show two characteristic
examples for the intensity dependence.

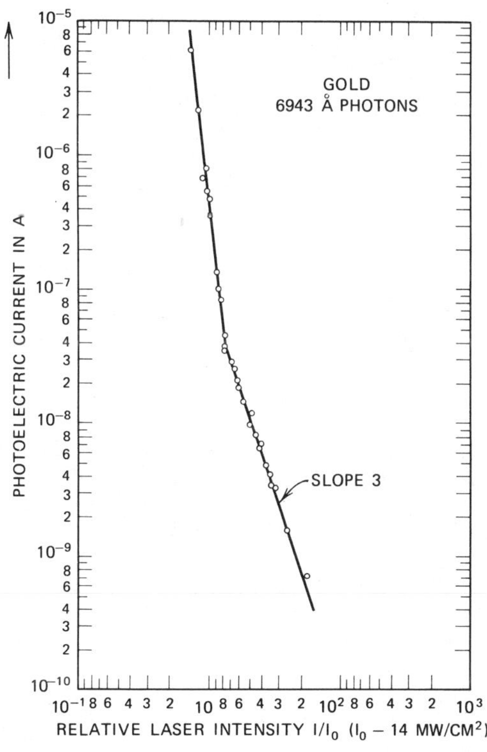

FIGURE 2

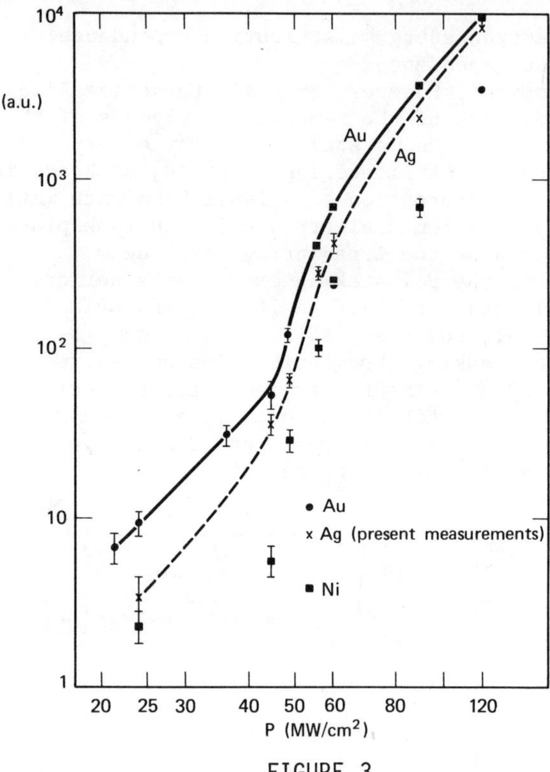

FIGURE 3

It can be concluded that these experiments have shown unambiguously the existence of the pure MPE of metals. The effect
might be observed however, using nanosec laser plulses up to
several Mw/cm² laser intensities only, because of the occurrence
of the thermionic emission at higher intensities. .Therefore the
behavior of the MPE at the theoretically predicted interesting
critical laser intensities could not be investigated owing to
the masking effect of thermionic emission.

The theoretical work of Bunkin and Prokhorov [28] showed how
to bypass this intensity limit for the investigations in the high
intensity range. Reducing the laser pulse duration and at the
same time increasing the light intensity the thermal processes
of the crystal lattice become negligible in comparison with the
multiphoton processes. In the next part we describe the experiments performed starting from these possibilities.

I.   INTENSITY DEPENDENCE OF MULTIPHOTON PHOTOEFFECT
AT HIGH LASER INTENSITIES

The theoretical calculations listed before predict a departure from the n_o-th power law for the intensity dependence of the multiphoton photocurrent at a given intensity value, as we have seen in Fig. 1. The main task of the experimental investigations is to find the intensity value at which this deviation takes place - if it exists; i.e. to decide which of the theories is correct.

Treating these simple considerations, however, we must keep in mind another interesting effect. Recently Anisimov et al [29,30] and Kantorovich [31,32] pointed out that if the low heat capacity of the conduction electrons themselves is taken into account, by using intensive ultrashort light pulses, the fast heating of only the conduction electrons may occur without simultaneously heating the whole crystal lattice. In this way the fast distortion of the Fermi distribution of electrons may occur permitting the occurrence of all multiphoton electron emission processes of lower order than the original n_o (see Fig. 4(a)).

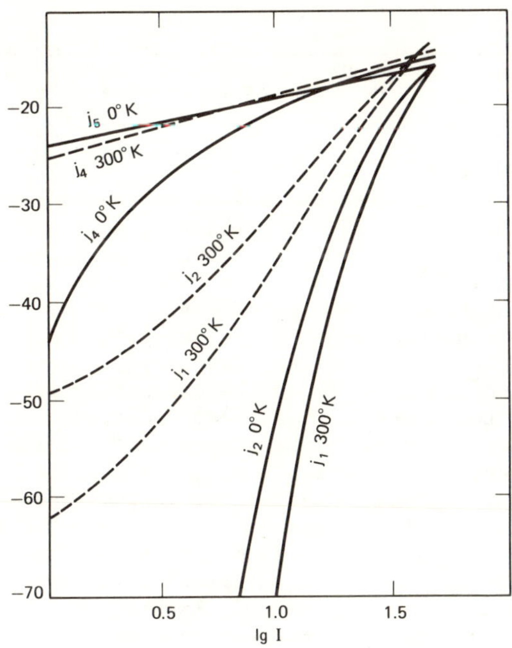

FIGURE 4(a)

Summing up these different order partial multiphoton currents, the resulting current j depends on a power of I higher than n_o, i.e., the relation $j \propto I^{n'}$ will hold with $n' \geqslant n_o$ above a given I value depending on the duration of the laser pulse and the initial temperature of the metal. Finally, for further increase of I only the fast thermionic emission of the electrons takes place with a very high n value (Fig. 4(b)).

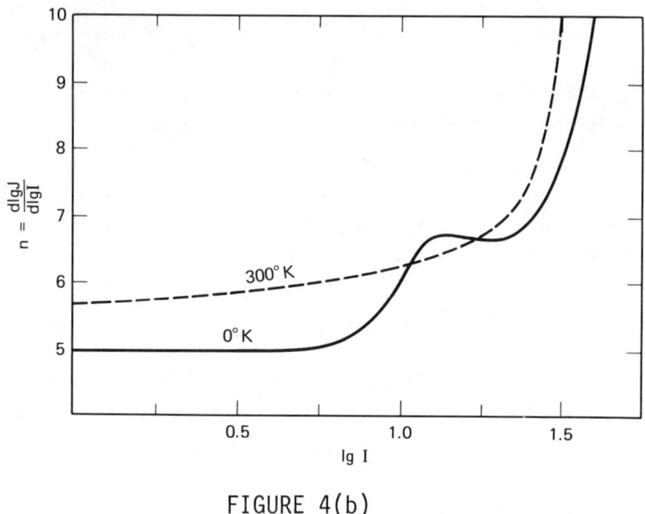

FIGURE 4(b)

It is mentioned here that a very interesting method has been elaborated by Körmendi [33] both theoretically and experimentally for the distortion of the Fermi distribution using a field emission tip in a high outer static electric field illuminated by laser pulses. The new Fermi distribution can be determined by observing the electrons emitted by the "photon enhanced" field emission process in a very sensitive energy analyzer system.

Similar phenomena occur in the experiments [34] performed to study the "photon assisted" classical Richardson emission.

After these remarks let us return to the chronology of the intensity dependence experiments of the photoeffect.

As a first experimental step we, in Budapest [35], have proven that the earlier mentioned predictions of Bunkin and Prokhorov [28] using ultrashort picosec duration laser pulses enable the pure multiphoton photoeffect up to 100 MW/cm^2 to be observed without any background of classical thermoemission for $n_o = 2,3,4,5$ combining different cathodes and lasers. Grazing light incidence was used which further reduced the heating, reducing the light absorption by the Au mirror-like cathode. After this, increasing the laser intensity, we performed the first experiment [36] demonstrating the existence of the predicted deviation from the n-th power law. Using ultrashort pulse trains of a mode-locked Nd laser we found a deviation at an $E_{\perp crit}$ value for which the tentative qualitative estimation gave the $10^{6.6}$ V/cm value, which is somewhat less than the theoretically predicted $10^{7.3}$ V/cm in [15] and much higher than the 10^5 V/cm value expected in [16,17].

It turned out in the course of these experiments that the characteristics of the photoemission were strongly dependent on the nature of the ultrashort pulses: namely quite different yield and n was found for regular gauss-shaped bandwidth-limited pulses and for pulses having irregular subpicosecond substructure and broad irregular spectrum. It is well known that in an Nd:glass laser the regular pulses may originate in the first part of the mode-locked train produced by a Q-switch of low initial absorption, while the pulses are irregular in the descending part of this train, or if using a Q-switch of high initial absorption the obtained laser pulses are already irregular both in the ascending and descending parts. Typical values for the duration of the regular pulses are of 6 picosec and for the irregular ones 50 picosec, the latter having substructures of 10^{-13} sec and a broadened spectrum.

We performed systematic investigations both with these regular and irregular pulses when they formed trains and when single pulses were selected from the trains. For the sake of completeness experiments were carried out using a single selected regular Nd:YAG laser pulses of 30 picosec duration, too.

In the experiment [37] using pulse trains produced by a Q-switch of low absorption the respective photocurrent signals of the linear detector and the multiphoton cathode were delayed with respect to each other and were photographed on the same oscilloscope (Fig. 5). It can be seen that owing to the non-linear response the envelope of the multiphoton signal train is shorter than that of the linear detector. Figure 6 shows the intensity dependence curve obtained by plotting in log-log scale the values of Fig. 5. It can be seen from the figure (Fig. 6) that the curve has three distinct parts: in the ascending part of the mode-locking train where the laser pulses are regular, the slope shows excellent agreement with the theoretically expected $j \propto I^5$ dependence thereby indicating the $n_0 = 5$ order multiphoton photoeffect; towards the higher maximum intensity values the slope decreases in accordance with the theoretically predicted decrease [15] and with our former result [36]; the third part of the curve, however, where the laser pulses are irregular, exhibits an unexpected steep slope in the descending part of the train.

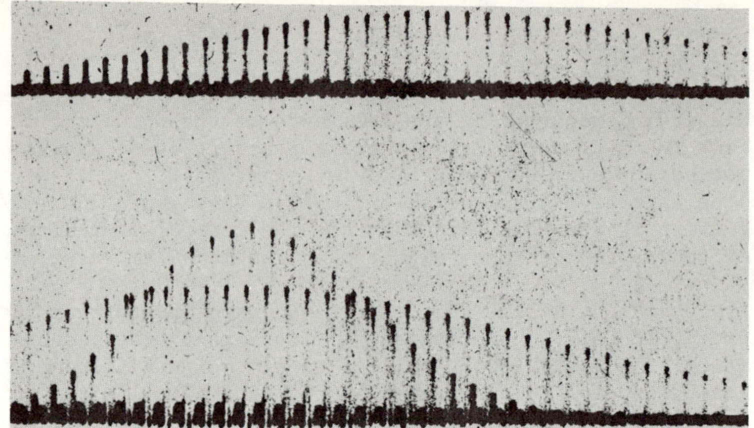

FIGURE   5

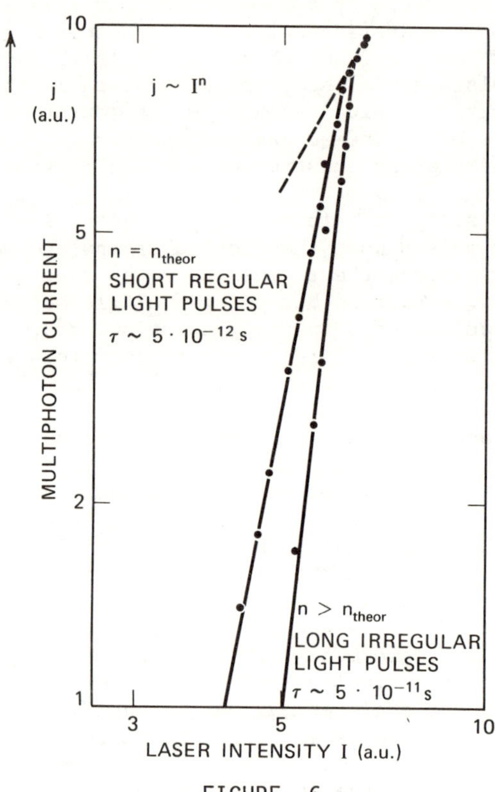

FIGURE   6

The slope values (i.e. the orders of nonlinearity n) obtained for the relation $j = f(I)$ by increasing the Q-switch absorption or in other words by spoiling the regularity of the laser pulses are shown in Fig. 7.

The experiment was repeated using single electro-optically selected regular as well as irregular pulses [38] and the results exactly corresponded to the expectations based on the results summarized in the former figures (Figs. 6 and 7). We therefore conclude that the results of these experiments coincide with the theoretical predictions of the multiphoton photoeffect, but only for the case of short and regular bandwidth-limited laser pulses.

Experiments performed with longer, but also single bandwidth-limited 30 psec duration Nd:YAG laser pulses confirmed this conclusion; Bloembergen et al. [39] who carried out an investigation on a tungsten cathode found the same to be true. The single selected pulse of an Nd:YAG oscillator was amplified and its time shape and space dimensions were determined in an indirect way generating the second harmonic of the laser frequency. The laser field strength values used (evaluated from the peak intensity value of the light pulse) were under $10^{6.2}$ V/cm, and the result gave a theoretical value of $n_o = 4$ indicating pure MPE (Fig. 8).

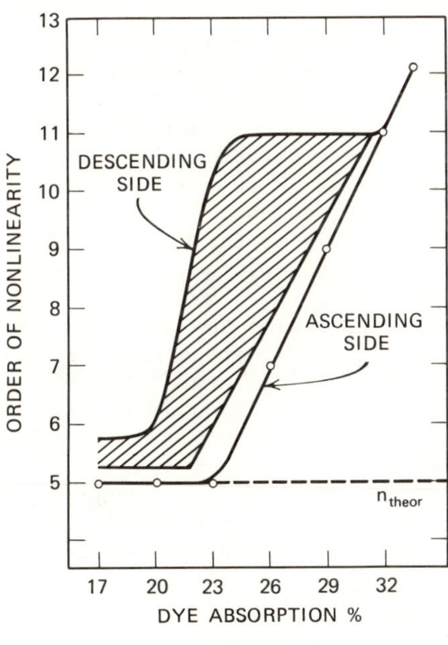

FIGURE 7

A similar experiment was performed by us in a cooperative program with the Saclay group in France [40] at the same time. The selected single transverse mode Nd:YAG laser pulse was amplified and the pulse shape and duration were controlled by a fast streak camera showing a Gaussian-shaped pulse with a duration of 30 psec. The laser beam reached the Au cathode under grazing incidence and the total electric vector E_O of the light was perpendicular to the cathode surface: $E_\perp = E_O$. The results are shown in Fig. 9. Up to several GW/cm^2 intensity values the $j \propto E^{2n_O} \propto I^{n_O}$ relation again holds with the theoretically predicted $n_O = 5$. Around $E_\perp = 10^{6.48 \pm 0.48}$ V/cm a deviation from this power dependence was observed.

The intensity scales used in [39], Fig. 8, and in [40], Fig. 9, are not equivalent, those used in [40] being the average and those in [39] the peak light intensity values,i.e., in [39] the peak intensity values of Fig. 8 must be reduced to obtain the average intensity values plotted in Fig. 9. To confirm that the observed MPE effect is a surface one, we carried out very sensitive polarization measurements [40] in the intensity range where the n_O-th power law is valid. The direction of the polarization of the light incident on the cathode is obtained by insertion of a Glan-Thompson prism in the path of the light beam. This direction could be varied by rotating the prism. The geometrical configurations and the results are presented in Fig.10, which shows the theoretically predicted dependences.

Summarizing these results we may conclude that the emission process observed in our experiments using bandwidth limited laser pulses is purely that of the surface multiphoton photoeffect of gold. Its experimental behavior towards the higher intensities supports the recent theoretical predictions. On the other hand, the explanation of the origin of the observed high n values in the case of irregular, non bandwidth limited laser pulses must be elsewhere.

Finally we remark that using the same experimental setup the same intensity dependence have been carried out in Saclay for noble gases (see Dr. Mainfray's lecture at this Conference). No deviation was observed from the n_O power law even in the case when Keldysh's γ was less than one ($\gamma \sim 0.3$), while in the microwave ionization experiments also at $\gamma < 1$, the deviation existed (see Dr. Bayfield's lecture at this Conference). These unexpected results are not yet understood theoretically.

Therefore experiments are in progress also in Saclay for metals when $\gamma < 1$ using CO_2 lasers.

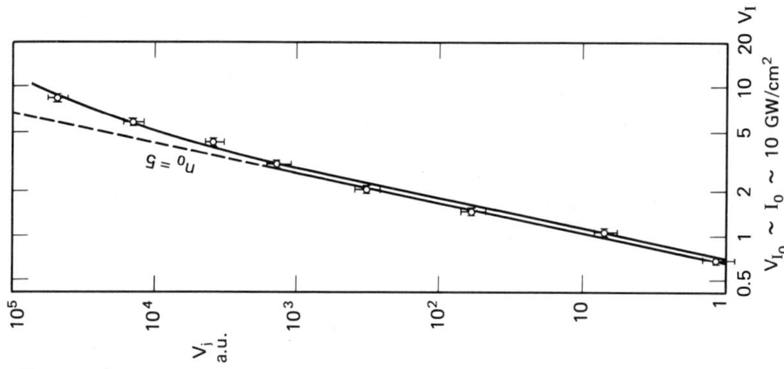

FIGURE 9

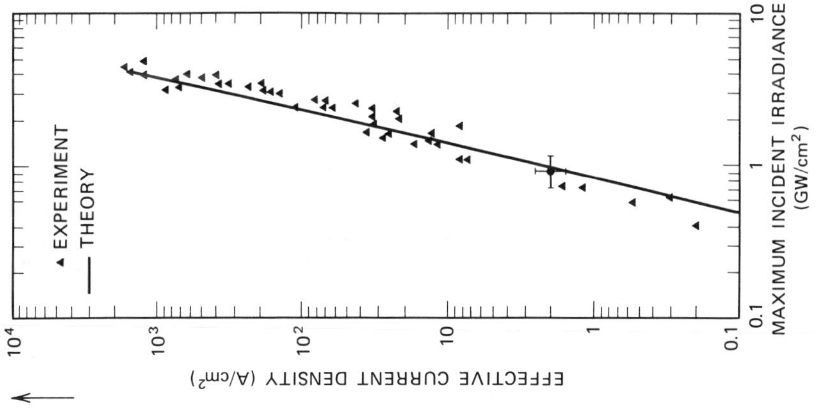

FIGURE 8

V_j ; $|E_\perp(\alpha)|^{2n_0}$; $|E_{||}(\alpha)|^{2n_0}$
a.u.
$E_\perp(\alpha)$; $E_{||}(\alpha)$
a.u.
10^5
10^4
$|E_{||}(\alpha)|^{2n_0}$
10^3
$|E_\perp(\alpha)|^{2n_0}$
V_j
(Exp.)
10^2
$E_{||}(\alpha)$
$E_\perp(\alpha)$
10
$|E_{||}(\alpha)|$
1
0°   30°   60°   90°   120°   150°
α
E_0
POLARIZER
α
GOLD TARGET
$E_\perp$
45°
$E_\perp(\alpha) = E_0 \cos\left(\frac{\pi}{4} - \alpha\right) \sin\alpha$
$E_{||}(\alpha) = E_0 \cos\left(\frac{\pi}{4} - \alpha\right) \cos\alpha$

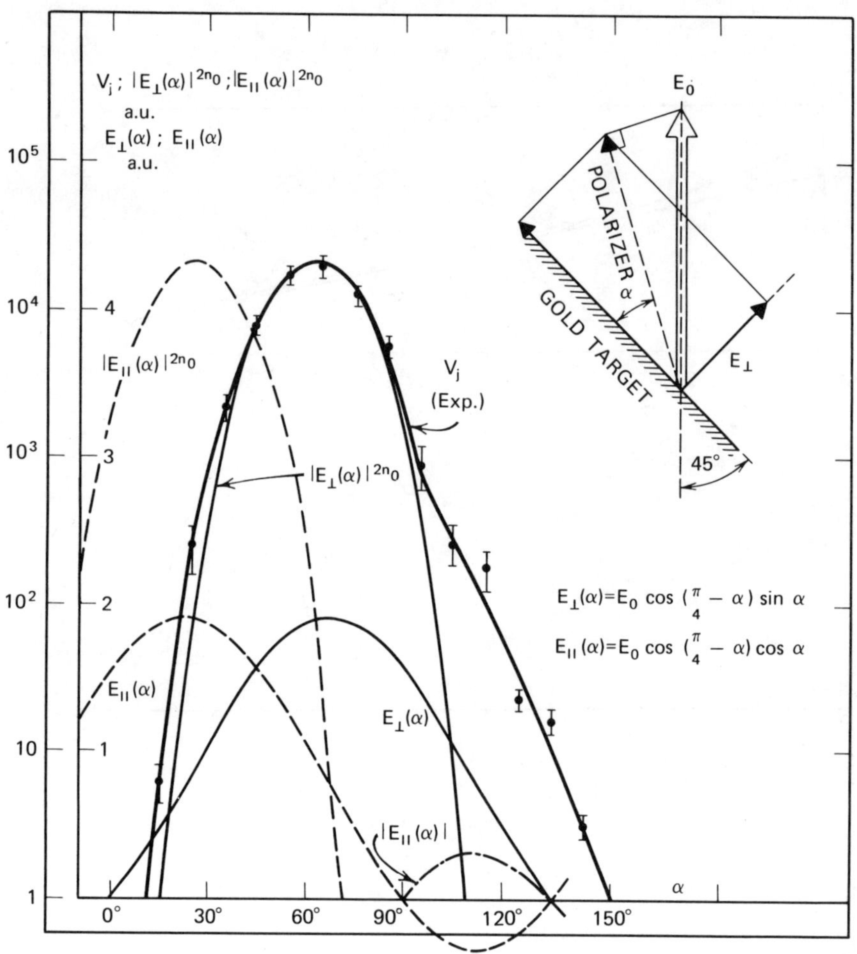

FIGURE   10

II. MULTIPHOTON RESONANCE EFFECTS IN THE
MULTIPHOTON PHOTOEFFECT OF METALS

Considering that the work function A of our metal is characterized by two levels, namely the Fermi level and the vacuum level, the question arises whether resonance effects might occur by varying the laser frequency. The question seems to be natural, keeping in mind that the work function of gold is $A = 4.8$ eV, which is very near to the 4 photon energy value of the Nd laser (1.17 eV). It should be noted of course that at first sight, the resonance, if any, may not be very sharp owing to the $kT \sim 10^{-2}$ eV energy spread of the Fermi electrons at room temperatures. We have seen, however, that with bandwidth limited laser pulses the intensity dependence relations corresponded to the situation, as if the electron gas were of nearly zero temperature. On the other hand, when irregular non bandwidth limited pulses of broadened spectrum were used an unexpectedly strong intensity dependence, i.e. high n values, was observed. It is quite natural to imagine that when broadening the spectrum, a bypass from the n-1 photon interaction to the n photon one occurs accompanied by a resonance phenomenon in which case very strange intensity dependence relations may hold - as is well known from the case of gases.

We decided to investigate this question experimentally; Dr. Horvath will present the problem in detail at the Conference. The experiments [41] were performed in Saclay using a tunable picosecond laser oscillator with an amplifying system and by automatic recording of the tuned wavelength and the duration and time shape of the laser pulse. The description of this setup is published elsewhere. The spectral bandwidth and the duration of the bandwidth limited laser pulse were 1.5 Å and 15 psec, respectively, the tuning range was 80 Å.

Two experiments were performed: in the first one the gold cathode was situated in a static vacuum bulb at 10^{-8} torr, baked and degassed during two days; in the second the gold cathode was polished again and was closed in a vessel with dynamic vacuum of less than 10^{-9} torr, but baked for only about 8 hours.

The experimental results are summarized in Fig. 11 where the exponent n of the $j \propto I^n$ intensity dependence relation is plotted against the wavelength λ. It can be seen that an unexpectedly sharp resonance exists at the wavelength $\lambda_o = 10598$ Å with a width of $\Delta\lambda \sim 7$ Å. It can also be seen that the place of the resonance does not depend on the different treatment of the cathode, but a certain damping of the resonance was found in the second experiment.

At wavelength less than $\lambda_o = 10598$ Å, $n = 4$ order and at those higher than λ_o, $n = 5$ order MPE occurs.

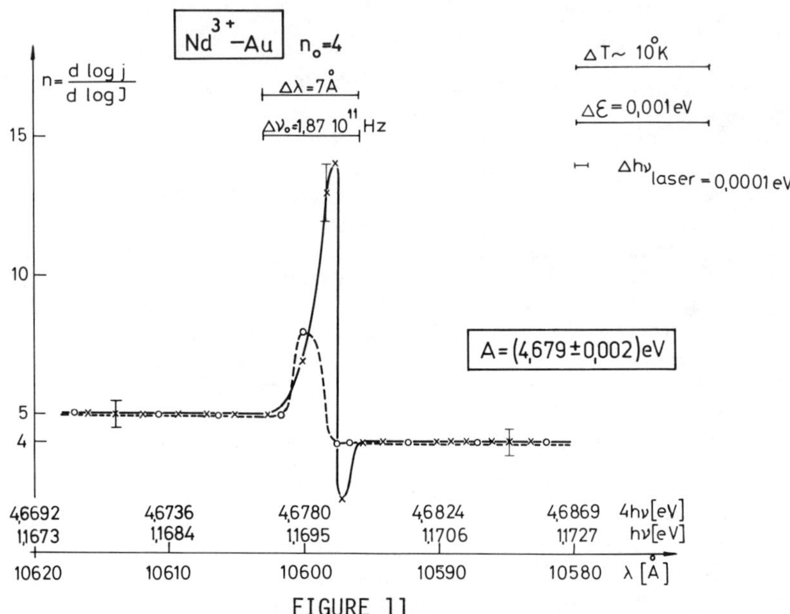

FIGURE 11

At the time of these experiments extensive theoretical works
were continued in Moscow by Anisimov, Kantorovich and others [31,
32,42,43] in connection with the theoretical aspects of these
questions also bearing in mind the temperature effects.

Using perturbation theory the dependence of n = d log j/d log I
on $\omega = 2\pi\nu$ was calculated [42] in a wide ω-range for Au cathode
and nonosec ruby laser pulses. Bumps were found on this n(ω) dependence curve at $\hbar\omega = A/n$ when n was integer. At high intensities the bumps disappeared due to heating.

In other papers [31,32] the emission characteristics were
calculated for Ag cathode and picosec Nd laser pulses. The order
of nonlinearity n was determined in the function of the laser intensity, laser frequency and the angles of incidence and polarization, taking into account the initial cathode temperature.
Figure 12 shows the frequency dependence of n where the parameters are the laser intensity I and the initial cathode temperature T_o. It can be seen that sharp resonance occurs at $\hbar\omega \sim A/4$
just as in our experimental curve in Fig. 11. The great difference, however, is that the half width values of the experimental
curves are many orders lower than those of the theoretical ones.

Quite similar results were found in the calculation where
the j(I) and n(ω) dependences were determined when an Au cathode
was irradiated simultaneously by two laser beams of frequencies
ω_1 and ω_2; the obtained resonance curve had even wider half width.

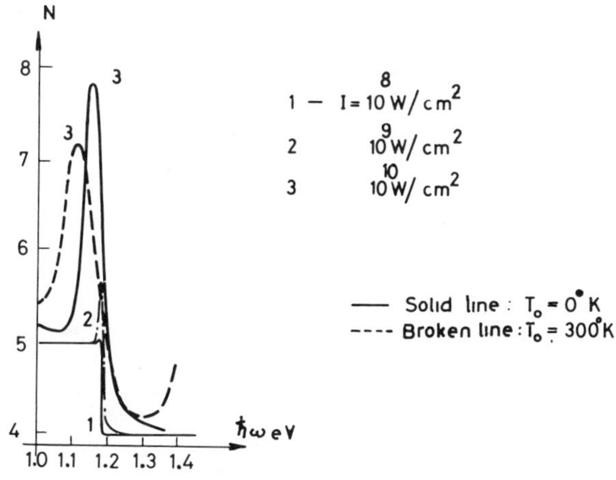

FIGURE 12

It can be concluded that the preliminary experimental resonance curves coincide more with those theoretical ones for which
the initial cathode temperature $T_{o} \sim 0$, just as it was stated in
the case of the intensity dependence. It must be emphasized again
that all of these results were obtained with regular bandwidth
limited laser pulses.

Although it is clearly too early to make definite statements,
it is certain that the experimental and theoretical discovery of
the resonance in the case of solid state systems may open a quite
new field for the research, both from the point of view of the
theory of multiphoton resonances and from that of solid state
physics. By this method the work function determination may be
improved by several orders.

III. INTERACTION OF LASER PHOTONS AND FREE ELECTRONS

We have seen in the preceeding sections that when calculating the probability of the multiphoton photoemission in most of
the theoretical works the Volkov state was used as the final
state for the photoelectron. It is well known that the Volkov
state is obtained from the exact solution of the Dirac equation
or the Klein Gordon equation, when the only potential in which
the electron is moving is that of the plane e.m. field of the
laser beam. The most complete review of the problems of the
photon-electron interaction in an intense laser beam is given by
Eberly [44]. Among the many interesting theoretical results
treated, the most important one for us is the optically induced

energy level structure of the electron in a laser beam predicted
first by Eberly [45] himself and later by others [44,46]. This
states that an electron with initial energy E_i will have a discrete energy set $E_n = E_i \pm nh\nu$ in a laser beam, with $n = 0 \pm 1,2,..$.
For the multiphoton photoemission $E_i = nh\nu - A$, i.e., $E_i < h\nu$, which
ought to be the maximum electron energy without considering the
Volkov state treatment.

Experimental evidence shows, however, that instead of E_i, unexpectedly high electron energies may be observed after the ionization or photoemission. Thus we published the observation of
~ 10 eV electron energies in our former photoeffect experiments
[23]. Later in the multiphoton gas ionization experiments of the
Saclay group, as high as 30 eV energy values for the emitted
electrons were detected [47]. And just now, here in Rochester a
correct quantitative determination was carried out by Mandel and
Martin [48] showing both the expected E_i values and a 10 eV peak
in the energy distribution of the electrons of the multiphoton
ionized gases. The theoretical interpretation of these unexpected
high energies was not quite satisfying on the basis of energy gain
from the classical electron oscillation process in the laser field.

In this situation we decided to follow Eberly's suggestion
[45], which urged experimentalists to investigate the problem
directly, i.e. the possible absorption of n hν quanta from the
running laser beam in a simple process.

We have performed [49] the following preliminary experiment
(Fig. 13). A heated oxide cathode at $\sim 800°$K emitted electrons of
$kT \sim 10^{-2}$eV thermal energies with density $\sim 10^{10}$ electron/cm^3. In
front of the cathode a tungsten mesh was situated with a 4 cm
spacing. Between the cathode and the mesh a negative variable
potential was applied. When the polarity of the potential of the
mesh is negative and has a value of several volts, the electrons
with low thermal energies cannot penetrate the mesh. When, however, the electron cloud is shot by an intense laser pulse, according to the prediction of Eberly, there are electrons which
may absorb 1,2,...,n quanta. Since the quantum energy of Nd laser
light h$\nu = 1.17$ eV, these electrons acquire energy enough to pass
through the negative grid. Unfortunately no quantitative numerical value for the probability of the process was given in the
theoretical work. Nevertheless it is sure that it is very low.
Therefore to be able to detect these electrons, if any, we built
behind the grid, an electron multiplier with a gain value of
10^6 (EMI-6903) at a pressure of 10^{-8} torr. The emitted and multiplied electrons will charge the 1 pF capacity giving a signal.
From this value the number of interacting electrons may be roughly
determined.

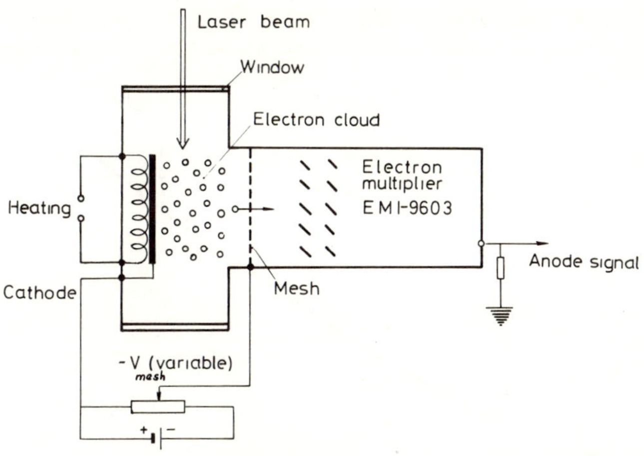

FIGURE 13

An I^n power law function is to be expected since this is the form generally taken by the n photon interaction. The experiment was performed in Saclay, where we used the same laser system as in the photoeffect experiments. We used a Nd:YAG laser of duration 30 psec and of maximum intensity of 10 GW/cm^2. The unfocused beam diameter was 0.5 cm. As for the grid potential we must keep in mind that owing to the contact potentials of the electrodes [50], even without any outer potential, a 4 V potential existed a priori on the mesh, with which we corrected always the measured values.

The experiment consisted of the determination of the multiplier anode signals in the function of the laser intensity with the parameter of the mesh potential. The results are shown in Fig. 14. It can be seen that the results show an I^n power law intensity dependence and the respective grid potentials correspond more or less to the nhν values expected for n photon interactions.

With the effect just detectable, we were able to determine only the intensity dependence and were not able to measure the energy distribution curve. Careful control measurements were performed to exclude the possibility of other secondary effects, (e.g. photoeffect from the walls and electrodes, etc.).

It follows already from simple qualitative considerations, that the effect is stronger than expected. Recently, Körmendi carried out a short numerical estimation for this problem in which he also predicted the existence of such a kind of effect, but with lower probability than was found by us. He attributes this to the fact that our electrons may not be considered as quite free ones due to the effect of the space charge field and to the grid potential. His results will be presented at this Conference.

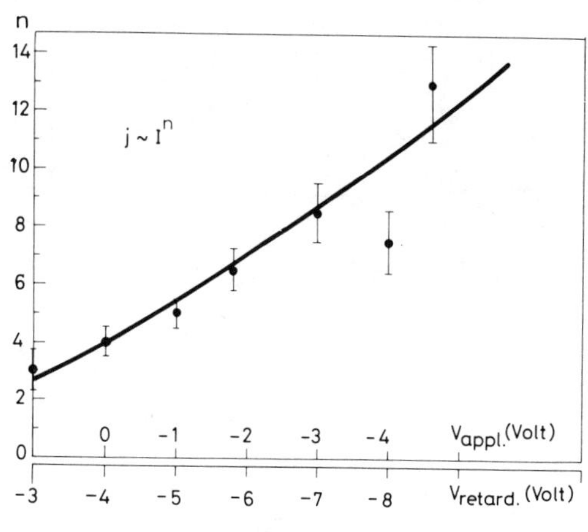

FIGURE 14

We may conclude that this very preliminary experiment shows the existence of the photon energy absorption by an electron in a strong laser beam. For the correct interpretation of the process, however, more serious experiments and theoretical estimations are required.

Further experiments are in progress and we hope that this preliminary experimental result will stimulate theoreticians to perform calculations on this subject.

REFERENCES

1.   P. P. Barashev, Phys. Stat. Sol. a9, 9 (1972).

2.   A. D. Gladun, P. P. Barashev, Usp. Fiz. Nauk 98, 493 (1969).

3.   Gy. Farkas, Proceedings of the I. Conference on Interaction of Electrons with Strong Electromagnetic Field. Invited papers, Budapest, 1973. p. 179.

4.   J. F. Ready, Effects of High-Power Laser Radiation, Academic Press, New York, London (1971).

5.   S. I. Anisimov, V. A. Benderskii, Gy. Farkas, Usp. Fiz. Nauk 122, 185 (1977).

6.   K. Mitchell, Proc. Roy. Soc. 146A, 442 (1934).

7.   R.E.B. Makinson, M. J. Buckingham, Proc. Roy. Phys. Soc. 64A, 135 (1951).

8.   R. L. Smith, Phys. Rev. 128, 2225 (1962).

9.   I. Adawi, Phys. Rev. 134A, 788 (1964).

10. M. E. Marinchuk, Phys. Rev. Lett. 34A, 97 (1971).

11. P. P. Barashev, Fiz. Tverd. Tela 12, 1973 (1970).

12. M. C. Teich, G. J. Wolga, Phys. Rev. 171, 809 (1968).

13. L. V. Keldysh, Zh. Eksp. Teor. Fix. 47, 1945 (1964).

14. F. V. Bunkin, M. V. Fedorov, Zh. Eksp. Teor. Fiz. 48, 1341 (1965).

15. A. P. Silin, Fix. Tverd. Tela 12, 3553 (1970).

16. A. M. Brodskii, Yu. Ya. Gurevich, Zh. Eksp. Teor. Fis. 60, 1453 (1971).

17. A. M. Brodskii, Yu. Ya. Gurevich, Teorija Elektronnoi Emissii iz Metallov, Nauka, Moscow, 1973.

18. D. M. Volkov, Z. für Phys. 94, 250 (1935).

19, M. C. Teich, J. M. Schroer, G. J. Wolga, Phys. Rev. Lett. 13, 611 (1964).

20. E. M. Logothetis, P. L. Hartman, Phys. Rev. Lett. 18, 581 (1967).

21. E. M. Logothetis, P. L. Hartman, Phys. Rev. 187, 460 (1969).

22. Gy. Farkas, Zs. Náray, P. Varga, Phys. Lett. 24A, 134 (1967).

23. Gy. Farkas, I. Kertész, Zs. Náray, P. Varga, Phys. Lett. 24A, 475 (1967).

24. Gy. Farkas, I. Kertész, Zs. Náray, P. Varga, Phys. Lett. 25A, 572 (1967).

25. Gy. Farkas, I. Kertész, Zs. Náray, Phys. Lett. 28A, 190 (1968).

26. L. J. Korshunov, V. A. Benderskii, V. I. Goldanskii, Yu. M. Zolotovskii, Pismo Zh. Teor. Fiz. 7, 55 (1968).

27. M. Louis-Jacquet, C. R. Ac. Sci. Paris, B273, 192 (1971).

28. F. V. Bunkin, A. M. Prokhorov, Zh. Eksp. Teor. Fiz. 52, 1610 (1967).

100   Gy. Farkas

29.   S. I. Anisimov, Ya.A. Imas, G. S. Romanov, Yu. V. Khodyko, Deisvie Izluchenia Bolshoi Moshnosti na Mettaly, Nauka, Moscow, 1970.

30.   S. I. Anisimov, N. A. Inogamov, Yu. P. Petrov, Phys. Lett. 45A, 449 (1976).

31.   I. Kantorovich, Zh. Techn. Fiz. 47, 660 (1977).

32.   I. Kantorovich, Pismo Zh. Techn. Fiz. 3, 230; 3, 280 (1977).

33.   F. F. Körmendi, Physica 75, 359 (1974).

34.   J. H. Bechtel, P. A. Franken, Phys. Rev. B11, 3159 (1975).

35.   Gy. Farkas, Z. Horváth, I. Kertész, G. Kiss, Nuovo Cim. Lett. 1, 314 (1971).

36.   Gy. Farkas, Z. Gy. Horváth, I. Kertész, Phys. Lett. 39A, 231 (1972).

37.   Gy. Farkas, Z. Gy. Horváth, Opt. Comm. 12, 392 (1974).

38.   Gy. Farkas, Z. Gy. Horváth, L. A. Lompré, G. Petite, Phys. Stat. Sol. a39, K25 (1977).

39.   J. H. Bechtel, W. L. Smith, N. Bloembergen, Opt. Comm. 12, 392 (1975).

40.   L. A. Lompré, J. Thébault, Gy. Farkas, Appl. Phys. Lett. 27, 110 (1975).

41.   Gy. Farkas, Z.Gy. Horváth, L. A. Lompré, G. Mainfray, C. Manus, J. Thébault, Contributed paper at this Conference, to be published.

42.   S. I. Anisimov, V. Fisher, I. Ponrovskaya, to be published.

43.   Yn. Petrov, N. Inogamov, to be published.

44.   J. H. Eberly, Progress in Optics VII, p. 361 (1969)(North Holland).

45.   J. H. Eberly, Proc. International Conf. on Optical Pumping and Atomic Line Shape, Warsaw, 1968, p. 311.

46.   A. M. Dukhne, G. L. Liudin, Usp. Fiz. Nauk 121, 157 (1977).

47.   L. A. Lompré, G. Mainfray, J. Thébault, Saclay Preprint to be published.

48.   E. A. Martin, L. Mandel. Appl. Opt. 15, 2378 (1976).

49.   Gy. Farkas, L. A. Lompré, G. Mainfray, C. Manus, J. Thébault, to be published.

50.   P.A.Redhead et al.,The Physical Basis of Ultrahigh Vacuum (1968) (Chapman and Hall).

Part 2

STRONG-FIELD RESONANCE FLUORESCENCE

Dressed-Atom Approach
to Resonance Fluorescence

C. COHEN-TANNOUDJI AND S. REYNAUD
Ecole Normale Supérieure et Collège de France
75231 Paris Cedex 05 - France

I. INTRODUCTION

Resonance fluorescence, which is the subject of this session, has been studied for a long time. The first quantum treatment of the scattering of resonance radiation by free atoms was given by Weisskopf and Wigner, in the early days of quantum mechanics [1].

The importance of this process in various fields such as spectroscopy, optical pumping, lasers,... is obvious and does not require further discussion. In the last few years, the interest in the problem of resonance fluorescence has been renewed by the development of tunable laser sources which made it possible to irradiate atomic beams with intense monochromatic laser waves and to study the characteristics of the fluorescence light. For example, the fluorescence spectrum, which is monochromatic at very low laser intensities, as predicted by lowest order QED for elastic Rayleigh scattering, exhibits more complex structures at higher intensities when absorption and induced emission predominate over spontaneous emission. Some of these experiments, which have been initiated by the work of Schuda, Stroud and Hercher [2], here in Rochester, will be discussed in subsequent papers [3].

From the theoretical point of view, a lot of papers have been devoted to this problem, and it would be impossible here to review all of them [4]. Let's just mention the publication of Mollow [5], who, in 1969, presented a complete and correct treatment of the problem, starting from the Bloch equations for the atomic density matrix driven by a c-number applied field, and using the quantum regression theorem for evaluating the correlation function of the atomic dipole moment. Perhaps the theoretical activity in this field can be interpreted as an attempt to build some simple physical pictures of resonance fluoresecence at high intensities in terms of photons. Actually this problem is not so simple and, before entering into any calculations, it seems interesting to point out some of these difficulties.

Let us first introduce some important physical parameters. It's well known that an atom, irradiated by a resonant monochromatic wave, oscillates between the ground state g and the excited state e with a frequency ω_1, which is the Rabi nutation frequency, and which is equal to the product of the atomic dipole moment d and the electric field amplitude E. ω_1 characterizes the strength of absorption and stimulated emission processes.

Γ, the natural width of the excited state, is the spontaneous emission rate. In intense fields, when $\omega_1 \gg \Gamma$, each atom can oscillate back and forth between e and g several times before spontaneously emitting a fluorescence photon.

T is the transit time of atoms through the laser beam and is usually much longer than the radiative lifetime Γ^{-1} of the excited state e.

From the preceding considerations, it appears first that one cannot analyze the situation in terms of a single fluorescence process. In intense fields, when each atom spends half of its time in e, there is on the average, for each atom, a *sequence* of several ($\sim \Gamma T/2 \gg 1$) fluorescence processes, which cannot be considered as independent, as a consequence of the coherent character of the laser driving field.

Secondly, we have clearly a *non equilibrium* situation. Any "steady-state" which can be eventually reached by the system is actually a dynamical equilibrium: photons are constantly transferred, through fluorescence processes, from the laser mode to the empty modes of the electromagnetic field.

Finally, and this is perhaps the most difficult point, one must not forget that, in quantum theory, the corpuscular and wave aspects of light are *complementary*. There is not a unique physical description of the sequence of fluoresecence processes which can be applied to all possible experiments.

Suppose for example we are interested in the *time aspect* of the problem, more precisely in the probability $p(\theta)$ for having 2 successive fluorescence photons emitted by the same atom separated by a time interval θ. This can be achieved by measuring, with a broad-band photomultiplier, the intesity correlations of the fluorescence light emitted by a very dilute atomic beam (for a theoretical analysis of this problem, see references 6 and 7). Note also that, throughout this paper, we restrict ourselves to very dilute atomic systems so that we can ignore any cooperative effects such as those discussed in reference 8. Once we have detected one fluorescence photon, the atom is certainly in the ground state because of the "reduction of the wave packet". In order to be able to emit a second photon, it must be reexcited in the upper state by the laser light. It is therefore not surprising that $p(\theta)$ is given by the Rabi transient describing the excitation of the atom from the ground state. Note in particular that $p(\theta) \to 0$ when $\theta \to 0$, showing an "antibunching" of the fluorescence photons emitted by a single atom.

One can also be interested in the *frequency distribution* of
the fluorescence light, and from now we will only consider this
type of problem. In such experiments, the fluorescence photons
are sent into an interferometric device, such as a high finesse
Fabry-Perot etalon, inside which they are kept for such a long
time that we lose all information concerning their order of emission. This clearly shows the complementarity between time and
frequency which cannot be simultaneously determined. This means
also that, for each ensemble of fluorescence photons with frequencies ω_A, ω_B,...,ω_N, we have several indistinguishable sequences of fluorescence processes, differing by the order of
emission of photons, and that we *must* take into account possible
interferences between the corresponding quantum amplitudes. There
is another example of such a difficulty which is well known in
atomic physics: the paradox of spontaneous emission from an harmonic oscillator [*9*]. It is well known that the linewidth of the
spontaneously emitted radiation is independent of the initial excitation of the oscillator. Such a result can be derived quantum
mechanically only if one takes into account the interferences between the N! possible cascades through which the oscillator decays from its initial excited state N to the ground state O.

II.   THE DRESSED ATOM APPROACH

In this paper, we would like to present a dressed-atom approach to resonance fluorescence, discussed in detail in reference 10, and which, in our opinion, solves the previous difficulties and leads to simple interpretations for the fluorescence
and absorption spectra of atoms irradiated by intense resonant
laser beams.

Let's emphasize that such an approach does not lead to new
results which could not be derived from a c-number description of
the laser field. Actually, the c-number description may be shown
to be strictly equivalent to the quantum description provided that
the initial state of the field is a coherent one [*11*]. What we
would like to show here is that introducing from the beginning
the energy levels of the combined system [atom + laser mode] leads,
in the limit of high intensities, to simpler mathematical calculations and more transparent physical discussions.

We give now the general idea of this method. In a first
step, one neglects spontaneous emission and one considers only
the total isolated system "atom + laser mode interacting together".
We call such a system the dressed-atom or the atom dressed by
laser photons. One easily determines the energy diagram of such
a system, which exhibits a quasiperiodicity associated with the
quantization of the radiation field.

Then, we introduce the coupling with the empty modes, responsible for the transfer of photons from the laser mode to the empty modes (Fig. 1). Resonance fluorescence can therefore be considered as spontaneous emission from the dressed atom. Similarly, a sequence of fluorescence processes corresponds to a radiative cascade of the dressed atom downwards its energy diagram.

Due to the very broad frequency spectrum of the empty modes, it is always possible to describe the spontaneous decay of the dressed atom by a master equation. In the limit of high intensities, more precisely in the limit of well resolved spectral lines, we will see that this equation takes a much simpler form.

Finally, it would be in principle necessary to introduce the coupling of the laser mode with the lasing atoms and with the cavity losses, in order to describe the injection of photons into the laser mode and the fluctuations of the laser light.

We will suppose here that the laser fluctuations are negligible and we will forget this coupling. We will describe the laser beam as a free propagating wave corresponding to a single mode of the radiation field, initially excited in a coherent state, with a Poisson distribution $p_o(n)$ for the photon number. The width, Δn, of this distribution is very large in absolute value, but very small compared to the mean number of photons $\bar{n}$ (quasi classical state):

$$1 \ll \Delta n \simeq \sqrt{\bar{n}} \ll \bar{n} \tag{1}$$

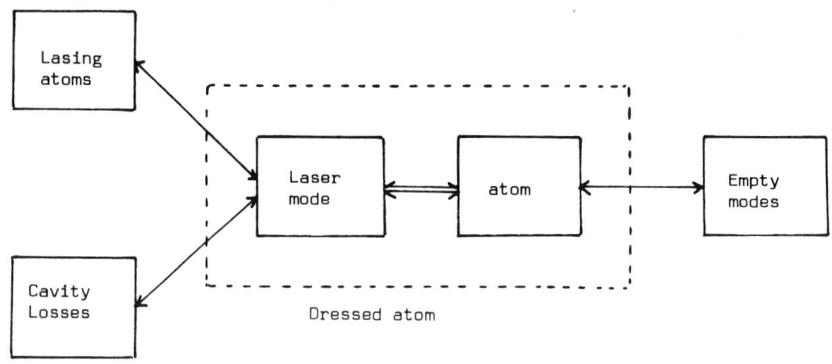

FIGURE 1.   Various couplings appearing in the dressed atom
            approach.

Furthermore, $\bar{n}$ has no real physical meaning: we can always let $\bar{n}$ and the quantization volume V tend to ∞, the ratio $\bar{n}/V$ remaining constant and related to the electromagnetic energy density experienced by the atom. This is why it will be justified to consider the dressed atom energy diagram as periodic over a very large range Δn, and to neglect the variation with n of any matrix element when n varies within Δn.

III. APPLICATION TO A TWO-LEVEL ATOM

We now show how this method works in the simple case of a two-level atom.

A. Energy Diagram

The unperturbed states of the dressed atom are labelled by two quantum numbers: e, g for the atom; n for the number of photons in the laser mode. They are bunched in two-dimensional multiplicities E_n, E_{n-1} separated by ω_L (laser frequency)(Fig.2a). The splitting between the two states $|g,n+1\rangle$ and $|e,n\rangle$ of E_n is the detuning δ between the atomic and laser frequencies ω_0 and ω_L:

$$\delta = \omega_0 - \omega_L \tag{2}$$

An atom in g can absorb one laser photon and jump to e. This means that the two states $|g,n+1\rangle$ and $|e,n\rangle$ are coupled by the laser mode-atom interaction Hamiltonian V, the corresponding matrix element being

$$\langle e,n|V|g,\ n+1\rangle = \frac{\omega_1}{2} \tag{3}$$

Since δ and ω_1 are small compared to ω_L, one can neglect all other couplings between different multiplicities, which is equivalent to making the rotating wave approximation. Thus, one is led to a series of independent two-level problems. One immediately finds that the two perturbed states associated with E_n (Fig. 2b) are separated by a splitting

$$\omega_{12} = \sqrt{\omega_1^2 + \delta^2} \tag{4}$$

and are given by

$$|1,n\rangle = \cos\phi|e,n\rangle + \sin\phi|g,\ n+1\rangle$$
$$|2,n\rangle = -\sin\phi|e,n\rangle + \cos\phi|g,\ n+1\rangle \tag{5}$$

where the angle ϕ is defined by

$$tg\ 2\phi = \omega_1/\delta \tag{6}$$

For the following discussion, it will be useful to know the matrix elements of the atomic dipole operator D. This operator does not act on the number n of laser photons and therefore the only non-zero matrix elements of D in the unperturbed basis are

$$\langle g,n|D|e,n\rangle = \langle g|D|e\rangle = d \tag{7}$$

From the expansion (5) of the dressed atom states, one deduces that D only couples states belonging to two adjacent multiplicities. We will call d_{ji} these matrix elements

$$d_{ji} = \langle j,\ n-1|D|i,n\rangle \tag{8}$$

One finds immediately the matrix elements corresponding to the various allowed Bohr frequencies (arrows of Fig. 2b):

$$d_{21} = d\cos^2\phi \qquad \text{(frequency } \omega_L + \omega_{12})$$
$$d_{12} = -d\sin^2\phi \qquad \text{(frequency } \omega_L - \omega_{12}) \tag{9}$$
$$d_{11} = -d_{22} = d\sin\phi\cos\phi \qquad \text{(frequency } \omega_L)$$

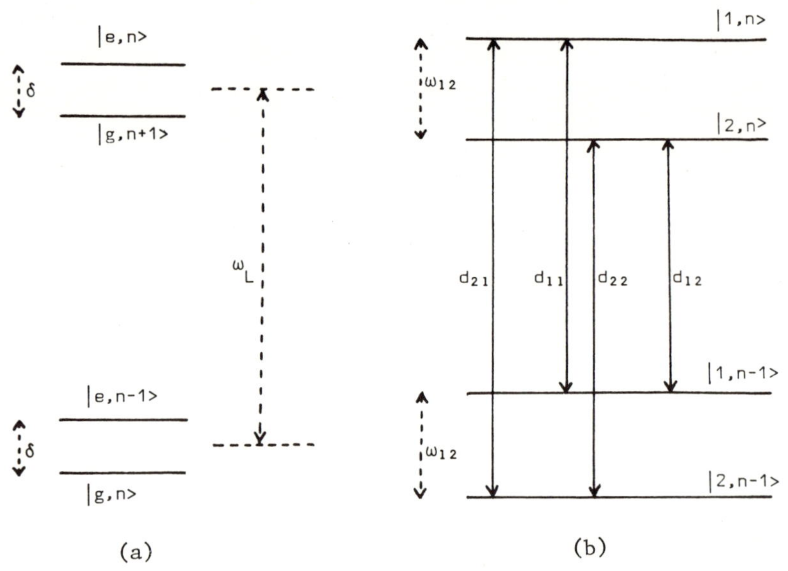

FIGURE 2. (a) Unperturbed states of the total system "atom + laser mode";(b) perturbed states of the same system.

B. Secular Approximation

Let us introduce now spontaneous emission, characterized by Γ. We will not discuss the general case, but restrict ourselves to the limit of well resolved lines:

$$\omega_{12} \gg \Gamma \tag{10}$$

From Eq. (4), this condition means either intense fields ($\omega_1 \gg \Gamma$) or large detunings ($|\delta| \gg \Gamma$) or both.

In such a case, the master equation describing spontaneous emission can be considerably simplified. Any coupling, which is of the order of Γ, between two density matrix elements evolving at different frequencies, differing at least by ω_{12}, can be neglected. This is the so called secular approximation which is the starting point of an expansion in powers of Γ/ω_{12} and which leads to independent sets of equations only coupling the elements of the dressed atom density matrix σ evolving at the same Bohr frequency.

C. Evolution of the Populations

As a first illustration of this discussion, let us consider the set of equations coupling the elements of σ evolving at frequency 0, i.e. the diagonal elements of σ which represent the populations $\Pi_{i,n}$ of the dressed atom energy levels:

$$\Pi_{i,n} = \langle i,n|\sigma|i,n\rangle \tag{11}$$

In these equations, important parameters appear which are the transition rates Γ_{ji} (Γ_{ji} is the transition rate from $|i,n\rangle$ to $|j,n-1\rangle$) and which are simply related to the dipole matrix elements introduced above:

$$\Gamma_{ji} = |\langle j,n-1|D|i,n\rangle|^2 = d_{ji}^2 \tag{12}$$

The evolution equations of the populations $\Pi_{i,n}$ can then be written

$$\dot{\Pi}_{1,n} = -(\Gamma_{11}+\Gamma_{21})\Pi_{1,n} + \Gamma_{11}\Pi_{1,n+1} + \Gamma_{12}\Pi_{2,n+1}$$

$$\dot{\Pi}_{2,n} = -(\Gamma_{12}+\Gamma_{22})\Pi_{2,n} + \Gamma_{21}\Pi_{1,n+1} + \Gamma_{22}\Pi_{2,n+1} \tag{13}$$

These equations have an obvious physical meaning in terms of transition rates: for example, the population $\Pi_{1,n}$ decreases because of transitions from $|1,n\rangle$ to the lower levels with a total rate

$$\Gamma_1 = \Gamma_{11} + \Gamma_{21} \tag{14}$$

and increases because of transitions from upper levels $|1,n+1\rangle$ (transition rate Γ_{11}) and $|2,n+1\rangle$ (transition rate Γ_{12}) (Fig. 3):

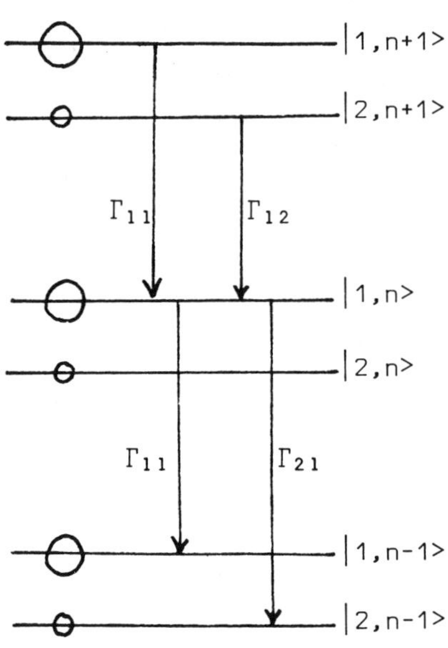

FIGURE 3.   Evolution of the population $\Pi_{1,n}$.

Because of the quasi-classical character of the laser mode state (condition 1), all matrix elements can be considered as periodic within the width Δn of the photon distribution $p_0(n)$. Thus $\Pi_{i,n+1}$ and $\Pi_{i,n}$ are practically equal and they can be written as

$$\Pi_{i,n+1} \simeq \Pi_{i,n} \simeq p_0(n)\, \Pi_i \tag{15}$$

where Π_i is a reduced population.

One deduces from Eqs. (13) and (15) that the reduced populations Π_i obey the simple equations

$$\dot{\Pi}_1 = -\Gamma_{21}\, \Pi_1 + \Gamma_{12}\, \Pi_2$$

$$\dot{\Pi}_2 = -\Gamma_{12}\, \Pi_2 + \Gamma_{21}\, \Pi_1 \tag{16}$$

One easily derives the following results for the evolution of the populations Π_i. Starting from the initial values $\Pi_i(o)$ obtained by expanding on the dressed atom states the initial state of the laser mode-atom system, the populations exhibit a transient behavior, on a time scale of the order of Γ^{-1}, and then reach a steady state $\Pi_i(\infty)$, or more precisely a dynamical equilibrium, where they do not vary any more.

The populations $\Pi_i(\infty)$ are determined by the two equations

$$\Pi_1(\infty) + \Pi_2(\infty) = 1 \tag{17}$$

$$\Gamma_{21}\,\Pi_1(\infty) = \Gamma_{12}\,\Pi_2(\infty) \tag{18}$$

The first one is the normalization condition. The second is the detailed balance condition, obtained from Eq. (16), and expressing that the number of transitions from $|2,n+1\rangle$ to $|1,n\rangle$ compensates the number of transitions from $|1,n\rangle$ to $|2,n-1\rangle$.

Solving these two equations leads to

$$\Pi_1(\infty) = \frac{\Gamma_{12}}{\Gamma_{12} + \Gamma_{21}} = \frac{\sin^4\phi}{\sin^4\phi + \cos^4\phi}$$
$$\Pi_2(\infty) = \frac{\Gamma_{21}}{\Gamma_{12} + \Gamma_{21}} = \frac{\cos^4\phi}{\sin^4\phi + \cos^4\phi} \tag{19}$$

D. Positions and Weights of the Various Components of the Fluorescence Spectrum

The positions of these components are given by the allowed Bohr frequencies of the atomic dipole moment which, according to the previous discussion, are $\omega_L - \omega_{12}$, $\omega_L, \omega_L + \omega_{12}$. So, we have a triplet of three well resolved lines since $\omega_{12} \gg \Gamma$.

It is clear that the total number of photons emitted on the component $\omega_L + \omega_{ij}$ is equal to the total number of transitions $|i,n\rangle \to |j,n-1\rangle$, corresponding to this frequency, and occurring during the transit time T of atoms through the laser beam. Since T is larger than Γ^{-1}, one can consider only the dynamical equilibrium regime. It follows that the weight $G_F(\omega_L + \omega_{ij})$ of the $\omega_L + \omega_{ij}$ component is given by:

$$G_F(\omega_L + \omega_{ij}) = T\,\Gamma_{ji}\,\Pi_i(\infty) \tag{20}$$

From the detailed balance condition (18), one deduces

$$G_F(\omega_L + \omega_{12}) = G_F(\omega_L - \omega_{12}) \qquad (21)$$

We have therefore a close connection between the detailed balance and the symmetry of the spectrum.

Since we know the Γ_{ji} and the Π_i, we can easily write analytical expressions for the weights of the two sidebands and for the weight $G_F(\omega_L)$ of the central component:

$$G_F(\omega_L + \omega_{12}) = G_F(\omega_L - \omega_{12}) = \Gamma T \frac{\sin^4\phi \, \cos^4\phi}{\sin^4\phi + \cos^4\phi} \qquad (22)$$

$$G_F(\omega_L) = T(\Gamma_{11} \Pi_1(\infty) + \Gamma_{22} \Pi_2(\infty)) = \Gamma T \sin^2\phi \, \cos^2\phi$$

They coincide with the now well known results concerning two level atoms at the limit of well resolved spectral lines.

E.   Widths of the Components

In order to obtain the width of the lateral components at $\omega_L \pm \omega_{12}$, we consider now the evolution of the off diagonal element of the density matrix connecting $|1,n\rangle$ and $|2,n-1\rangle$ and which we note σ^+_{12n}

$$\sigma^+_{12n} = \langle 1,n|\sigma|2,n-1\rangle \qquad (23)$$

The evolution equation of σ^+_{12n} can be written:

$$\dot\sigma^+_{12n} = -i(\omega_L + \omega_{12})\sigma^+_{12n} - \frac{1}{2}(\Gamma_1 + \Gamma_2)\sigma^+_{12n} + d_{11}d_{22} \, \sigma^+_{12 \, n+1} \qquad (24)$$

Three terms appear in this equation: the first one describes the free evolution of σ^+_{12n} at the Bohr frequency $\omega_L + \omega_{12}$; the second one describes the damping of σ^+_{12n} by spontaneous emission with a rate equal to the half sum of the total decay rates Γ_1 and Γ_2 from the two levels $|1,n\rangle$ and $|2,n-1\rangle$. Finally, one must not forget the coupling of σ^+_{12n} with another off diagonal element of σ, $\sigma^+_{12 \, n+1}$, which connects $|1,n+1\rangle$ and $|2,n\rangle$ and evolves at the same Bohr frequency, the coupling coefficient being the product of the two dipole matrix elements d_{11} and d_{22} (Fig. 4).

As above, we can use the periodicity property for replacing $\sigma^+_{12 \, n+1}$ by σ^+_{12n} in Eq. (24), which gives the damping rate L_{12} of σ^+_{12n}

$$L_{12} = \frac{1}{2}(\Gamma_1 + \Gamma_2) - d_{11}d_{22} \qquad (25)$$

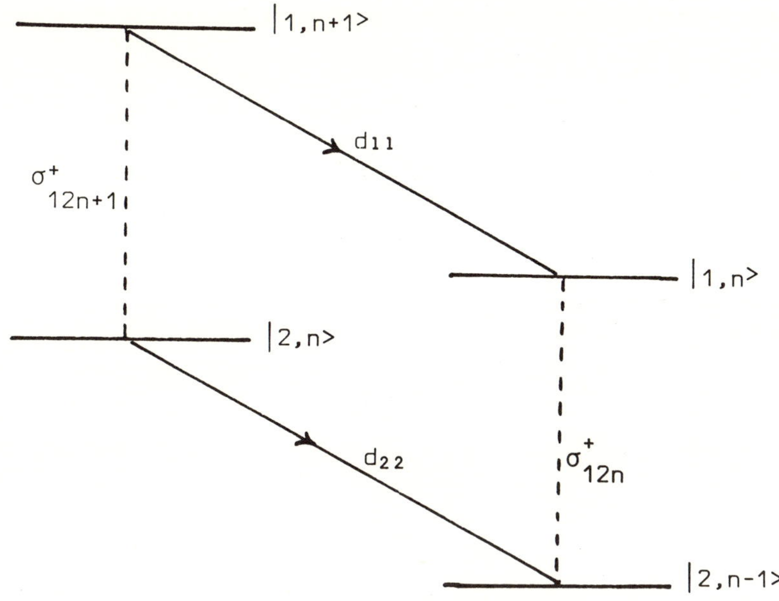

FIGURE 4. Coupling of off-diagonal density matrix elements
evolving at the same Bohr frequency.

This rate is also the damping rate of the component of the mean
dipole moment oscillating at frequency $\omega_L + \omega_{12}$, so that L_{12} is
also the width of the lateral components. One therefore concludes
that the width of the line emitted on the transition $|1,n> \to |2,n-1>$
is not simply given by the half sum of the natural widths Γ_1 and
Γ_2 of the two involved levels. Because of the periodicity of the
energy diagram, there is a phase transfer in the radiative cascade which is responsible for a correction equal to $-d_{11}d_{22}$. The
explicit expression of L_{12} in terms of Γ and ϕ is:

$$L_{12} = \Gamma(\tfrac{1}{2} + \cos^2\phi \, \sin^2\phi) \qquad (26)$$

The problem of the central component is a little more complicated. There are now two off diagonal elements

$$\sigma^+_{iin} = <i,n|\sigma|i,n-1> \qquad \text{with } i = 1,2 \qquad (27)$$

which connect the same multiplicities E_n and E_{n-1} and which
evolve at the same frequency ω_L. One can show that, except for
the free evolution terms, the σ^+_{iin} and the populations $\Pi_{i,n}$

obey the same equations. But the time evolution of the populations is given by the superposition of a transient regime and of
a steady state one. It follows that the central component is
actually the superposition of two lines: a δ-function corresponding to the coherent scattering due to the undamped oscillation of
the mean dipole moment driven at ω_L by the laser wave and an inelastic central component. Simple calculations (10) give the
weights G_{el} and $G_{inel}(\omega_L)$ of the elastic and inelastic central
components and the width L_c of the inelastic one:

$$
\begin{aligned}
G_{el} &= T(d_{11}\,\Pi_1(\infty) + d_{22}\,\Pi_2(\infty))^2 \\[2mm]
&= \Gamma T \cos^2\phi\,\sin^2\phi\,\left[\frac{\cos^4\phi - \sin^4\phi}{\cos^4\phi + \sin^4\phi}\right]
\end{aligned}
\tag{28}
$$

$$
\begin{aligned}
G_{inel}(\omega_L) &= G_F(\omega_L) - G_{el} \\[2mm]
&= \Gamma T\,\frac{4\,\cos^6\phi\,\sin^6\phi}{(\cos^4\phi + \sin^4\phi)^2}
\end{aligned}
\tag{29}
$$

$$
L_c = \Gamma_{12} + \Gamma_{21} = \Gamma(\sin^4\phi + \cos^4\phi)
\tag{30}
$$

F. Absorption Spectrum

We now show how this dressed atom approach provides a
straightforward interpretation of very recent experimental results [12].

Atoms are always irradiated by an intense laser beam at ω_L.
Instead of looking at the fluorescence light emitted by these
atoms, one measures the absorption of a second weak probe beam ω.
ω_L is fixed and ω is varied. One can say that, in this experiment, one measures the absorption spectrum of the dressed atom.

Since the perturbation introduced by the weak probe beam can
be neglected, the energy levels $|i,n\rangle$ $|j,n-1\rangle$... of the dressed
atom and their populations Π_i Π_j... are the same as before. To
the transition $|i,n\rangle \rightarrow |j,n-1\rangle$ of the dressed atom corresponds,
in the absorption spectrum, a component centered at $\omega_L + \omega_{ij}$, with
a width L_{ij}, and a weight $(\Pi_j - \Pi_i)\,\Gamma_{ji}T$ determined by the difference between absorption and stimulated emission processes.

This is the main difference between fluorescence and absorption spectra. An absorption signal is proportional to the *difference of populations* between the two involved levels; a spontaneous emission signal only depends on the population of the
upper state.

We therefore arrive at the following conclusions:

(i) Because of the periodicity property, the two levels $|i,n>$ and $|i,n-1>$ (with $i = 1,2$) have the same populations. So, the central component at ω_L disappears since it corresponds to transitions between two equally populated levels.

(ii) If Π_2 is larger than Π_1, the lateral component at $\omega_L + \omega_{12}$ (transition $|i,n> \rightarrow |2,n-1>$) is absorbing since the lower level $|2,n-1>$ is the most populated (Fig. 5). But, then, the second lateral component at $\omega_L - \omega_{12}$ (transition $|2,n> \rightarrow |1,n-1>$ is *amplifying* since it is now the upper level $|2,n>$ which is the most populated.

(iii)Finally, at resonance, one easily finds that $\Pi_1 = \Pi_2$ so that all levels are equally populated and all components disappear in the absorption spectrum.

Let's recall that all these results are only valid to 0th order in Γ/ω_{12}. Higher order corrections to the secular approximation, which is used here, would introduce smaller signals, which do not vanish at resonance or near ω_L.

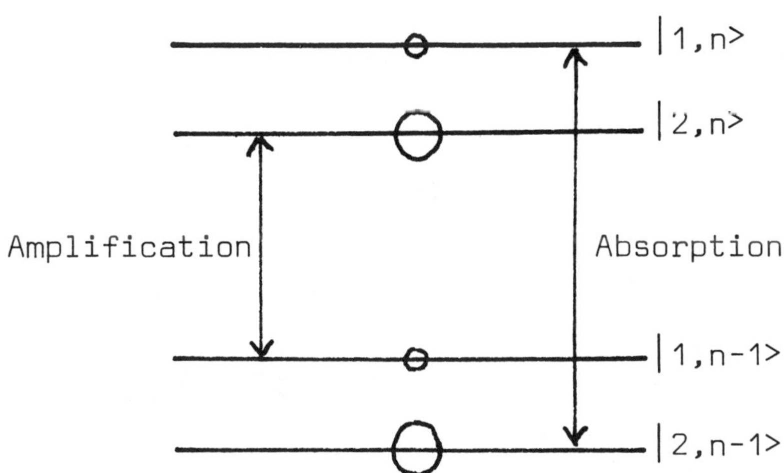

FIGURE 5.   If $\Pi_2 > \Pi_1$, the $\omega_L + \omega_{12}$ component is absorbing, whereas the $\omega_L - \omega_{12}$ component is amplifying.

It appears clearly from the previous discussion that good amplification requires a detuning between the laser and atomic frequencies ω_L and ω_o. One can easily compute the optimum conditions for such an amplification.

The maximum of the amplifying line is obtained by dividing the weight of this line by its width. Let us call G the ratio of this maximum amplification to the maximum absorption of free atoms. One obtains

$$G = \frac{(\Pi_2 - \Pi_1)\, \Gamma_{12}\, T}{L_{12}}\; \frac{\Gamma/2}{\Gamma T}$$

$$= \frac{\sin^4\phi\,(\cos^2\phi - \sin^2\phi)}{(\cos^4\phi + \sin^4\phi)(1 + 2\,\sin^2\phi\,\cos^2\phi)} \tag{31}$$

The value of the detuning which maximizes this amplification is given by

$$\frac{|\omega_L - \omega_o|}{\omega_1} = 0.334 \tag{32}$$

and it corresponds to a maximum amplification

$$G_{max} = 4.64\% \ . \tag{33}$$

IV.   CONCLUSION

To conclude, let us mention some further applications of the dressed atom approach.

First, this method can be directly applied to multilevel systems. Similar mathematical expressions having the same physical meaning can be derived for the characteristics of the various components of fluorescence and absorption spectra. We have already used it to study several problems such as the modification of the Raman effect at very high intensities [13]; the simultaneous saturation of two atomic transitions sharing a common level [14], a situation which occurs frequently in stepwise excitation experiments (in that case, we have to consider atoms dressed by two types of photons); polarization effects related to Zeeman degeneracy and which could be observed by exciting atoms with a given polarization and by observing the fluorescence spectrum with a different one [15].

Second, this method is very well suited to the study of the effect of collisions in the presence of resonant fields [*16,17, 18*]. One has to add in the master equation new terms describing the transition rates induced by collisions. The detailed balance condition, giving the dynamical equilibrium, depends now on both radiative and collisional rates. But. on the other hand, the weights of the lines only depend on radiative rates. This provides very simple interpretations for the asymmetries which appear in the fluorescence spectrum and which are due to collisions.

Finally, one can easily introduce the Doppler effect in the theory by plotting energy diagrams giving the dressed atom energy levels versus the atomic velocity. We have shown that these diagrams can provide very simple interpretations for the various saturation signals observed in laser spectroscopy. We are presently investigating some new effects suggested by such an approach [*18*].

REFERENCES

1.   V. Weisskopf and E. Wigner, Z. Phys. <u>63</u>, 54 (1930).

2.   F. Schuda, C. R. Stroud Jr. and M. Hercher, J. Phys. <u>B7</u>, L198 (1974).

3.   See subsequent papers of H. Walther and S. Ezekiel, including: H. Walther, Proc. of the 2nd Laser Spectroscopy Conf., June 75, ed. S. Haroche, J. C. Pebay-Peyroula, T. W. Hansch and S. H. Harris, 358-69 (Springer-Verlag, Berlin 1975); F. Y. Wu, R. E. Grove and S. Ezekiel, Phys. Rev. Lett. <u>35</u>, 1426 (1975); W. Hartig, W. Rasmussen, R. Schieder and H. Walther, Z. Phys. <u>A278</u>, 205 (1976); R. E. Grove, F. Y. Wu and S. Ezekiel, Phys. Rev.A <u>15</u>, 227-33 (1977).

4.   See subsequent papers of L. Mandel and H. J. Kimble, B. R. Mollow, C. R. Stroud, Jr. and H. R. Gray, D. F. Walls, W. A. McClean and S. Swain. See also references given in 10.

5.   B. R. Mollow, Phys. Rev. <u>188</u>, 1969 (1969).

6.   H. J. Carmichael and D. F. Walls, J. Phys. <u>B9</u>, L43 (1976); J. Phys. <u>B9</u>, 1199 (1976).

7.   C. Cohen-Tannoudji, Frontiers in Laser Spectroscopy, Les Houches 1975, Session XXVII, ed. R. Balian, S. Haroche and S. Liberman (North Holland, Amsterdam, 1977).

8.   R. Bonifacio and L. A. Lugiato, Opt. Com. <u>19</u>, 172 (1976).

9.   N. Kroll, in Quantum Optics and Electronics, Les Houches
      1964, ed. C. De Witt, A. Blandin and C. Cohen-Tannoudji
      (Gordon and Breach, 1965).

10.   C. Cohen-Tannoudji and S. Reynaud, J. Phys. $\underline{B10}$, 345 (1977);
      S. Reynaud,Thèse de 3ème cycle (Paris, 1977).

11.   B. R. Mollow, Phys. Rev. A $\underline{12}$, 1919 (1975).

12.   F. Y. Wu and S. Ezekiel, M. Ducloy, B. R. Mollow, Phys. Rev.
      Lett. $\underline{38}$, 1077 (1977). See also, A. M. Bonch-Breuvich,
      V. A. Khodovoi and N. A. Chiger', Zh. Eksp. Teor. Fiz. $\underline{67}$,
      2069 (1974) [Sov. Phys. JETP 40, 1027 (1975)]; S. L. McCall,
      Phys. Rev. A $\underline{9}$, 1515 (1974); H. M. Gibbs, S. L. McCall and
      T. N. C. Venkatesan, Phys. Rev. Lett. $\underline{36}$, 1135 (1976).

13.   C. Cohen-Tannoudji and S. Reynaud, J. Phys. $\underline{B10}$, 365 (1977).

14.   C. Cohen-Tannoudji and S. Reynaud to be published in J. Phys.
      $\underline{B10}$ (1977).

15.   C. Cohen-Tannoudji and S. Reynaud, J. de Phys. Lettres, $\underline{38}$,
      L173 (1977).

16.   J. L. Carlsten, A. Szöke and M. G. Raymer, Phys. Rev. A
      $\underline{15}$, 1029 (1977); E. Courtens and A. Szöke, Phys. Rev. A $\underline{15}$,
      1588 (1977).

17.   B. R. Mollow, Phys. Rev. A $\underline{15}$, 1023 (1977).

18.   C. Cohen-Tannoudji, F. Hoffbeck and S. Reynaud, to be
      published.

Resonance Fluorescence
under Finite Bandwidth Excitation

L. MANDEL AND H. J. KIMBLE*
Department of Physics and Astronomy
University of Rochester
Rochester, New York

I. INTRODUCTION

Until the last year or so it has been customary in all treatments of resonance fluorescence to assume that the exciting field
is completely monochromatic and of definite intensity. In practice, lasers are generally used for the experiments [1], so that
the assumption of fixed intensity is reasonable. However, even
a laser beam has a finite bandwidth, and the neglect of this bandwidth may not be justified when extremely sharp atomic transitions
are being investigated. We have recently succeeded in incorporat-
ing the effect of finite bandwidth excitation into the treatment
of resonance fluorescence [2], and have shown that it modifies
all aspects of the phenomenon somewhat, but that the effect is
greatest on the spectrum and greatest under off-resonant excitation. Although there have been other treatments of the same
problem [3], they have tended to concentrate on the case of resonant excitation, when some of the most interesting effects are
absent.

We treat the atom as a two-level quantum system with energy
level spacing $\hbar\omega_o$. Its natural dynamical variables are then
either the three Pauli spin operators [4] $\hat{R}_1(t)$, $\hat{R}_2(t)$, $\hat{R}_3(t)$, or
alternatively the atomic raising and lowering operators $\hat{b}^\dagger(t)$
and $\hat{b}(t)$. The energy of the atomic system located at the origin
$\underline{r} = o$ in an electromagnetic field, in the presence of an electric
dipole interaction, is then given by

$$\hat{H} = \frac{1}{2} \int \left[\epsilon_o \hat{E}^2(\underline{r},t) + \frac{1}{\mu_o} \hat{B}^2(\underline{r},t) \right] d\underline{r} +$$

$$+ \hbar\omega_o \hat{R}_3(t) + 2\omega_o \underline{\mu} \cdot \hat{\underline{A}}(0,t)\hat{R}_2(t), \tag{1}$$

---

*This work was supported by the National Science Foundation.
119

in which the last term is the interaction term. μ is the transition dipole moment, and the other variables have their usual meaning.

We shall be interested in the electromagnetic field seen by a photodetector at position $\underline{r}$ some distance from the atom, which requires that we integrate the Heisenberg equations of motion for the electromagnetic field. Fortunately, these equations are simply Maxwell's equations, which can be readily combined into an inhomogeneous wave equation, and the solution can then be written down directly with the help of the well-known retarded Green function. We find for the positive frequency part of the electric vector $\hat{\underline{E}}^{(+)}(\underline{r},t)$ in the far-field [5]

$$\hat{\underline{E}}^{(+)}(\underline{r},t) = \frac{\omega_o^2}{4\pi\varepsilon_o c^2}\left[\frac{\underline{\mu}}{r}-\frac{\underline{\mu}\cdot\underline{r}}{r^3}\,\underline{r}\right]\hat{b}\left(t-\frac{r}{c}\right)+\hat{\underline{E}}^{(+)}_{\text{free}}(\underline{r},t), \tag{2}$$

in which the first term represents the contribution of the atomic source, and the second term is the complementary function, which is the free field or the solution of the homogeneous wave equation.

The field itself is of course not measured in the optical domain; we are generally interested only in the expectation of certain products of field operators. Among these are the average light intensity $I(\underline{r},t)$, which is the expectation value of $\hat{\underline{E}}^{(-)}(\underline{r},t)\cdot\hat{\underline{E}}^{(+)}(\underline{r},t)$ and is proportional to the probability of photodetection at time t [6,7], the intensity correlation $\Gamma^{(2,2)}(\underline{r},t,\tau)$, which is the expectation value of the fourth order product of fields in normal order and in time order and is proportional to the joint probability of photodetection at two times t and $t+\tau$ [6,7], and the spectrum, which is the Fourier transform of the second order correlation function $\Gamma^{(1,1)}(\underline{r},t,\tau)$ of the field [6,7]. We may readily show from Eq. (2), provided the photodetector is not directly exposed to the exciting field, that [5]

$$I(\underline{r},t) \equiv \langle\hat{\underline{E}}^{(-)}(\underline{r},t)\cdot\hat{\underline{E}}^{(+)}(\underline{r},t)\rangle = \left(\frac{\omega_o^2\mu\sin\psi}{4\pi\varepsilon_o c^2 r}\right)^2\left[\langle\hat{R}_3(t)\rangle+\frac{1}{2}\right] \tag{3}$$

$$\Gamma^{(2,2)}(\underline{r},t,\tau) \equiv \langle \hat{E}_i^{(-)}(\underline{r},t)\hat{E}_j^{(-)}(\underline{r},t+\tau)\hat{E}_j^{(+)}(\underline{r},t+\tau)\hat{E}_i^{(+)}(\underline{r},t)\rangle$$

$$= \left(\frac{\omega_o^2\mu\sin\psi}{4\pi\varepsilon_o c^2 r}\right)^4 \left[\langle\hat{b}^\dagger(t)\hat{R}_3(t+\tau)\hat{b}(t)\rangle + \frac{1}{2}\langle R_3(t)\rangle + \frac{1}{4}\right] \tag{4}$$

$$\Gamma^{(1,1)}(\underline{r},t,\tau) \equiv \langle \hat{\underline{E}}^{(-)}(\underline{r},t)\cdot\hat{\underline{E}}^{(+)}(\underline{r},t+\tau)\rangle$$

$$= \left(\frac{\omega_o^2\mu\sin\psi}{4\pi\varepsilon_o c^2 r}\right)^2 \langle\hat{b}^\dagger(t)b(t+\tau)\rangle \, , \tag{5}$$

where ψ is the angle between the $\underline{\mu}$ and $\underline{r}$ vectors, which we shall
take to be 90° in order to avoid complications associated with
Doppler shifts. The problem is therefore one of determining the
various expectations on the right of Eqs. (3) to (5) for a given
initial state of the system.

II. THE STATE OF THE EXCITING FIELD

We take the initial state, immediately prior to the interaction, as a product state in which the atom is in any pure quantum state, and in which the field is describable by an ensemble
of coherent states, such that for each member of the ensemble

$$\hat{A}_{\underline{-}\mathrm{Free}}^{(+)}(0,t)|\rangle = \underline{\varepsilon}\, A\, e^{-i\omega_1 t}\, e^{i\phi(t)}|\rangle \, . \tag{6}$$

Here $\underline{\varepsilon}$ is a unit polarization vector, A is a real amplitude, ω_1
is the frequency on which the exciting field (the laser) is
centered, and $\phi(t)$ is a time-dependent, slowly varying phase,
that is treated as a random process and generates the ensemble
of states. Subsequently we shall have to average over the ensemble, and we denote this average by $\langle\langle\,\rangle\rangle$.

In order to define the ensemble more explicitly, we make the
simplifying assumption that the phase $\phi(t)$ in effect performs a
random walk in time, and may be treated like the displacement of
an imaginary particle undergoing a one-dimensional Brownian motion.
This phase diffusion model is in accordance with the well-known
Langevin equation for the laser field [8]. The mean squared

phase excursion in a time interval τ is taken to be proportional to τ,

$$\langle\langle[\phi(t) - (t + \tau)]^2\rangle\rangle = 2\lambda|\tau| , \qquad (7)$$

with a coefficient λ that determines the laser linewidth. It follows directly from the Langevin equation that

$$\langle\langle\exp ix[\phi(t) - \phi(t + \tau)]\rangle\rangle = \exp(-\lambda|\tau|x^2), \qquad (8)$$

which shows that the phase excursion $\phi(t) - \phi(t+\tau)$ is a Gaussian variable, and that the corresponding laser spectrum is Lorentzian of width λ.

In calculating the ensemble average of terms in Eqs. (3) to (5) we shall at times encounter combinations like

$$\langle\langle \langle\hat{0}_A(t)\rangle \exp i[\phi(t) - \phi(t + \tau)]\rangle\rangle ,$$

in which $\hat{0}_A(t)$ is some atomic operator that is to be evaluated at time t. By virtue of the fact that the atom at time t is not affected by subsequent phase fluctuations, it follows that

$$\langle\langle \langle\hat{0}_A(t)\rangle \exp i[\phi(t) - \phi(t + \tau)]\rangle\rangle = \langle\langle \langle\hat{0}_A(t)\rangle \rangle\rangle\langle\langle\exp i[\phi(t)-\phi(t+\tau)]\rangle\rangle$$

$$\langle\langle \langle\hat{0}_A(t)\rangle \rangle\rangle \exp(-\lambda \tau) , \quad \tau \geqslant 0.$$

$$(9)$$

This factorization is not valid of course for $\tau < 0$, as the atom is certainly influenced by preceding phase fluctuations.

We have now specified all the conditions required for the evaluation of the three quantities defined by Eqs. (3) to (5). The problem is to integrate the coupled Heisenberg equations of motion for all the relevant dynamical variables, to form the products indicated and to calculate expectation values. We shall not to into the details of the calculation here, as they are given in Ref. [2]. Instead we wish to discuss some of the solutions and to indicate how they differ from the corresponding results under monochromatic excitation.

III. SOLUTION UNDER MONOCHROMATIC EXCITATION

We have shown earlier [5] that, under monochromatic excitation, the light intensity $I(r,\tau)$, the intensity correlation $\Gamma^{(2,2)}(r,t,\tau)$ and the amplitude correlation $\Gamma^{(1,1)}(r,t,\tau)$ are all derivable from the solution of a Volterra integral equation of form

$$F(\tau) = y(t,\tau) + \int_{o}^{\tau} d\tau' \, K(\tau-\tau')F(\tau'), \tag{10}$$

in which $F(\tau)$ stands for any one of the three functions $I(\underline{r},\tau)$, $\Gamma^{(2,2)}(\underline{r},t,\tau)$, $\tilde{\Gamma}^{(1,1)}(\underline{r},t,\tau)$, and $\tilde{\Gamma}^{(1,1)}(\underline{r},t,\tau)$ is related to $\Gamma^{(1,1)}(\underline{r},t,\tau)$ by a linear transformation. The kernel $K(\tau)$ is the same for all three functions and is given by

$$K(\tau) = -\frac{\Omega^2}{\beta(1+\theta^2)} e^{-\beta\tau}\left[\cos\beta\theta\tau + \theta\sin\beta\theta\tau - e^{-\beta\tau}\right]. \tag{11}$$

Here β is half the Einstein A-coefficient, θ is the relative detuning parameter, which is the ratio of $\gamma + \omega_1 - \omega_0$ to β (γ is the Lamb shift), and Ω is the Rabi frequency $2\omega_0\underline{\mu}\cdot\epsilon\underline{A}/\hbar$. The inhomogeneous term $y(t,\tau)$ in Eq. (10) is generally different for the different functions $I(\underline{r},\tau)$, $\Gamma^{(2,2)}(\underline{r},t,\tau)$, $\tilde{\Gamma}^{(1,1)}(\underline{r},t,\tau)$. The solution of Eq. (10) is of the general form

$$F(\tau) = C + \sum_{i=1}^{3} D_i \, e^{p_i\tau} \tag{12}$$

in which the C and D_i depend on β,θ,Ω,t and on which function is being described, and the p_i ($i = 1,2,3$) are the roots of the cubic equation

$$p^3 + 4\beta p^2 + (5\beta^2 + \beta^2\theta^2 + \Omega^2)p + 2\beta^3 + 2\beta^3\theta^2 + \Omega^2\beta = 0, \tag{13}$$

which is the same in all three cases. It can be shown that the following important factorization property holds for the two-time intensity correlation function [5] (we neglect transit times from atom to detector)

$$\Gamma^{(2,2)}(\underline{r},t,\tau) = I(\underline{r},t)I_G(\underline{r},\tau), \tag{14}$$

where $I_G(\underline{r},\tau)$ is the light intensity radiated by the atom at time τ in the case in which it starts in the lower (or ground) state. A special case of the factorization property (14) was first given by Carmichael and Walls [9], but it holds very generally for reasons that have been discussed [2,5]. The fact that the same kernel $K(\tau)$ and the same cubic equation (13) describe the τ-dependence of all three functions $I(\underline{r},\tau)$, $\Gamma^{(2,2)}(\underline{r},t,\tau)$ and $\tilde{\Gamma}^{(1,1)}(\underline{r},t,\tau)$ is no accident of course; it is a reflection of the fact that the same underlying mechanism – namely, the quantum fluctuations – is responsible for the behavior of all three functions.

IV. SOLUTION UNDER FINITE BANDWIDTH EXCITATION

When the exciting field is no longer monochromatic, but is of the form described in Section II, the solutions for all three functions $I(\underline{r},\tau)$, $\Gamma^{(2,2)}(\underline{r},t,\tau)$ and $\tilde{\Gamma}^{(1,1)}(\underline{r},t,\tau)$ are changed. The first two functions can again be treated collectively and shown to obey an equation of the form (10), in which the kernel $K(\tau)$ is the same for both functions, and is given by

$$K(\tau) = -\left(\frac{\Omega^2}{\beta}\right) \frac{e^{-(\beta+\lambda)\tau}[(1-\lambda/\beta)\cos\beta\theta\tau + \theta\sin\beta\theta\tau] - (1-\lambda/\beta)e^{-2\beta\tau}}{(1-\lambda/\beta)^2 + \theta^2} .$$

$$(15)$$

We note that $K(\tau)$ is an explicit function of the laser bandwidth λ, and reduces to Eq. (11) when $\lambda = 0$. The general solution again has the structure of Eq. (12), in which C and D_i are more complicated functions of $\beta,\theta,\Omega,\lambda,t$, and the coefficients p_i in the exponent are solutions of the more complicated cubic equation

$$p^3 + (4\beta + 2\lambda)p^2 + [(\beta+\lambda)(5\beta+\lambda) + \beta^2\theta^2 + \Omega^2]p +$$

$$+ 2\beta(\beta+\lambda)^2 + 2\beta^3\theta^2 + (\beta+\lambda)\Omega^2 = 0, \qquad (16)$$

that reduces to Eq. (13) when $\lambda=0$. The effect of the finite bandwidth excitation is therefore to modify somewhat the evolution of the light intensity and the intensity correlation function, but qualitatively the effect is not great. Moreover, it can be shown that the factorization property (14) is preserved [2].

However, when we come to the spectrum of the fluorescence, we encounter much larger effects. Although the second order correlation function $\Gamma^{(1,1)}(\underline{r},t,\tau)$ is also derivable from an integral equation of type (10), the kernel is quite different this time. It is a complex function of τ in general, and has the form

$$K(\tau) = -\frac{1}{2}\,\Omega^2 \left[\frac{e^{-(\beta+4\lambda+i\beta\theta)\tau} - e^{-(2\beta+\lambda)\tau}}{\beta - 3\lambda - i\beta\theta} + \frac{e^{-\beta(1-i\theta)\tau} - e^{-(2\beta+\lambda)\tau}}{\beta + \lambda + i\beta\theta}\right]. (17)$$

Although the general solution for $\tilde{\Gamma}^{(1,1)}(\underline{r},t,\tau)$ still has the formal structure of Eq. (12), the p_i's are now the roots of a different - and, in general, complex - cubic equation

$$p^3 + p^2[4\beta + 5\lambda] + p[5\beta^2 + \beta^2\theta^2 + \Omega^2 + \lambda(14\beta + 4\lambda - 4i\beta\theta)]$$

$$+ 2\beta^3 + \beta\Omega^2 + 2\beta^3\theta^2 + \lambda(9\beta^2 + 2\Omega^2 + 4\lambda\beta + \beta^2\theta^2 - 8i\beta^2\theta - 4i\beta\lambda\theta) = 0.$$

$$(18)$$

Because of the complex kernal $K(\tau)$ and the complex cubic equation (18), the second order correlation function $\Gamma^{(1,1)}(\underline{r},t,\tau)$ is now intrinsically complex, whereas it was real (apart from oscillatory phase factors) under monochromatic excitation. As a consequence we find that the spectral density of the fluorescent light, which is the Fourier transform of $\Gamma^{(1,1)}(\underline{r},t,\tau)$ with respect to τ, is now asymmetric in general. We may note in passing that the imaginary contributions in Eqs. (17) and (18) are all associated with the detuning parameter θ, and vanish when $\theta = 0$, so that the asymmetries are absent when the excitation is on-resonance. This explains why they were not encountered in some other treatments [3].

The general solution for the spectrum $\Phi(\underline{r},\omega)$ of the light radiated by the atom at frequency ω in the steady state is obtained by letting $t \to \infty$ in $\Gamma^{(1,1)}(\underline{r},t,\tau)$, and taking the Fourier transform with respect to τ. We then arrive at the expression

$$\Phi(\underline{r},\omega) = \left(\frac{\omega_o^2 \mu \sin\psi}{4\pi\epsilon_o c^2 r}\right)^2 \left\{\frac{\frac{1}{2}\Omega^2/\beta^2}{\frac{1}{2}(\Omega^2/\beta^2)(1+\lambda/\beta)+(1+\lambda/\beta)^2+\theta^2}\right.$$

$$\times \left\{\frac{\frac{1}{2}(\beta+3\lambda+i\beta\theta)(\beta+\lambda-i\beta\theta)}{[(\beta+3\lambda+i\beta\theta)(\beta-\lambda-i\beta\theta)+\frac{1}{2}\Omega^2(1+\lambda/\beta)][\lambda+i(\omega_1-\omega)]}\right.$$

$$-\frac{1}{4}\Omega^2 \sum_{\substack{i=1 \\ i\neq j\neq k}}^{3} \frac{(p_i+2\beta)-(\lambda/\beta)(p_i-3\beta+\lambda)-2(\lambda/\beta\Omega^2)(p_i+\beta+4\lambda+i\beta\theta)(p_i+2\beta+\lambda)^2}{(p_i+\lambda)(p_i-p_j)(p_i-p_k)[-p_i+i(\omega_1-\omega)]}$$

$$\left. + \text{c.c.}\right\}, \quad (19)$$

in which the first term corresponds to elastic light scattering, and is the dominant contribution in a very weak field, as in Mollow's original formula [10], whereas the other terms represent inelastic scattering contributions.

The behavior of the solution (19) under several different conditions is illustrated in Figs. 1 and 2. Figure 1 shows the spectrum radiated by the driving atom in a relatively weak field, which is detuned 5 natural linewidths from resonance, for several different bandwidths of the laser. Under (almost) monochromatic excitation, the spectrum is dominated by the elastic scattering peak centered at the laser frequency ω_1 , but as the bandwidth λ is increased, the elastic scattering contribution falls sharply

and a peak develops at the atomic frequency. We note that even
for a bandwidth λ that is a small fraction of the natural linewidth β, there is a very significant effect. Figure 2 shows the
fluorescence spectrum in a moderately strong field that is detuned
three linewidths from resonance. Under (almost) monochromatic
excitation, the spectrum has the form of the familiar symmetric
Stark triplet. As the laser bandwidth λ is increased an asymmetry
develops, and by the time the laser spectrum has been broadened
to one or two natural linewidths, the satellite peak that is re-
mote from the atomic frequency disappears, and the Stark triplet
reduces to a doublet. Again we note that the asymmetry is not insignificant, even when λ is a fraction of a natural linewidth.

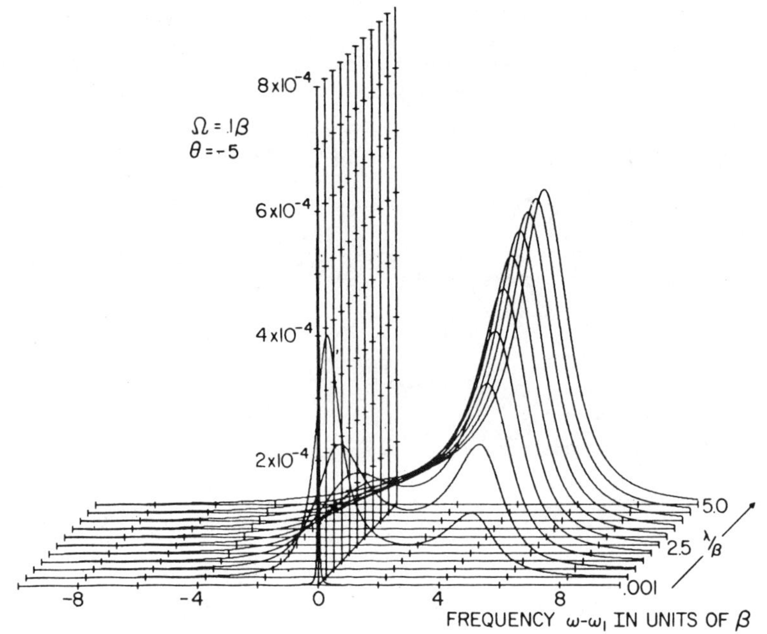

FIGURE 1.   (Reproduced from The Physical Review, Ref. 2)
             The spectral density of the fluorescence emitted by a
             driven atom in a moderately weak field ($\Omega = .1\beta$) with
             detuning of 5 natural linewidths from resonance, for
             various laser bandwidths λ.

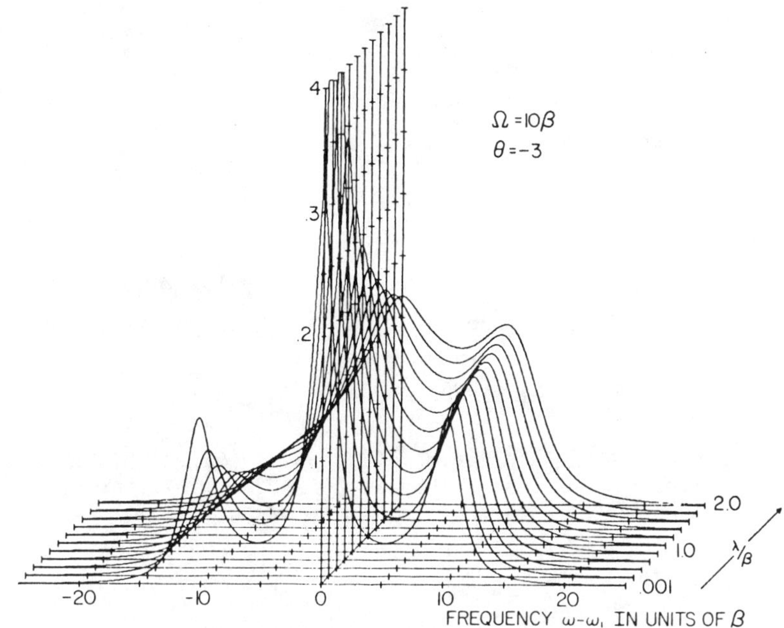

FIGURE 2.   (Reproduced from the Physical Review, Ref. 2)
The spectral density of the fluorescence emitted by a
driven atom in a moderately strong field ($\Omega = 10\beta$) with
detuning of 3 natural linewidths from resonance, for
various laser bandwidths λ.

It is possible that these effects played a role in some of
the early measurements of the fluorescence spectrum [1], in which
asymmetries were observed, although it is likely that other effects were present and perhaps more important. In any case, we
see that the finite bandwidth of the exciting field is reflected
more in the spectrum of the resonance fluorescence than in the
behavior of the light intensity, as is of course to be expected,
and that the effects can be significant.

REFERENCES

1.   See for example, F. Schuda, C.R. Stroud,Jr., and M. Hercher,
J. Phys. B7, L198 (1974); F.Y. Wu, R.E.Grove and S. Ezekiel,
Phys. Rev. Lett. 35, 1426 (1975); R.E.Grove, F.Y. Wu and
S. Ezekiel, Phys. Rev. A15, 227 (1977); H. Walther, in *Laser
Spectroscopy*, edited by S.Haroche et al. (Springer-Verlag,
Berlin, 1975),p.358; W.Hartig, W. Rasmussen, R. Schieder and

H. Walther, Z. Physik A278, 205 (1977); H.M.Gibbs and T.N.C. Vankatesan, Opt. Commun. 17, 87 (1976).

2.   H. J. Kimble and L. Mandel, Phys. Rev. A15, 689 (1977).

3.   J. H. Eberly, Phys. Rev. Lett. 37, 1387 (1976); G.S. Agarwal, Phys. Rev. Lett. 37, 1383 (1976); P. Avan and C. Cohen-Tannoudji, J. Phys. B10, 155 (1977).

4.   We label all Hilbert space operators by the caret ^ .

5.   H. J. Kimble and L. Mandel, Phys. Rev. A13, 2123 (1976).

6.   R. J. Glauber, Phys. Rev. 130, 2529 (1963) and 131, 2766 (1963).

7.   See for example, L. Mandel and E. Wolf, Rev. Mod. Phys. 37, 231 (1965.

8.   H. Risken, Zeits, f. Physik 191, 186 (1965); H. Risken and H. D. Vollmer, Zeits. F. Physik 201, 323 (1967); R. D. Hempstead and M. Lax, Phys. Rev. 161, 350 (1967).

9.   H. J. Carmichael and D. F. Walls, J. Phys. B8, 177 (1975) and 9, 1199 (1976).

10.   B. R. Mollow, Phys. Rev. 188, 1969 (1969).

Atomic Fluorescence
with Monochromatic Excitation

H. WALTHER

Sektion Physik, Universität München and Projektgruppe
für Laserforschung der Max-Planck-Gesellschaft,
Garching, W. Germany

I. INTRODUCTION

The interaction of monochromatic laser radiation with atoms
is characterized by a correlation time of the light wave being
much longer than the respective excited state lifetime of the
particles. Therefore, each atom undergoes many absorption and
emission processes during the correlation time of the laser wave.
In the two level situation this interaction can be described by
Bloch-type equations. Many phenomena in laser physics have been
explained in this way, the representation being especially useful
in connection with the description of coherent transient effects
[1,2].

The optical Bloch equations include the spontaneous decay of
the excited states as relaxation of the average dipole moment.
This gives, in general, an adequate description. However, when
the frequency distribution of the spontaneously emitted light is
investigated not only the mean variation of the average dipole is
important but also its fluctuations around the mean change [3].
Such investigations are of special interest since the fluctuations of the electric dipole play the essential role and are not
only a source of noise as in the case when the absorption or amplification of a light wave is considered. Therefore, the study
of the frequency distribution of the spontaneously emitted radiation has received large theoretical and experimental attention
in recent years. Particularly at higher laser intensities the
investigations can be used to check the validity of various
theoretical approaches and give a test of the quantum electrody-
namic theory in the optical domain. (For a survey on the theoretical work see e.g. [3,4,5]). The frequency distribution observed in the fluorescence can also be understood as a Stark
effect induced by the incoming laser radiation. The observed
spectrum is a result of the dynamic Stark splitting of the upper
and lower states determined by the Rabi nutation frequency between the two levels.

When the spontaneously emitted radiation is observed without frequency selection the above mentioned fluctuations of the electric dipole are of no importance and the experiments are again well described by Bloch-type equations. An example of this kind is the study of the level crossing resonance which appears on the total fluorescent light when a static field is scanned around the value corresponding to a crossing between two excited sublevels (level crossing or Hanle effect [6,7]). In the case of monochromatic excitation the density matrix description of this effect involves not only the nondiagonal elements which describe the coherent superposition of the sublevels but also the nondiagonal elements connecting the fundamental state and the excited state. Under monochromatic excitation a coherent oscillation between the excited and fundamental states is built up, being analogous to the Rabi nutation in magnetic resonance. The nondiagonal elements of the optical transition are, of course, of no importance in the case of broad band excitation since the short coherence time of the radiation does not allow a coherent mixing of excited and fundamental states. Therefore, the level crossing experiment under monochromatic excitation gives some additional information which is not contained in the classical experiment.

A review of our fluorescence experiments performed with monochromatic excitation and an outlook on future experiments in connection with the above mentioned problems will be presented here. The work discussed was performed in collaboration with W. Hartig, J. Häger, W. Rasmussen and R. Schieder.

II. LEVEL CROSSING EXPERIMENT

The first results on the level crossing experiments with monochromatic excitation have been published some time ago [8,9]. The $6s^2$ $^1S_0$–$6s6p$ $^1P_1$ resonance transition in the BaI spectrum at 5535 Å was investigated. In order to achieve excitation of the atoms by the same laser frequency the Doppler effect had to be eliminated. Therefore the light was scattered by the free atoms of a well collimated atomic beam (collimation ratio about 1:500). This corresponds to a residual Doppler width of about 3 MHz. The absorption width of the Ba atomic beam was therefore essentially determined by the natural width of the $^1P_1$ level being 20 MHz. For the measurements the laser was tuned to the Ba^{138} line to avoid the complication of hyperfine splitting (Fig. 1). The dye laser used in the experiment has been described earlier [11]. The spectral line width of the free running laser is less than 1 MHz, with a frequency drift of less than 1 MHz/min. The experimental geometry was chosen so that the magnetic field was

directed parallel to the atomic beam. The direction of the exciting light, the direction of the magnetic field and the direction of observation were taken mutually perpendicular. The laser
light was linearly polarized perpendicular to the magnetic field.
The part of the fluorescent light polarized perpendicular to the
magnetic field was observed.

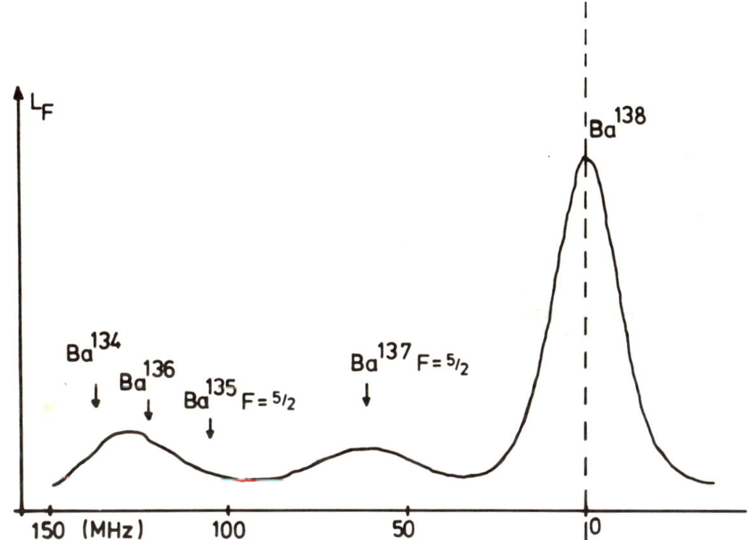

FIGURE 1.   Part of the spectrum of the $6s^2$ $^1S_0$-6s6p $^1P_1$ transi-
            tion of Ba at 5535 Å. For the level crossing experi-
            pents the laser was tuned to the wavelength indi-
            cated by the broken line.

     In the standard level crossing experiment this experimental
geometry gives a variation of the fluorescent light described by
a Lorentzian with a minimum at zero magnetic field and maximal
values at high positive and negative magnetic field values. In
the case of monochromatic excitation the signal decreases at
high magnetic fields being a result of the magnetic scanning
effect caused by the Zeeman splitting of the levels. The signal
around zero field is mainly determined by the interference of
the transition amplitudes. At low laser intensity the halfwidth
of the dip at zero magnetic field is 45% of that observed in the
standard level crossing experiment [8] (see Fig. 2).

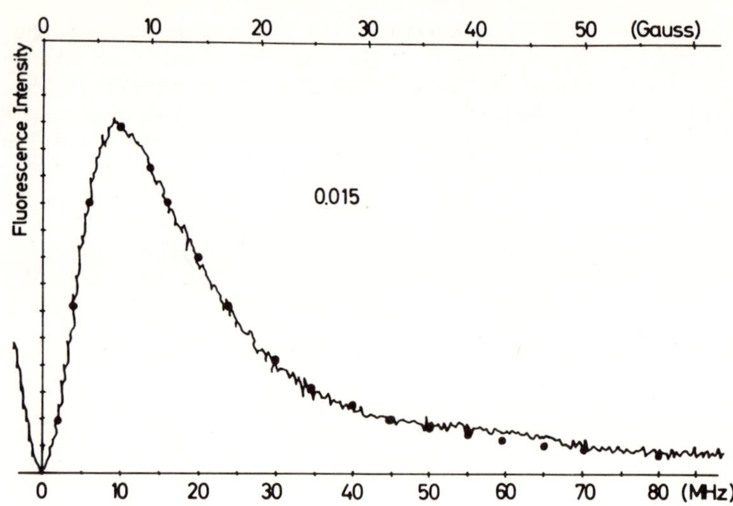

FIGURE 2.   Level crossing signal.   For this measurement the
parameter $4v^2/\gamma^2$ was equal to 0.015.   The solid dots
are the theoretical values which follow from Eq. (1).

The power dependence of the signal shape has been calculated
by Avan and Cohen-Tannoudji [12] using generalized Bloch equations.
The result for our experimental geometry and for the case of a
$J = 0 \rightarrow J = 1$ transition is given by

$$L \propto \frac{\Omega^2 v^2}{\left[\left(\frac{\gamma}{2}\right)^2 + \Omega^2\right]^2 + 4v^2 + \left[5\left(\frac{\gamma}{2}\right)^2 + \Omega^2\right] v^2} \tag{1}$$

where $\hbar\Omega$ is the Zeeman energy and γ the decay constant of the
excited state.   The parameter v is given by $v^2 = 3E^2 e^2 f_{ge}/16m\omega$,
where E is the laser electric field amplitude, ω is the resonance
frequency of the optical transition at zero field and f_{ge} is the
oscillator strength of the optical transition.
The signal curves obtained in our measurements are shown in
Figs. 2 and 3 for two laser intensities.   The parameters given on
the figures are the values $4v^2/\gamma^2$ being proportional to the laser
intensity.   The solid points are the theoretical values given by
Eq. (1).   The signal curves are normalized, to account for the
different intensities.
A good agreement between experiment and theory is obtained
at low laser intensity.   At higher intensity, however, the

deviation between theor and experiment is considerable. The reason is that in this case the Zeeman components of the other Ba
isotopes are brought into near resonance of the laser frequency.
These extra lines are therefore excited and give an appreciable
contribution to the fluorescence intensity. Due to power broadening this contribution is more significant at higher laser intensities (see Ref. [10]). When this contribution is subtracted
from the measured signal the curve indicated by crosses and the
broken line is obtained (Fig. 3). This is in good agreement with
the theoretical result which is again given by the solid points.
(For a detailed comparison between theory and experiment see
[10]).

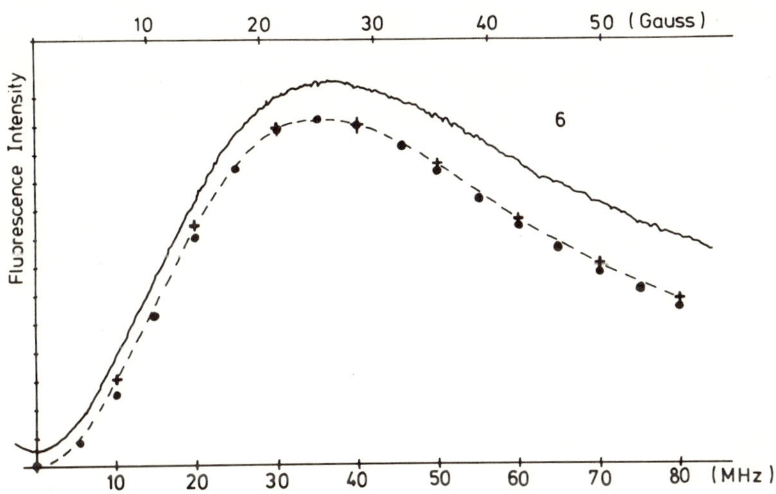

FIGURE 3.   Level crossing signal for high laser intensity. For
this measurement the parameter $4v^2/\gamma^2$ was equal to 6.
The broken curve was obtained from the measured signal
by subtracting the influence of the odd Ba isotopes.

It should me mentioned that recently Avan and Cohen-Tannoudji
have extended their study of the level crossing signal to the case
of a fluctuating intense quasi-monochromatic laser beam [13].
Assuming that the laser intensity is so high that several Rabi
nutations occur during the correlation time of the laser light
and that the atoms are sensitive to the laser fluctuations during
their lifetime (spectral width of the laser larger or comparable
to the decay constant) the level crossing signal was calculated.
These conditions do not allow a treatment by means of Bloch equations or by rate equations which are used in the case of broad

band excitation. A light wave was assumed which has short-time fluctuations resulting from an erratic motion of the phase and long-time fluctuations which may be caused by a slow phase diffusion or a slow amplitude variation. The short-time fluctuations were considered by a perturbation treatment and the slow variations were assumed to be adiabatically followed by the atom. Avan and Cohen-Tannoudji have shown in this way that the level crossing signal is quite sensitive to the fast phase fluctuations and that a detailed study of the signal shape of the level crossing signal could give an interesting insight into higher order correlations of the light beam.

III. FREQUENCY DISTRIBUTION OF SPONTANEOUSLY EMITTED FLUORESCENCE

In order to study the frequency distribution of the spontaneously emitted light following monochromatic laser excitation the Doppler broadening has to be almost completely excluded; as in the level crossing experiment discussed above. Hence, again, the laser light has to be scattered by the free atoms of a strongly collimated atomic beam. In order to measure the frequency distribution the fluorescent light has to be investigated by means of a highly resolving spectrometer. The experiments on this topic published so far are, in the order of their publication, [14, 9, 15, 18, 5, 16].

The essential results of the experimental studies are that in the case of weak monochromatic excitation a linewidth is observed which is smaller than the natural width [18,5]; however, so far it has not yet been demonstrated that this spectral distribution is identical with the laser linewidth as predicted by theory. In our experiment the transition $6s^2$ ^1S_o - $6s6p$ $^1P_1$ of Ba^{138}, mentioned above, was investigated. It has been chosen since the natural linewidth is 20 MHz which is twice as large as the linewidth observed in the experiment. This width is determined by the collimation ratio of the atomic beam, the acceptance angle of observation, the laser linewidth and the finesse of the Fabry Perot interferometer.

The frequency distribution of the fluorescence radiation at high intensities consists of a strong central component at the wavelength of the incoming laser radiation and two weaker side components having 1/3 of the intensity of the central line. All the experimental investigations published so far confirm this structure. However, there is not yet complete agreement between the measurements by different authors when linearly and circularly polarized light is used for excitation [1,16,5].

In all experiments with high laser intensities the $^2P_{3/2}$, $F = 3 \rightarrow {}^2S_{1/2}$, $F' = 2$ hyperfine transition of the sodium D_2-line was used. This transition is suitable as the upper $F = 3$ level can only decay into the $F' = 2$ level of the ground state from which the excitation is performed. Therefore, multiple excitations are possible and, in addition, no hyperfine pumping can occur. The transition has, of course, the disadvantage that it deviates from the two-level system usually considered in the theoretical treatments since the two hyperfine levels are degenerate. Furthermore, the other hyperfine levels of $^2P_{3/2}$ are so close that their influence cannot be neglected at high laser intensities, as they overlap with the $F = 3$ level due to power broadening. As a consequence, spontaneous decay to the $^2S_{1/2}$, $F' = 1$ can occur besides the decay to the $F' = 2$ level. Since the hyperfine splitting between $F' = 1$ and $F' = 2$ is 1,772 MHz a reexcitation of those atoms is impossible.

The degeneracy of the hyperfine levels has an influence on the frequency distribution of the fluorescent light since the Rabi nutation frequency, which determines the separation of the side components, is proportional to the transition probability; therefore the different Zeeman substates contribute in a different way to the spectrum. In order to get a good signal the number of involved Zeeman states has to be reduced. This was done in our experiment [5] by using circularly polarized light for excitation. Due to optical pumping during the first excitation processes when the atoms enter the laser beam [17] almost all are pumped into the $m_{F'} = 2$ or $m_{F'} = -2$ sublevels of the $F' = 2$ hyperfine level for righthanded or lefthanded circular polarization, respectively. Finally, only the transition $^2S_{1/2}$, $m_{F'} = \pm 2 \rightarrow {}^2P_{3/2}$, $m_F = \pm 3$ is excited and no other Zeeman transition has to be taken into account. In addition, no coupling with other $^2P_{3/2}$-hyperfine levels is possible since there is no other allowed transition starting from $^2S_{1/2}$, $F' = 2$, $m_F = \pm 2$ which may be populated by the laser light. As a consequence the two level system is closely approached.

In our experimental set-up the interaction region between the atomic beam and the laser radiation was situated within the Fabry Perot interferometer used to analyze the frequency spectrum of the fluorescence. In such an arrangement the observed signal intensity is enhanced by a factor approximately equal to the finesse of the Fabry-Perot [19]. A confocal interferometer was used with a mirror radius of curvature of 25 cm, corresponding to a free spectral range of 300 MHz. The finesse was about 100. The interferometer could be tuned by a piezoelectric variation of the cavity length. Radiation transmitted by the interferometer was passed through an arrangement of two circular apertures and a

lens, to limit the effective field of view and observed on a
photomultiplier.

As the laser beam has a Gaussian spatial intensity distribution it must be expanded in order to obtain an almost constant
intensity over the part of the interaction region observed by the
Fabry Perot. The diameter of this region was about 0.5 mm which
has to be compared to the Gaussian diameter of the expanded laser
beam of about 3 mm. The accurate alignment of the Fabry Perot interferometer with respect to the axis of the atomic beam is quite
important for the experiment. (For details see Ref. [5]). Since
the fluorescent intensity was sufficiently high the spectrum of
the fluorescent light could be directly displayed by an oscilloscope, the sweep voltage of which was used to drive the piezoelectric spacers of the interferometer. The axis of the Fabry
Perot was adjusted with respect to the atomic beam until the linewidth of the fluorescence spectrum was minimized.

Some of the results obtained in our measurements are shown
in Figs. 4 and 5. They were all obtained using circularly polarized laser radiation. For the measurement in Fig. 4 the laser
frequency was tuned to the center of the hyperfine transition.
It is evident that the positions of the side maxima vary with
laser power. At maximal power the signal intensity of the side
components is about one third of that of the central peak. This
is in very good agreement with the theory of Mollow [20] and
others [3]. For a detailed comparison between our measurements
and the theoretical results see Ref. [5].

The variation of the fluorescence spectrum for different detunings of the laser frequency with respect to the transition
frequency is shown in Fig. 5.

The separation of the two side maxima from the central component is plotted in Figs. 6 and 7 as a function of laser power
and laser detuning. The solid curves represent the fit of the
corresponding functional dependence of the Rabi nutation frequency to the experimental points. It follows that the curves
in Figs. 6 and 7 are in reasonable agreement with the results
expected from the measured laser power and from the estimated result for the expectation value of the electric dipole moment of
the investigated transition.

As mentioned above, our results obtained using linearly
polarized excitation are not in agreement with those obtained by
Wu et al. [15] and by Grove et al. [16]. Both observe a symmetric three peak structure whereas our result [5] suggests an asymmetry towards smaller frequencies. This is the direction where
the $^2S_{1/2}$, F' = 2 - $^2P_{3/2}$, F = 2 hyperfine transition also has its
absorption. The separation between these transitions is about
60 MHz which is comparable to the power broadening expected at
the laser powers used in the measurements. The effect of power

can directly be seen in the experiment since the fluorescence resulting from the decay $^2P_{3/2}$,F = 2 – $^2S_{1/2}$,F' = 1 can be observed
even when the laser is tuned to the $^2P_{3/2}$,F = 3 hyperfine level.
(The decay to the $^2S_{1/2}$,F = 1 level cannot be seen when circularly
polarized light is used for excitation.) Therefore, it can be
expected that the deviation from the two-level case is responsible
for the asymmetry observed for the excitation with linearly polarized light. Calculations by Hopf and Milonni [21] who take the
other hyperfine levels into account seem to confirm this result.
     Grove et al. [16] showed that by observing a region of the
atomic beam where the laser intensity has a gradient may cause a
similar asymmetry. Hence, misalignment may be responsible for
the observed deviation. At present no final conclusion on this
discrepancy can be made and further investigations have to be
performed.

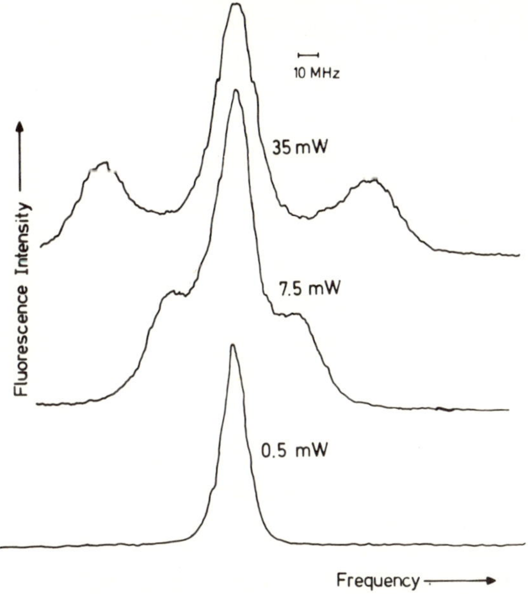

FIGURE 4.   Spectra of the fluorescent light of the $^2S_{1/2}$, F' = 2 –
            $^2P_{3/2}$, F = 3 hyperfine transition of Na for different
            laser powers.

For this purpose we started some time ago to study the fluorescent light by means of a photon correlation experiment. It has
been shown independently by Cohen-Tannoudji [3] and by Carmichael
and Walls [22,23] that the study of the intensity correlation of
the fluorescence provides another useful approach for the

determination of the linewidth and splitting frequencies. These parameters may be obtained from a Fourier transform of the intensity correlation function of the fluorescence (Fig. 8).

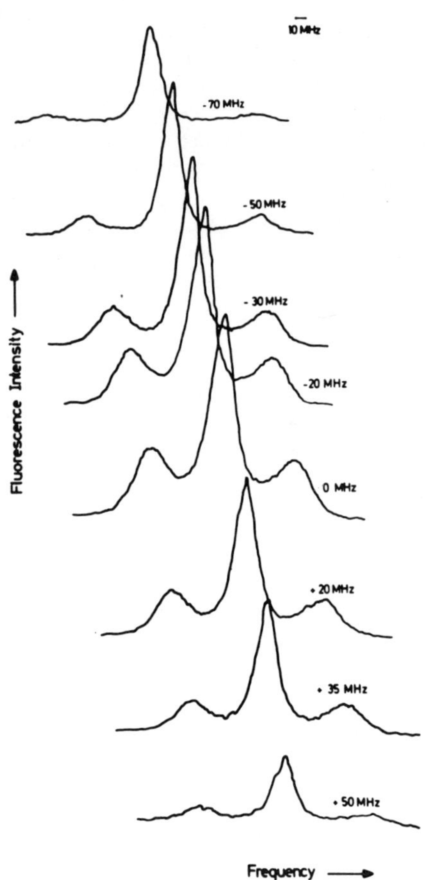

FIGURE 5. Variation of the fluorescence spectrum for different detunings of the laser frequency with respect to the transition frequency. The laser power was for all measurements 30 mW. The figure is taken from Ref. [5].

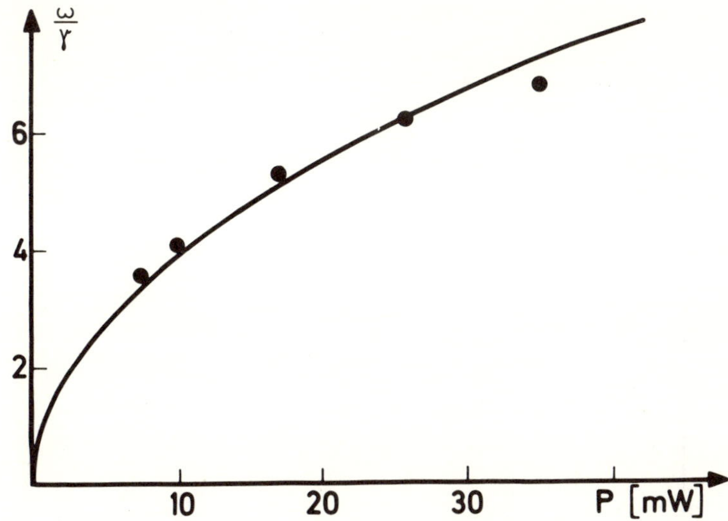

FIGURE 6.   Separation of the side maxima from the central com-
             ponent as a function of the laser power. The solid
             points are the experimental results. The figure is
             taken from Ref. [5].

Furthermore, if one could measure the correlation of single
atoms one would be able to observe the antibunching effect which
is particularly significant since it is only predicted by quantum
electrodynamical theory. The antibunching effect gives a zero
intensity correlation for a zero time interfal (Fig. 9). This behavior is different from the photon-bunching in the Hanbury-Brown
and Twiss experiment where the intensity correlation has a maximum at zero and decreases for larger time intervals. The antibunching can be explained by the fact that after the first photon
is detected the atom must be in the lower state and in order that
a second photon can be emitted the atom has to be raised into the
excited state. The increase of the intensity correlation with
time reflects the evolution of the atom. The experiments we have
performed so far demonstrate that the observation of the single
atom correlation should be possible, however, a measuring time
of about 2 hours seems to be necessary to achieve a sufficient
signal to noise ratio. During this time the laser must remain
amplitude stable to $\simeq$ 5% and frequency stable to $\simeq$ 1 MHz (1 part
in 10^9).

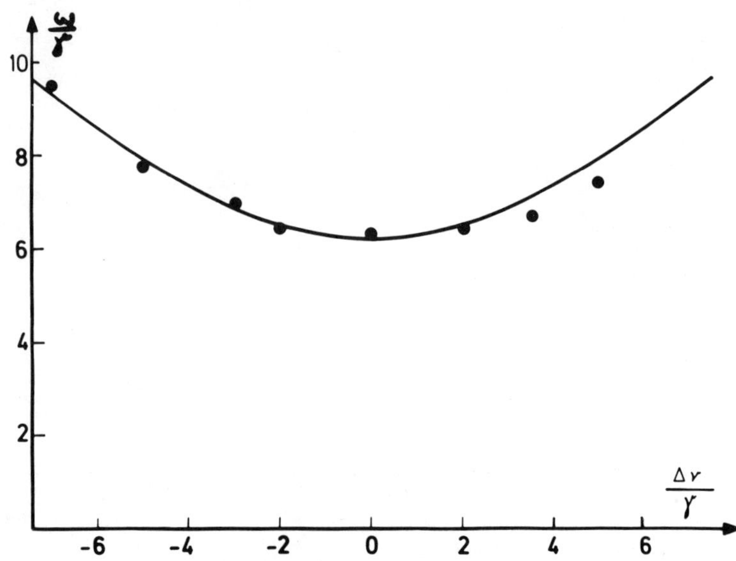

FIGURE 7.   Position of the side maxima with respect to the main component as a function of the laser frequency. The points are obtained from an evaluation of the results shown in Fig. 5. The figure is taken from Ref. [5].

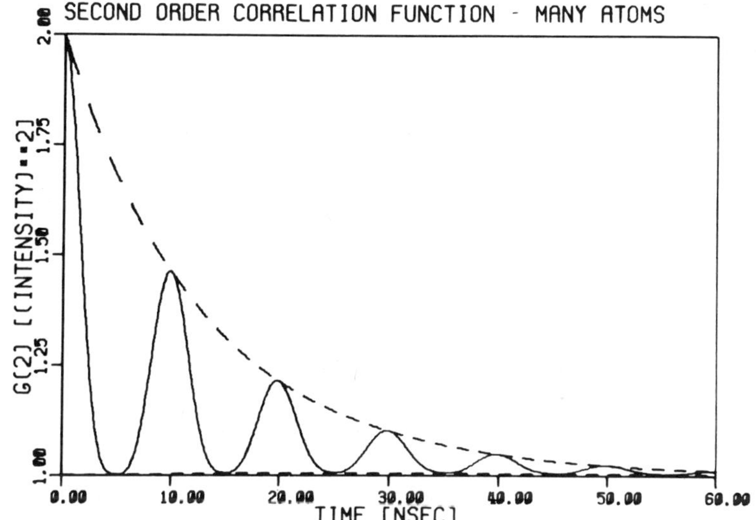

FIGURE 8.   Second order correlation function (intensity cor-
            relation) for many atoms. The figure was calculated
            using the results of Carmichael and Walls [23].

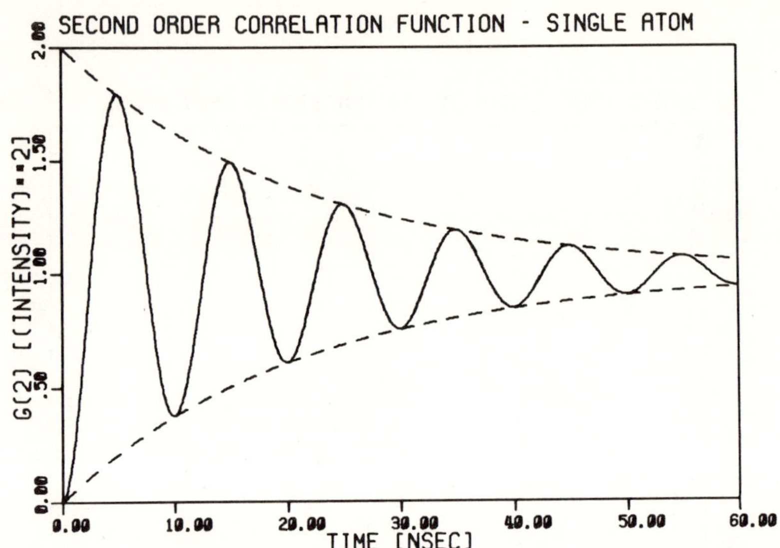

FIGURE 9.   Second order correlation function (intensity correlation) for a single atom. The figure was calculated using the results of Carmichael and Walls [23].

REFERENCES

1.   H. Walther, in Laser Spectroscopy of Atoms and Molecules, H. Walther, ed., (Springer-Verlag, Berlin, Heidelberg, New York, 1976).

2.   L. Allen, J. H. Eberly, Optical Resonance and Two-Level Atoms, (Wiley, New York, 1975).

3.   C. Cohen-Tannoudji, in Frontiers in Laser Spectroscopy, R. Balian, S. Haroche, S. Liberman, eds. (North-Holland Publishing Company, Amsterdam, New York, Oxford, 1977).

4.   H. J. Kimble, L. Mandel, Phys. Rev. A13, 2123 (1976).

5.   W. Hartig, W. Rasmussen, R. Schieder, H. Walther, Z. Physik A278, 205 (1976).

6.   W. Hanle, Z. Physik 30, 93 (1924).

7.   P. Franken, Phys. Rev. 121, 508 (1961).

8.   W. Rasmussen, R. Schieder, H. Walther, Opt. Comm. 12, 315
     (1974).

9.   H. Walther, in Laser Spectroscopy, S. Haroche, J. C. Pebay-
     Peyroula, T. W. Hänsch, S. E. Harris, eds., (Springer-
     Verlag, Berlin, Heidelber, New York, 1975).

10. J. Häger, Ph.D. Thesis, University of Köln (1975) and
     J. Häger and H. Walther, to be published.

11. H. Walther, Physica Scripta 9, 297 (1974).

12. P. Avan, C. Cohen-Tannoudji, J. Physique 36, L85 (1975).

13. P. Avan, C. Cohen-Tannoudji, J. Phys. B: Atom. Molec. Phys.,
     10, 171 (1977).

14. F. Schuda, C. R. Stroud Jr., M. Hercher, J. Phys. B7, L198
     (1974).

15. F. Y. Wu, R. E. Grove, S. Ezekiel, Phys. Rev. Lett. 35,
     1426 (1975).

16. R. E. Grove, F. Y. Wu, S. Ezekiel, Phys. Rev. A15, 227
     (1977).

17. R. Schieder, H. Walther, Z. Physik 270, 55 (1974).

18. H. M. Gibbs. T. N. C. Venkatesan, Opt. Commun. 17, 87 (1976).

19. A. Kastler, Appl. Opt. 1, 17 (1962). See also L. D. Vil'ner
     S. G. Rautian, S. A. Khaikin, Opt. a. Spectr. 12, 240 (1962).

20. B. R. Mollow, Phys. Rev. 188, 1969 (1969).

21. F. A. Hopf, P. W. Milonni, Bull. Am. Phys. Soc. 20, 1449
     (1975).

22. H. J. Carmichael, D. F. Walls, J. Phys. B: Atom. Molec.
     Phys. 9, L43 (1976).

23. H. J. Carmichael, D. F. Walls, J. Phys. B: Atom. Molec.
     Phys. 9, 1199 (1976).

Two-Level Atoms in an Intense Monochromatic Field: A Review of Recent Experimental Investigations

S. EZEKIEL AND F. Y. WU
Research Laboratory of Electronics
Massachusetts Institute of Technology
Cambridge, Massachusetts

The advent of tunable lasers is at last enabling researchers
to study in great detail the interaction of intense monochromatic
radiation with two-level atoms. Since such studies are very basic
to our understanding of atom-field interaction, they must be performed as carefully as possible in order to test theoretical predictions with high accuracy.

Stroud [1] and his group at the University of Rochester performed the first measurement in this area, namely the spectrum of
resonance fluorescence in intense fields. A number of other groups
then began to investigate this spectrum in more detail [2-5] as
well as other features of atom-field interaction.

In this paper we shall briefly review recent measurements
that have been performed on carefully prepared two-level atoms
interacting with intense monochromatic fields at optical frequencies.

I. EXPERIMENTAL PREREQUISITES

To perform precise measurements in this area, a number of
prerequisites must be satisfied. An atomic system is needed in
which only two nondegenerate energy levels can participate in the
interaction [5]. Doppler and collisional broadening of the transition must be kept much smaller than the natural broadening. The
monochromatic light source must be precisely tunable across the
atomic transition and its intensity should be sufficient to

---

*Supported by the Joint Services Electronics Program.

saturate the transition strongly. The spectral width of the
light source [6] must be much smaller than the natural width.
Finally, since the behavior of the atoms depends on the field intensity,only atoms in a region of uniform field may be examined.

II.   PREPARATION OF TWO-LEVEL ATOMS

The majority of experiments performed so far have been conducted on sodium. The sodium D_2 line is a good choice because
it has a large matrix element, can be prepared as a two-level
system, and the transition frequency is conveniently located
within the tuning range of dye lasers.

The preparation of an atomic beam of sodium in one magnetic
sublevel has been suggested by Abate [7]. In the experiments
discussed here a rather simple preparation scheme is used to put
almost all the population of the $F = 2$ ground state of sodium in
the $F = 2$, $m_F = 2$ magnetic sublevel [5]. This can be accomplished
by optical pumping by subjecting a highly collimated beam of
sodium to intense σ^+ circularly polarized light resonant with the
$3\ ^2S_{1/2}$ $(F = 2) - 3\ ^2P_{3/2}$ $(F' = 3)$ transition. The $F = 2$ atoms are
pumped into the $F = 2$, $m_F = 2$ sublevel from which the only allowed
$\Delta m = +1$ transition is to the $F' = 3$, $m_F' = 3$; these two-levels therefore comprise a two-level system. In order to prevent the redistribution of magnetic sublevel populations by stray fields in
the interaction region, a weak magnetic field $(0.7\,G)$ is applied
parallel to the laser beam [5].

It should be noted that the effect of atomic population in
the $3\ ^2S_{1/2}$ $(F = 1)$ ground state is negligible in our case because
of the large separation 1772 MHz between the $F = 1$ and $F = 2$ levels.

III.   SPECTRUM OF RESONANCE FLUORESCENCE

This experiment measures the spectrum of the light spontaneously emitted by a two-level atom resonantly excited by monochromatic light. Prepared two-level atoms interact with a resonant circularly polarized laser beam of variable intensity.
Fluorescence emitted orthogonal to the laser and atomic beams is
collimated by apertures and analyzed by a scanning Fabry-Perot
interferometer. The apertures also restrict the field of view
to the central, most uniformly illuminated region of the sodium
beam.

In the first theoretical treatment of the problem, Weisskopf
[8] showed that if the incident field intensity is well below
saturation, the resulting fluorescence does not exhibit the

natural width Γ, but is monochromatic and has the same frequency
as the incident light. Several groups [3,4,9,10] have verified
that when the field is weak, the fluorescence spectrum is a single
line whose width is less than natural width and can be explained
by experimental broadening mechanisms.

The case of strong-field excitation has been considered by
many authors recently [11-29]. The spectrum which is now generally accepted was first calculated by Mollow [15] and consists
of three peaks; two side peaks are separated from the central
peak by the Rabi frequency Ω. When the field is exactly resonant
and very intense ($\Omega \gg \Gamma$), the peaks have the height ratio 1:3:1
and widths $3/2\,\Gamma$, Γ, $3/2\,\Gamma$. There is also a delta-function component which is negligible in the on-resonance case but appre-
ciable for moderate detunings. The peak heights and widths change
when the excitation is detuned from resonance, but the entire
spectrum remains symmetric about the incident radiation frequency.
Examples of on- and off-resonance theoretical spectra are shown
in Fig. 1a,b,c.

The strong-field spectrum has been measured by groups led by
Stroud [1] at the University of Rochester, Walther [2,4] at the
University of Cologne, and Ezekiel [3,5] at the Massachusetts
Institute of Technology. The agreement with theory is very good,
as is illustrated in Fig. 1d,e,f from Ref. 5. The smooth curves
superimposed on the data are convolutions of the theoretical
spectra with the instrumental lineshape of the apparatus.

The experimental results also show that, under some conditions, the observed spectra are asymmetric, i.e., one side peak
is smaller than the other. Walther [4] has presented results
which indicate that use of linearly polarized light leads to
asymmetry due to the possibility of exciting transitions to another hyperfine level. However, Grove, Wu, and Ezekiel [5] demonstrated that asymmetric spectra may be produced by observing
atoms in a nonuniform field, regardless of polarization. Some
calculations predict asymmetries due to finite observation times
[30] or effects of laser jitter [31,32], but they appear not to
account for the observed asymmetry. Recently we have considered
another effect that can give asymmetry for atoms in a nonuniform
field; the effect is created by a combination of residual Doppler
shifts in the atomic beam and the spread of Rabi frequencies due
to the nonuniform field. Nevertheless, the resulting asymmetry
is not sufficiently large to account completely for the experimental spectra.

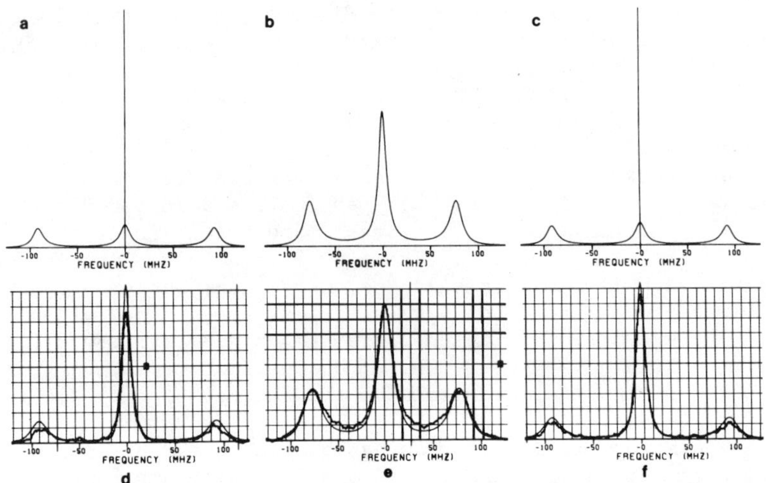

FIGURE 1.   Theoretical and experimental spectra of resonance
fluorescence from strongly driven two-level sodium
atoms. (a), (b), and (c) are theoretical spectra for
a detuning from resonance of -50 MHz, 0, and +50 MHz,
respectively, for a Rabi frequency $\Omega = 78$ MHz and
natural width $\Gamma = 10$ MHz. (d), (e), and (f) are experimental spectra and convolutions (smooth curve) of
theoretical spectra with instrumental lineshape for a
detuning of -50 MHz, 0, and +50 MHz, respectively.
Peak laser intensity is 640 mW/cm^2. (Taken from Ref.5.)

IV. ABSORPTION LINESHAPE IN THE PRESENCE
OF A STRONG FIELD

The simple Lorentzian absorption spectrum of a two-level
system, as measured by a nonsaturating, tunable probe field, is
greatly modified by the presence of a strong, near-resonant,
fixed-frequency field [29,33-37]. Previous experiments with rf
[38] and millimeter-wave radiation [39] have given evidence of
probe amplification by the saturated two-level system. The experiment [40] described here is designed to measure the absorption spectrum, as it is altered by the strong field.

For *this* measurement, the prepared two-level atoms interact simultaneously with a "driving field" and a "probe beam." The driving field is fixed at or near resonance and is very intense, while the probe beam strength is kept well below saturation. The probe beam is tuned across the resonance frequency and its absorption recorded. The probe beam is focused to approximately one-tenth the diameter of the driving field at the interaction region so that only atoms in a uniform field region are probed.

When the driving field at frequency ω is exactly resonant and very intense, peak absorption is greatly reduced compared to its unsaturated value. At frequencies differing from resonance (ω_0) by less than the Rabi frequency Ω, the probe is amplified, but all spectral features are very small. However, as seen in Fig. 2, when the driving field is detuned above resonance, the spectrum has one large absorption peak at ω_0' shifted down from the original resonance frequency, and one large gain peak at $2\omega - \omega_0'$. The shift $\omega_0' - \omega$ is known as the "light shift" [41] and is caused by the strong, nonresonant field.

With an atomic beam density which gives 9.4% absorption in the absence of a driving field, the measured peak amplification in Fig. 2 is 0.7%. This agrees with calculations [37], which predict that in the limit of high intensities, the peak gain occurs when the detuning $\Delta\omega$ is equal to $\Omega/3$, and is approximately 5% of the probe field absorption in the absence of the saturating field.

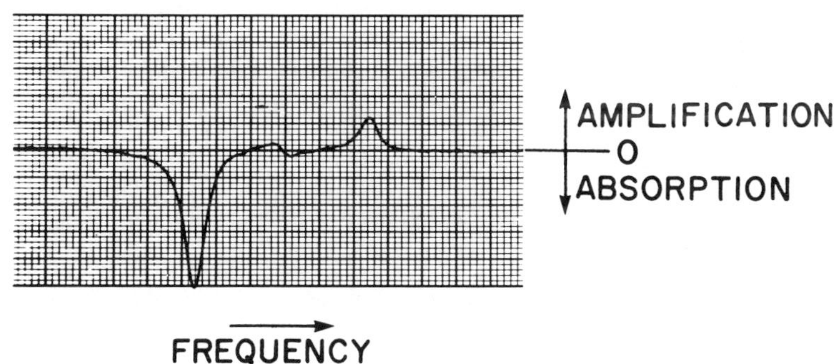

FIGURE 2.   Absorption spectrum of a two-level sodium atom driven by a strong field (ω) at a detuning from resonance of 28 MHz. Peak intensity 560 mW/cm^2. Peak absorption is at the level shifted frequency ω_0' and peak amplification is at $2\omega - \omega_0'$. (Taken from Ref. 40.)

V. POWER BROADENING AND SATURATION

Studies of the absorption spectrum of a two-level system as
a function of intensity can provide quantitative verification of
the familiar phenomena of saturation and power broadening. In
this experiment, which differs from the previous one in that there
is no fixed-frequency driving field, the fluorescence from atoms
in the most uniform region of the probe beam is monitored and recorded as a function of probe frequency and probe intensity.
Figures 3a-3e are a progression of observed lineshapes as the intensity of the probe is increased. At low intensities ($0.2\mathrm{mW/cm}^2$,
in Fig. 3a) the measured linewidth is 11 MHz, caused by the 10 MHz
natural width and residual Doppler broadening. In Fig. 3e, the
intensity was 150 $\mathrm{mW/cm}^2$, approximately 23 times the saturation
intensity, $I_s = 6.4$ $\mathrm{mW/cm}^2$. The lineshape, replotted as the dotted
curve in Fig. 3f, has a 47 MHz width, consistent with the 49.4
MHz width predicted by $\Delta\nu = \Gamma\sqrt{1 + I/I_s}$.

The measured power-broadened line is slightly asymmetric,
as can be seen by comparison with a Lorentzian (solid line) in
Fig. 3f. The asymmetry may be due to atomic recoil; the effect
of optical pumping caused by imperfect circular polarization
would result in asymmetry of the opposite sense (see Fig. 1b of
Ref. 3).

VI. OPTICAL AUTLER-TOWNES EFFECT

The splitting of an energy level in the presence of a strong
driving field, known as the dynamic Stark splitting or Autler-
Townes effect [42], has also been investigated at optical frequencies. The dynamic Stark splitting can be examined by probing
either one of the energy levels of a strongly driven two-level
atom. The probing is accomplished by using a second optical
field which can induce transitions to a third level.

The splitting of an optical transition was first observed
in saturated absorption of a neon line in a laser cavity [43].
Other groups have also obtained line splittings with saturation
techniques [44,45]. Picqué and Pinard [46] and others [47,48]
have performed such experiments using a sodium atomic beam and
two dye lasers, eliminating problems caused by velocity distribution and the limited tuning range of gas lasers. However, the
sodium atoms in these studies had not been prepared as two-level
systems and consequently the results cannot be easily interpreted [49].

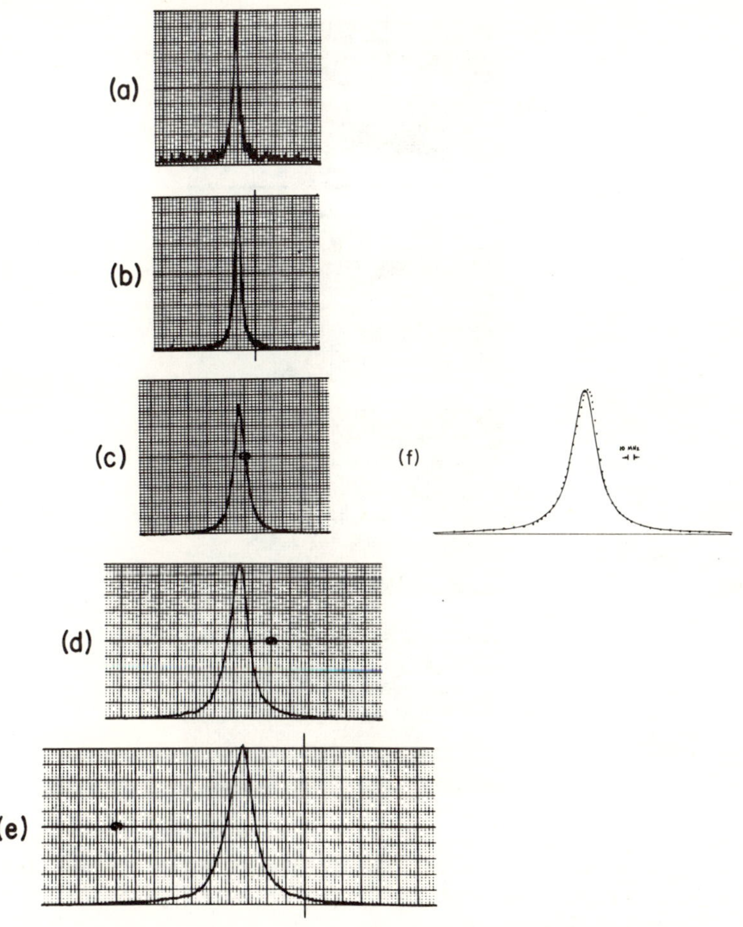

FIGURE 3.   Power Broadening and Saturation of a Two-Level
            Sodium Atom. Horizontal scale: 5.8 MHz/small division.
            Intensity and vertical scale (in photoelectron counts
            per sec per large division); (a) $I = 0.03\,I_s$, 200
            cts/sec, (b) $0.25\,I_s$, 1000 cts/sec, (c) $3\,I_s$, 5000
            cts/sec, (d) $12\,I_s$, 5000 cts/sec, (e) $23\,I_s$, 5000
            cts/sec. (f) Power-broadened lineshape with $I = 23\,I_s$.
            Theoretical Lorentzian-solid curve, measured line-
            shape-dotted curve.

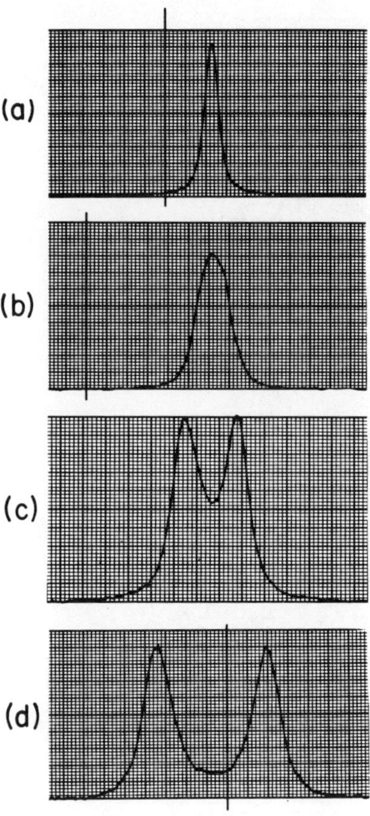

FIGURE 4. Optical Autler-Townes Splitting in Sodium. Horizontal
scale: 8.3 MHz/large division. 5890 Å laser intensity; (a) 1 mW/cm^2, (b) 16 mW/cm^2, (c) 73 mW/cm^2,
(d) 410 mW/cm^2.

We have recently performed the optical Autler-Townes experiment using a sodium beam prepared as a two-level system described earlier. Circularly polarized dye laser beams at 5890 Å and 5688 Å corresponding to $3\ ^2S_{1/2}$ $(F = 2)$-$3\ ^2P_{3/2}$ $(F = 3)$ and $3\ ^2P_{3/2}$ $(F = 3)$-$3\ ^4D_{5/2}$ $(F = 4)$ transitions, respectively, are propagated collinearly and intersect the atomic beam at right angles. The fluorescence at 5688 Å is monitored through an interference filter. The results obtained with the $3\ ^2S_{1/2}$ $(F = 2, m_F = 2)$-$3\ ^2P_{3/2}$ $(F = 3, m_F = 3)$-$3\ ^4D_{5/2}$ $(F = 4, m_F = 4)$ transition are shown in Fig. 4, as a function of the 5688 Å laser frequency, for several values of 5890 Å laser intensity. In the low-power case (Fig. 4a), the linewidth is 7 MHz, consisting of 3.3 MHz natural width of the $3\ ^4D_{5/2}$ state and residual Doppler broadening [50]. As the intensity of the 5890 Å laser increases, the structure of the intermediate level $3\ ^2P_{3/2}$ $(F = 3, m_F = 3)$ is altered; the weak 5688 Å probe laser beam reveals the dynamic Stark splitting (Figs. 4c,d). The observed splitting is approximately the Rabi frequency, or 45 MHz in Fig. 4d. The lineshapes are symmetric when the 5890 Å laser is on resonance; as in an earlier experiment [40], it was necessary to compensate for atomic recoil caused by the intense 5890 Å laser beam. The results are in agreement with theory [51,52].

VII. FUTURE MEASUREMENTS

One of the more exciting measurements yet to be made is that of photon correlation in resonance fluorescence where antibunching phenomena are expected to be observed [53]. This would imply that a two-level atom can emit only one photon at any one time and should provide yet another test of QED [28].

REFERENCES

1.   F. Schuda, C. R. Stroud, Jr., and M. Hercher, J. Phys. B7, L198 (1974).

2.   H. Walther, in Proceedings of the Second Laser Spectroscopy Conference, Megève, France, 1975 (Springer-Verlag, Berlin, 1975).

3.   F. Y. Wu, R. E. Grove, and S. Ezekiel, Phys. Rev. Lett. 35, 1426 (1975).

4.   W. Hartig, W. Rasmussen, R. Schieder, and H. Walther, Z. Physik A278, 205 (1976).

5.   R. E. Grove, F. Y. Wu, and S. Ezekiel, Phys. Rev. A15, 227 (1977).

6.   See, for example, F. Y. Wu and S. Ezekiel, Laser Focus 13, 78 (1977); R. L. Barger, J. B. West, and T. C. English, App. Phys. Lett. 27, 31 (1975).

7.   J. A. Abate, Opt. Commun. 10, 269 (1974).

8.   V. Weisskopf, Ann. Phys. Leipzig 9, 23 (1931).

9.   H. M. Gibbs and T. N. C. Venkatesan, Opt. Commun. 17, 87 (1976).

10. P. Eisenberger, P. M. Platzman, and H. Winick, Phys. Rev. Lett. 36, 623 (1976).

11. S. G. Rautian and I. I. Sobel'man, Zh. Eksp. Teor. Fiz. 41, 456 (1961) [Sov. Phys. JETP 14, 328 (1962)].

12. P. A. Apanesevich, Optics and Spectroscopy 16, 387 (1964).

13. M. C. Newstein, Phys. Rev. 167, 89 (1968).

14. V. A. Morozov, Optics and Spectroscopy 26, 62 (1969).

15. B. R. Mollow, Phys. Rev. 188, 1969 (1969).

16. M. L. Ter-Mikaelyan and A. O. Melikyan, Sov. Phys. JETP 31, 153 (1970).

17. C. R. Stroud, Jr., Phys. Rev. A3, 1044 (1971); and Coherence and Quantum Optics, Mandel and Wolf (Eds.) (Plenum Pub., New York, 1973), p. 537.

18. G. Oliver, E. Ressayre, and A. Tallet, Lettere al Nuovo Cimento 2, 777 (1971).

19. R. Gush and H. P. Gush, Phys. Rev. A6, 129 (1972).

20. E. V. Baklanov, Zh. Eksp. Teor. Fiz. 65, 2203 (1973) [Sov. Phys. JETP 38, 1100 (1974)].

21. G. S. Agarwal, Quantum Optics, Springer Tracts in Modern Physics (1974), p.108.

22. M. E. Smithers and H. S. Freedhoff, J. Phys. B7, L432 (1974); J. Phys. B8, L209 (1975).

23. H. J. Carmichael and D. F. Walls, J. Phys. B8, L77 (1975).

24. S. S. Hassan and R. K. Bullough, J. Phys. B8, L147 (1975).

25. C. Cohen-Tannoudji, Proceedings of the Second Laser Spectroscopy Conference, Megève, France, 1975 (Springer-Verlag, Berlin, 1975).

26.   S. Swain, J. Phys. B8, L437 (1975).

27.   B. R. Mollow, Phys. Rev. A12, 1919 (1975).

28.   H. J. Kimble and L. Mandel, Phys. Rev. Lett. 34, 1485 (1975);
      Opt. Commun. 14, 167 (1975); Phys. Rev. 13, 2123 (1976).

29.   C. Cohen-Tannoudji and S. Reynaud, J. Phys. B10, 345 (1977).

30.   B. Renaud, R. M. Whitley, and C. R. Stroud, Jr., J. Phys.
      B10, 19 (1977).

31.   G. S. Agarwal, Phys. Rev. Lett. 37, 1383 (1976); J. H.
      Eberly, Phys. Rev. Lett. 37, 1387 (1976).

32.   P. Aran and C. Cohen-Tannoudji, J. Phys. B10, 155 (1977);
      H. J. Kimble and L. Mandel, Phys. Rev. A15, 689 (1977).

33.   S. G. Rautian and I. I. Sobel'man, Zh. Eksp. Teor. Fiz. 41,
      456 (1961) [Sov. Phys. JETP 14, 328 (1962)].

34.   B. R. Mollow, Phys. Rev. A5, 2217 (1972).

35.   S. Haroche and F. Hartmann, Phys. Rev. A6, 1280 (1972).

36.   M. Sargent III and P. E. Toschek, Appl. Phys. 11, 107 (1976).

37.   M. Ducloy, in Proceedings of the Symposium on Resonant Light
      Scattering, Massachusetts Institute of Technology, Cambridge,
      MA, April, 1976 (unpublished).

38.   A. M. Bonch-Bruerich, V. A. Khodovoi, and N. A. Chigir',
      Zh. Eksp. Teor. Fiz. 67, 2069 (1974) [Sov. Phys. JETP 40,
      1027 (1975)].

39.   B. Senitzky, G. Gould, and S. Cutler, Phys. Rev. 130, 1460
      (1963); B. Senitzky and S. Cutler, Microwave J. 7, 62 (1964).

40.   F. Y. Wu, S. Ezekiel, M. Ducloy and B. R. Mollow, Phys. Rev.
      Lett. 38, 1077 (1977).

41.   J. P. Barrat and C. Cohen-Tannoudji, J. Phys. Rad. 22, 329,
      443 (1961); W. Happer, Rev. Mod. Phys. 44, 169 (1972).

42.   S. H. Autler and C. H. Townes, Phys. Rev. 100 , 703 (1955).

43.   A. Schabert, R. Keil, and P. Toschek, Appl. Phys. 6, 181
      (1975); Opt. Commun. 13, 265 (1975).

44.   Ph. Cahuzac and R. Vetter, Phys. Rev. A14, 270 (1976).

45.   C. Delsart and J.-C. Keller, J. Phys. B9, 2769 (1976).

46.   J. L. Picqué and J. Pinard, J. Phys. B9, L77 (1976).

47.   S. E. Moody and M. Lambropoulos, Phys. Rev. A15, 1497
      (1977).

48.   P. Liao and J. Bjorkholm (to be published).

49.   W. A. McClean and S. Swain, J. Phys. B$\underline{10}$, L143 (1977); S. Swain and W. A. McClean (to be published).

50.   Experiments using counter-propagating laser beams to eliminate this Doppler broadening are now under way.

51.   B. R. Mollow, Phys. Rev. A$\underline{5}$, 1522 (1972); R. M. Whiteley and C. R. Stroud, Jr., Phys. Rev. A$\underline{14}$, 1498 (1976).

52.   Recent measurements of the Autler-Townes splitting were presented by H. R. Gray and C. R. Stroud, Jr., in Fourth Rochester Conference on Coherence and Quantum Optics, June 8-10, 1977.

53.   H. J. Carmichael and D. F. Walls, J. Phys. B$\underline{9}$, L43 (1976).

Part 3

ATOMIC MULTIPHOTON PROCESSES

Multiphoton Ionization of Atomic Hydrogen

E. KARULE
Physics Institute
Latvian SSR Academy of Sciences,
Riga, Salapils, USSR

I.   INTRODUCTION

The hydrogen atom has always interested theoreticians. First,
as the simplest of atoms, it is of interest in its own right. As
the wave functions of the hydrogen atom are well known we expect
that rather good approximations can be obtained. Second, the results of the calculations for atomic hydrogen provide the establishment of reasonable approximations for a large number of atomic
systems: alkali atoms, negative ions and so on. Multiphoton ionization from the lower levels of hydrogen atoms has been extensively investigated theoretically though there are few experimental data available till now [1].

II.   TRANSITION RATE FOR MULTIPHOTON IONIZATION

Unlike single-photon processes which are independent of the
intensity and polarization of the light, multiphoton processes
strongly depend on the intensity and polarization of the incident radiation. If we assume that the light is linearly polarized, then for the multiphoton ionization rate we have

$$\frac{Q^{(N)}}{I^{N-1}} = 2\pi^2 \alpha a_o^2 \omega I_o^{1-N} \sum_\ell \left| T^{(N)}(n\ell_1,\ E\ell|\omega)\right|^2 \tag{1}$$

where $Q^{(N)}$ is the total cross section for N-photon ionization, I
is the radiation intensity in W/cm^2, α is the fine structure constant, a_o is the Bohr radius, ω is the photon energy (it is assumed that the energy of all photons is the same), $I_o = ce^2/2\pi a_o^4$
$= 14.038 \times 10^{16}$ W/cm^2, $T^{(N)}$ is the transition amplitude, $n\ell_1$ are
the atomic quantum numbers in the initial state, E is the energy
and ℓ is the orbital angular momentum of the ejected electron.
In calculating the ionization rate, the main difficulty is
in the calculation of the radial parts of transition matrix

elements. If the intensity of the radiation field is less than or about 10^7 V/cm, then perturbation theory may be used. We restrict our review to the papers in which the multiphoton ionization of hydrogen atom is considered in the framework of perturbation theory and the ionization process is correctly described in the off-resonance regions.

Let us consider the interaction between the atom and the radiation field as dipole interaction. Then in the lowest non-vanishing order of the nonrelativistic perturbation theory, radial parts of transition matrix elements may be written in the form

$$T_{rad}^{(N)}(n\ell_1, E\ell) = \int_0^\infty R_{E\ell}(r_1) r_1^3 \prod_{j=1}^{N-1} \int_0^\infty G_{L_j}(r_j, r_{j+1}; \Omega_j)$$

$$\times R_{n\ell_1}(r_N) r_j^3 dr_1 dr_j \qquad (2)$$

where $\Omega_j = E_{n\ell_1} + j\omega$. The selection rules for orbital angular momenta are

$$L_1 - \ell_1 = \pm 1 \quad , \quad L_j - L_{j-1} = \pm 1 \quad , \quad \ell - L_{N-1} = \pm 1,$$

Ω is the energy and L is the orbital angular momentum of the electron in virtual state, $R_{n\ell_1}$ and $R_{E\ell}$ are the radial parts of hydrogenic initial and final state wavefunctions. G_L is the Coulomb Green's function.

III. REPRESENTATIONS OF COULOMB GREEN'S FUNCTION

The hydrogen atom wave functions are known exactly. Therefore in calculating multiphoton transition amplitudes (2), the approach of various authors differs mainly in the choice of Green's function representation and in the technique of calculations.

The most popular in multiphoton calculations is the Green's function eigenfunction expansion

$$G_L(r_1, r_2; \Omega) = - \underset{n}{S} \frac{R_{nL}(r_1) R_{nL}(r_2)}{E_n - \Omega} \qquad (3)$$

where S means summation over discrete spectrum and integration over continuum (a. u. are used).

Zernik [2-4] who was the first to calculate the two-photon ionization rate for atomic hydrogen used expansion (3). To avoid

explicitly summing over intermediate states he reduced the problem
of obtaining transition amplitudes to the solution of the first-
order inhomogeneous differential equation using the methods proposed by Schwartz and Tieman [5,6]. Gontier and Trahin [7] have
generalized this method for multiphoton ionization. Using their
technique one must solve a set of first-order differential equations. The solution to the first equation gives the two-photon
result and is used to construct the inhomogeneous part of the
second (which gives the three-photon result) and so on. In the
two-photon case a similar method is used also by Chang and Tang
[8]. Bebb and Gold [9] summed the series (3) directly, but they
have not taken into account the continuum and only approximately
have they taken into account the contribution from those intermediate states, which are far from one of the atomic states.
Therefore their results for 2-12 photon ionization differs from
those of Gontier and Trahin [10] and [11] by several orders of
magnitude.

The Coulomb Green's function may be expressed also as the
product of Whittaker's functions

$$G_L(r_1,r_2;\Omega) = - \frac{\Gamma(1+L-p)\,p}{\Gamma(2L+2)r_1r_2}\, M_{p,L+1/2}\left(\frac{2r_<}{p}\right) W_{p,L+1/2}\left(\frac{2r_>}{p}\right) \quad (4)$$

where $p = (-2\Omega)^{-\frac{1}{2}}$. Whittaker's functions are defined as in [12].
The representation (4) is used by Zon et al [13] to calculate
three-photon ionization rate from 1s and 2s states and by
Laplanche et al [14] in two- and three-photon calculations for
ionization from the ground state. Using expressions (3) and (4)
for the Green's function one may calculate numerically transition amplitudes, but it is impossible in such a way to obtain
expressions in a closed form. When considering two-photon processes it is convenient to use the integral representation of
the Green's function

$$G_L(r_1,r_2;\Omega) = 2(r_1r_2)^{-1/2} \int_0^\infty dt\, \exp\left(-\frac{r_1+r_2}{p}\cosh t\right)$$

$$(\coth t/2)^{2p}\, I_{2L+1}\left(\frac{2r_1r_2}{p}\sinh t\right)$$

$$\Omega < 0 \qquad (5)$$

Expression (5) has been applied to two-photon ionization calculations by Rapoport et al [15], Klarsfeld [16] and Arnous et al
[17]. They have obtained two-photon transition amplitudes for

162   E. Karule

ionization from s- and p-states in closed form. In expression
(5) the radial variables are not completely separated which prevents one from obtaining transition amplitudes in closed form
when the number of photons is more than two.

The most appropriate expansion of the Green's function for
multiphoton calculations is the Sturmian expansion given by
Hostler [18]

$$G_L(r_1,r_2;\Omega) = -p \sum_{m=L+1} \frac{S_{mL}(\frac{2r_1}{p})\, S_{mL}(\frac{2r_2}{p})}{m-p} \qquad \Omega < 0, \qquad (6)$$

where S_{mL} are Sturmian functions [19], which are the charge eigenfunctions. p serves as a principal quantum number of virtual
states. If p=m is an integer, then $S_{mL}(2r/m)$ differs from
$R_{mL}(2r/m)$ only by a constant factor. A method equivalent to the
Green's function Sturmian expansion was first used by Podolsky
[20], who calculated two-photon bound-bound transitions.

The radial variables in expression (6) are completely separated which enables one to obtain expressions for multiphoton
transition amplitudes in a closed form [21,22] and to do calculations up to the number of photons 16 [11] and even more if needed.
Calculations performed by various authors are shown in Table 1.

IV. TWO-PHOTON IONIZATION ABOVE ONE-PHOTON THRESHOLD

Most N-photon ionization calculations are carried out for
photon energy varying between the thresholds of N- and (N-1)-
photon ionization. Only Zernik and Klopfenstein [3] have performed two-photon ionization calculations above single-photon
ionization threshold earlier. They showed that, when both one-
photon and two-photon ionization are energetically possible, ionization cross section by linearly polarized light of wavelength
λ (in cm) is dependent on light intensity as follows

$$Q^{(2)} = 1.045 \times 10^{-2} n^{-5} \lambda^3 (g_1 + \xi I \lambda^3 g_2) \qquad (7)$$

where
$$\xi = 2I_o^{-1}\lambda_o^{-3} = 0.1504 \ W^{-1} cm^{-1}$$

g_1 is the usual Gaunt factor of order unity and g_2 is a similar
factor also of order unity. In paper [3] g_2 is calculated for
n=1,^,3,4,5. For ionization from the ground state g_2 is given
for 911 Å > λ > 455 Å.

TABLE 1. Multiphoton Ionization Calculations for Atomic Hydrogen.

Number of Photons	Initial State	Polarization of Light	Method	References
2	2s	ℓ	DEM	Zernik (1964)
2	2s,3s	ℓ	DEM	Zernik and Klopfenstein (1965)
2-12	1s,2s	ℓ	(3)	Bebb and Gold (1966)
2	1s	ℓ	DEM	Zernik (1968)
2-7	1s	ℓ	DEM	Gontier and Trahin (1968)
2	1s	ℓ	DEM	Chang and Tang (1969)
2	2s	ℓ	(5)	Rapoport, Zon and Manakov (1969)
2	1s	ℓ	(5)	Klarsfeld (1969)
2-8	1s	ℓ	DEM	Gontier and Trahin (1971)
2,3	2s	ℓ		
3	1s,2s	ℓ,c	(4)	Zon, Manakov and Rapoport (1971)
2-8	1s	c	DEM	Gontier and Trahin (1973)
2-4,6	2s	c		
2	1s,2s,2p	ℓ,c	(5)	Arnous, Klarsfeld and Wane (1973)
2-16	1s	ℓ	(6)	Karule (1975)
2	1s	ℓ	DEM	Chang and Poe (1976)
2,3	1s	ℓ,c	(4)	Laplanche et al (1976)
2	1s,2s,3s	ℓ	(6)	Khristenko and Vetchinkin (1976)
3	1s,2s	ℓ		
2	1s	ℓ,c	(6)	Karule (1977)

DEM   - differential equation method
(3)-(6) - representations of Green's function used
ℓ - linearly polarized light
c   - circularly polarized light

164   E. Karule

    It is also possible to obtain analytically in closed form transition amplitudes for two-photon ionization above the one-photon ionization threshold. As expressions (6) and (7) for the Green's function are valid only for $\Omega < 0$, we continued analytically the transition amplitudes to the region where $\Omega > 0$. First we derived, using the Green's function Sturmian expansion, the following expression for the transition amplitudes for two-photon ionization from the ground state

$$T_{rad}^{(2)}(1s, E\ell) = 3/2^{\ell} C_k k^{\ell} \omega^{-5} (T_1 + T_2) \tag{8}$$

where k is the wave number of the ejected electron, C_k is the normalization factor of the Coulomb wave function, and T_1 and T_2 are given by

$$T_1 = -\frac{1}{2} z_3^{-q}(1+z_3)^{\ell-3} {}_2F_1(-q+\ell+1, \ell-3; \ 2\ell+2; \ 1-z_3), \tag{9}$$

where $q = (-2E)^{-\frac{1}{2}} = (1-4\omega)^{-\frac{1}{2}}$ and $z_3 = q + 1/q - 1$.

$$T_2 = 2^3 q^{\ell+1} p^{-3}(2-p)^{-1}(z_2/z_1)^2 z_2^{-q}$$

$$\times \sum_{n=0}^{4} \frac{M_n}{n! z_2^n} F_1(2-p; \ q+n, \ -q-n+4; \ 3-p; \ \frac{1}{z_1 z_2}, \frac{z_2}{z_1}) \tag{10}$$

where

$$M_n = \sum_{m=0}^{3-\ell} \frac{(-m-\ell-1)_n \ (\ell-3)_m \ (q+\ell+1)_m}{m! \ (2\ell+2)_m} \tag{11}$$

$$p = (1-2\omega)^{-\frac{1}{2}} \qquad z_1 = \frac{1+p}{1-p} \qquad z_2 = \frac{q+p}{q-p} \quad .$$

As $\ell = 0, 2$, T_1 given by (9) is a simple algebraic function which is valid below as well as above the one-photon ionization threshold. The convergence of sums in expression (10) for T_2 depends on the convergence of Appell's hypergeometric series F_1 [12]. The series are convergent if the moduli of both variables are less than unity. For the first argument $1/z_1 z_2$ it is valid below as well as above threshold, but $|z_2/z_1| < 1$ only below the threshold. To continue analytically Appell's functions to the region above the single-photon threshold one must obtain an expansion in series of z_1/z_2 instead of z_2/z_1. Therefore we first expressed the Appell's functions as series of Gauss hypergeometric functions and next used the transformation formulae for them. Thus we obtained

$$T_2 = 2^3 q^{\ell+1} p^{-3} \left\{ z_2^{p-q} (-z_1)^{-p} \, B(2-p,\ p-q+2) \right.$$

$$\times \sum_{n=0}^{4} \frac{(q-3)_n \, M_n}{n! \, (q-p-1)_n \, z_2^n} \; {}_2F_1\left(2-p, q+n;\ q-p+n-1;\ \frac{1}{z_2^2}\right) -$$

$$- \left(\frac{z_1}{z_2}\right)^2 (-z_1)^{-q} \sum_{n=0}^{4} \frac{M_n}{n! \, (-z_1)^n} \sum_{x=0}^{\infty} \frac{(q+n)_x}{x! \, (z_1 z_2)^x}$$

$$\left. \times \sum_{\nu=0}^{\infty} \frac{(-q-n+4)_\nu}{\nu! \, (p-q-n-x+\nu+2)} \left(\frac{z_1}{z_2}\right)^\nu \right\} \tag{12}$$

where

$$B(2-p,\ p-q+2) = \frac{\Gamma(2-p)\ \Gamma(p-q+2)}{\Gamma(-q+4)} \quad .$$

M_n is defined by expression (11). T_2 given by expression (12)
converges above the threshold and diverges below it. Calculations of the two-photon ionization rate using expressions (8)-
(12) have been carried out for radiation wavelengths $0 < \lambda < 1823\text{Å}$
[23].

V.   RESULTS AND CONCLUSIONS

Two-photon transition rates computed by various authors below one-photon ionization threshold are given in Table 2. There
is very good agreement among the results of papers [8,14,23 and
25]. The comparatively large discrepancy between these and the
results of [10,22] seems to have resulted from the use of experimental value for R_∞ instead of theoretical when calculating wavelength λ [in Å] $= 10^8/2R_\infty\omega$ [in a.u.]. The two-photon transition
rate above the threshold for linearly and circularly polarized
light is given in Figs. 1 and 2. For comparison we evaluated the
transition rate also using the Gaunt factor g_2, given in [3].The
two calculations are in good agreement (Table 3).
The variables of Appell's functions in expression (10) are
the same as in [17] and very similar to those in [15] so it is
easy to continue them analytically above the one-photon threshold
too. For multiphoton transition amplitudes, expressions valid
above the threshold may be obtained in closed form if the summation over intermediate states with positive energy is completed
by the technique proposed in [23]. Numerically the same may be

done by generalizing the method developed by Zernik and
Klopfenstein [3].

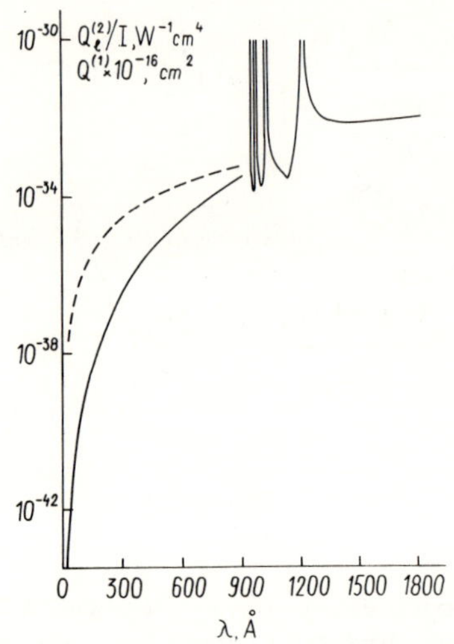

FIGURE 1.   Solid line: two-photon ionization rate for atomic
hydrogen in the ground state when linearly polarized
light is used. Dashed line: photoionization cross
section.

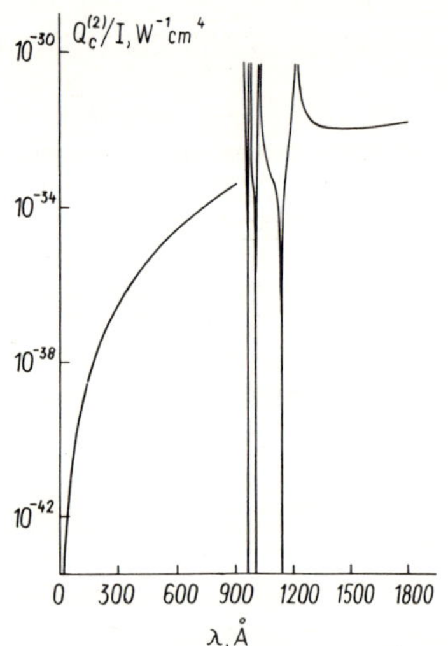

FIGURE 2.   Two-photon ionization rate for atomic hydrogen in the
ground state when circularly polarized light is used.

TABLE 2. $Q_\ell^{(2)}/I$ 10^{34}, $W^{-1}cm^4$ for atomic hydrogen in the ground state when linearly polarized light is used.

λ, Å	Chang and Tang (1969)	Gontier and Trahin (1971)	Khristenko and Vetchinkin (1976)	Laplanche et al (1976)	Chang and Poe (1976)	Karule (1977)
925	–	–	1.29	2.04	–	2.03
975	51.5	–	67.7	49.2	–	49.1
1019.45	–	–	–	–	55.2	55.3
1019.9	–	–	–	–	67.6	67.6
1020	67.5	55.2	55.2	70.8	70.8	70.9
1100	4.01	4.05	4.05	4.04	–	4.00
1199.35	–	–	–	–	580	580
1199.88	–	–	–	–	630	630
1200	630	580	580	642	642	642
1300	128	128	128	128	–	127
1400	84.5	84.5	84.5	84.7	–	84.5
1600	91.5	91.4	91.4	91.8	–	91.5
1700	103	102	102	103	–	103

TABLE 3. Two-photon ionization rate for atomic hydrogen in the ground state above single-photon ionization threshold (numbers in parentheses indicate powers of 10).

$\lambda,\overset{\circ}{A}$	$Q_\ell^{(2)}/I \times 10^{34}, \quad W^{-1}\ cm^4$		$Q_c^{(2)}/Q_\ell^{(2)}$
	[3]	[23]	[23]
20	–	1.61 (–10)	0.68
100	–	4.09 (–6)	0.74
200	–	3.02 (–4)	0.81
300	–	3.67 (–3)	0.88
400	–	2.15 (–2)	0.94
500	8.7 (–2)	8.49 (–2)	1.00
600	2.7 (–1)	2.62 (–1)	1.06
700	7.0 (–1)	6.83 (–1)	1.12
800	1.6	1.58	1.17
900	3.3	3.33	1.22

REFERENCES

1.   M. Lu Van, G. Mainfray, C. Manus and I. Tugov, Phys. Rev. A7, 91 (1973).

2.   W. Zernik, Phys. Rev. 135, A51 (1964).

3.   W. Zernik and R. W. Klopfenstein, J. Math. Phys. 6, 262 (1965).

4.   W. Zernik, Phys. Rev. 176, 420 (1968).

5.   C. Schwartz, Ann. Phys. 6, 156 (1959).

6.   C. Schwartz and J. J. Tieman, Ann. Phys. 6, 178 (1959).

7.   Y. Gontier and M. Trahin, Phys. Rev. 172, 83 (1968).

8.   T. N. Chang and C. L. Tang, Phys. Rev. 185, 42 (1969).

9.   H. B. Bebb and A. Gold, Phys. Rev. 143, 1 (1966).

10. Y. Gontier and M. Trahin, Phys. Rev. A5, 1896 (1971).

11. E. Karule in Atomnije Processi (Riga, Zinatne), p.5 (1975).

12. H. Bateman and A. Erdelyi, Higher Transcendental Functions, V1, New York (1953).

13. B. Zon, N. L. Manakov and L. P. Rapoport, Zh. Eksp. Teor. Fiz. 61, 968 (1971), Trans. Sov. Phys.-JETP 34, 515 (1972).

14. G. Laplanche, A. Durrieu, Y. Flank, M. Jaquen and A. Rachman, J. Phys. B: Atom. Molec. Phys. 9, 1263 (1976).

15. L. Rapoport, B. Zon and N. Manakov, Zh. Eksp. Teor. Fiz. 56, 400 (1969); Trans. Sov. Phys.-JETP 28, 480 (1969).

16. S. Klarsfeld, Nuovo Cimento Lett. 1, 682 (1969); 2, 548 (1969).

17. E. Arnous, S. Klarsfeld and S. Wane, Phys. Rev. A7, 1559 (1973).

18. L. C. Hostler, J. Math. Phys. 11, 2966 (1970).

19. M. Rotenberg, Adv. Atom. Molec. Phys. 6, 233 (1970).

20. B. Podolsky, Proc. Natl. Acad. Sci. US 14, 253 (1928).

21. E. Karule, J. Phys. B, Atom. Molec. Phys. 4, L67 (1971).

22. S. V. Khristenko and S. I. Vetchinkin, Opt. Spektrosk. 40, 417 (1976).

23. E. Karule, J. Phys. B, Atom. Molec. Phys. (in press).

24. Y. Gontier and M. Trahin, Phys. Rev. A7, 2069 (1973).

25. T. M. Chang and R. T. Poe, J. Phys. B, Atom. Molec. Phys. 9, L311 (1973).

L²-Floquet Treatment of Intensity-Dependent Multiphoton Ionization of the Hydrogen Atom

SHIH-I CHU*
Joint Institute for Laboratory Astrophysics
University of Colorado and National Bureau of Standards
Boulder, Colorado

and

WILLIAM P. REINHARDT[†]
Department of Chemistry, University of Colorado
and Joint Institute for Laboratory Astrophysics
University of Colorado and National Bureau of Standards
Boulder, Colorado

There is currently much interest in the problem of multi-photon ionization of atoms [1-5]. At low fields perturbative techniques [2] have been developed and the theory is well in hand for the H atom [2], although, for more complex atoms, experiment and theory are not in complete agreement [3]. At higher fields the "generalized" cross sections become intensity dependent: Gontier et al. [4] have used higher order perturbation theory and diagrammatic summation techniques to study these effects. However, within the framework of perturbation theory it is difficult to take into account the intensity dependent broadening and shifting of the atomic levels in a straightforward and self-consistent manner as rms field strengths approach 10^7 or 10^8 V/cm, becoming a major perturbation on the atomic levels. We are thus led to seek a non-perturbative method for field strengths which are within a few orders of magnitude of the internal field strengths. That is, we seek a technique appropriate to the intermediate regime where fields are too high for conventional multiphoton perturbation theory, yet not high enough to reverse the procedure [5], treating the Coulomb interaction as a small perturbation on

---

*Present address: Department of Physics, Yale University, New Haven, Connecticut.

[†]Camille and Henry Dreyfus Teacher-Scholar.

the motion of a charged particle in an intense ac field.

In the rf region of the spectrum the Floquet or "dressed atom" model [6] has been used in non-perturbative treatments of intense field level shifts and anti-crossings in the approximation that only finite (usually 2) numbers of atomic states are directly involved. It is the purpose of this paper to generalize the dressed atom model to include the complete set of bound and continuum states of the bare atom. This has the effect of giving each of the dressed levels an intensity dependent imaginary part (width) in addition to the usual field induced shifts. Proper interpretation of the frequency and intensity dependence of these complex energies gives rise to rates for multiphoton processes without prior specification of the specific number of photons involved in the process, and is equivalent to infinite order perturbation theory, self consistent in that the shifts and widths of all levels are simultaneously determined.

Corresponding to the time dependent Hamiltonian $H = H$ (atomic) $+ \vec{F} \cdot \vec{r} \cos \omega t$ describing the interaction of an atom with a monochromatic field, an equivalent time independent Hamiltonian H_F may be written in analogy with the semiclassical Floquet Hamiltonian of Shirley [6a]. The resulting block structure is shown in Fig. 1, where the $V_{\ell,\ell'}$'s are dipole coupling elements and the angular momentum blocks $\tilde{S}, \tilde{P}, \tilde{D}...$ represent the projection of the atomic electronic Hamiltonian onto states of total $L = 0, 1, 2...$ etc. Thus, for example, in the case of the H atom the $\tilde{S}$ block consists of the 1s,2s,3s,...ns... bound states and the entire ks Coulomb continuum. The Hamiltonian of Fig. 1 has no discrete spectrum, and the existence of Floquet solutions of the time dependent Schrödinger equation is not established [7]. However, writing the time evolution operator as

$$ e^{-iH_F t/\hbar} = \frac{1}{2\pi i} \int_C dz \frac{e^{-izt/\hbar}}{z - H_F} \tag{1} $$

gives the usual result [8] that the time dependence is dominated by poles of $(z-H_F)^{-1}$ near the real axis but on higher Riemann sheets, and that the complex energies of the poles are related to positions and widths of the shifted and broadened dressed states. These complex pole positions may be found directly from the analytically continued Floquet Hamiltonian, $H_F(\theta)$, obtained by the dilatation transformation [9] $r \to e^{i\theta}r$. This transformation effects an analytic continuation of $(z-H_F)^{-1}$ into the lower half plane on appropriate higher Riemann sheets, allowing the dressed states to be determined by solution of a non-Hermitian eigenproblem. In practice the atomic blocks $\tilde{S}, \tilde{P}, \tilde{D}, ...$ were discretized [10] by use of a finite subset of the complete discrete

Laguerre basis $r^{\ell+1}e^{-\lambda r/2}L_n^{2\ell+2}(\lambda r)$, $n = 0,1,2...$ which gives a Pollaczek [10] quadrature representation of the bound and continuum contributions to the spectral resolution of the hydrogenic Hamiltonian.

$$
\begin{array}{|c|c|c|c|c|}
\hline
A+4\omega I & B & O & O & O \\
\hline
B^T & A+2\omega I & B & O & O \\
\hline
O & B^T & A & B & O \\
\hline
O & O & B^T & A-2\omega I & B \\
\hline
O & O & O & B^T & A-4\omega I \\
\hline
\end{array}
$$

$$
A =
\begin{array}{|c|c|c|c|c|}
\hline
S & V_{sp} & O & O & O \\
\hline
V_{ps} & P-\omega I & V_{pd} & O & O \\
\hline
O & V_{dp} & D & V_{df} & O \\
\hline
O & O & V_{fd} & F-\omega I & V_{fg} \\
\hline
O & O & O & V_{gf} & G \\
\hline
\end{array}
$$

$$
B =
\begin{array}{|c|c|c|c|c|}
\hline
O & O & O & O & O \\
\hline
V_{ps} & O & V_{pd} & O & O \\
\hline
O & O & O & O & O \\
\hline
O & O & V_{fd} & O & V_{fg} \\
\hline
O & O & O & O & O \\
\hline
\end{array}
$$

FIGURE 1.   Structure of the Floquet Hamiltonian in the electric dipole approximation. The Hamiltonian is composed of *Floquet* blocks, of type A, which are in turn composed of *angular momentum* blocks $\tilde{S}$, $\tilde{P}$, $\tilde{D}$...; both types of blocks are coupled by the (dipole) coupling elements $V_{\ell,\ell\pm1}$.

As the techniques of dilatation transformation and L^2 discretization are unfamiliar within the context of intense field effects, we illustrate the methods with a calculation of the dc Stark width and shift of the 1s state of the H atom. In this case the dilatation transformation gives [11]

$$H_{F} \to H_{F}(\theta) \equiv H_{Atomic}(re^{i\theta}) + e^{i\theta} \vec{F} \cdot \vec{r} \qquad (2)$$

where

$$H_{Atomic}(re^{i\theta}) = e^{-2i\theta} \frac{1}{2}\left(-\frac{d^2}{dr^2} + \frac{\ell(\ell+1)}{r^2}\right) - \frac{e^{-i\theta}}{r}, \qquad (3)$$

and the Floquet matrix reduces to block A of Fig. 1. Convergence to the complex eigenvalue adiabatically connected to the exact 1s state at $F = 0$ is illustrated in Table I, where convergence as a function of basis set (and number of angular momentum blocks) is shown.

TABLE I.   Convergence[a] of the complex eigenvalue (in a.u.) representing the broadened and shifted "1s" state of the H atom as a function of the number of angular momentum blocks (N_ℓ) and number of basis functions (n) for a dc field $F = 0.1$ a.u. (5.1×10^8 V/cm).[b]

n	$N_\ell = 4$	$N_\ell = 6$
5	$-0.52728 - 0.7154 \times 10^{-2}$ i	$-0.52731 - 0.7259 \times 10^{-2}$ i
8	$-0.52739 - 0.7234 \times 10^{-2}$ i	$-0.52741 - 0.7273 \times 10^{-2}$ i
12	$-0.52740 - 0.7234 \times 10^{-2}$ i	$-0.52742 - 0.7270 \times 10^{-2}$ i

[a]See Ref. 11.

[b]Henneberger et al. [Phys. Rev. A10, 1494 (1974)] give $(-0.527 - 0.72 \times 10^{-2}$ i) for this pole position.

Application to multiphoton ionization of hydrogen follows the same procedure as in the dc Stark case, except that now convergence must be obtained with respect to the number of Floquet blocks, as well as with respect to basis size and number of angular momentum blocks. To the extent that enough Floquet blocks are included to obtain convergence, a completely non-perturbative

result is obtained in that all orders of perturbation theory are
included, and all relevant processes involving differing photon
numbers are simultaneously and self-consistently included. For
example, Table II gives ionization cross sections in the frequency
region where a single photon is sufficient for ionization. These
flux dependent cross sections were computed from the transition
rate implied by the imaginary part of the *dressed* 1s state, and
are, at low fields, intensity independent and identical to the
usual photoeffect cross section.

TABLE II. Intensity dependent ionization cross section (a_0^2)
for the dressed "1s" state as a function of frequency, ω and field strength F_{rms}. In atomic units
(a.u.) $\hbar = m_e = e^2 = 1$ and $\omega = 0.5$ is the zero field
ionization threshold. M_F is the number of Floquet
blocks needed to achieve convergence. Nonlinear
effects are seen to be quite small in this frequency region.

ω	F_{rms} [a]		
	10^{-4} ($M_F = 3$)	0.05 ($M_F = 4$)	0.20 ($M_F = 5$)
0.5	0.2250	0.2282	0.2534
0.6	0.1378	0.1384	0.1418
2.0	0.0044	0.0044	0.0044

[a] F_{rms} is given in a.u. (1 a.u. = 5.1×10^9 V/cm).

A more interesting frequency region is that dominated by
resonant two-photon ionization. Figure 2 shows the intensity dependence of the generalized cross section [2], $\hat{\sigma}_2 \equiv$ [Ionization
rate/(Intensity)^2], of the *bare* 1s state in the neighborhood of
the 1s $\rightarrow$ 2p and 1s $\rightarrow$ 3p one-photon resonances. The time evolution
of the *bare* 1s state was determined using Eq. (1), and corres-
ponds to a sudden turning on of the field at $t = 0$. At low fields
($F \lesssim 10^{-4}$ a.u.) $\hat{\sigma}_2$ is independent of field strength, except very
close to resonance, and to our working accuracy (convergence to
better than 1%) identical with perturbation results, as shown in
Table III. As the fields approach 1 or 2% of atomic field
strengths the intermediate 2p state broadens substantially and
the strong near resonant enhancement near $\omega = 0.375$ is quenched;

however, *off resonance $\hat{\sigma}_2$ remains essentially field independent.* We note that at the higher fields shown in Fig. 2 the definition of $\hat{\sigma}_2$ depends on the choice of time scale: this is discussed in detail in Ref. 13.

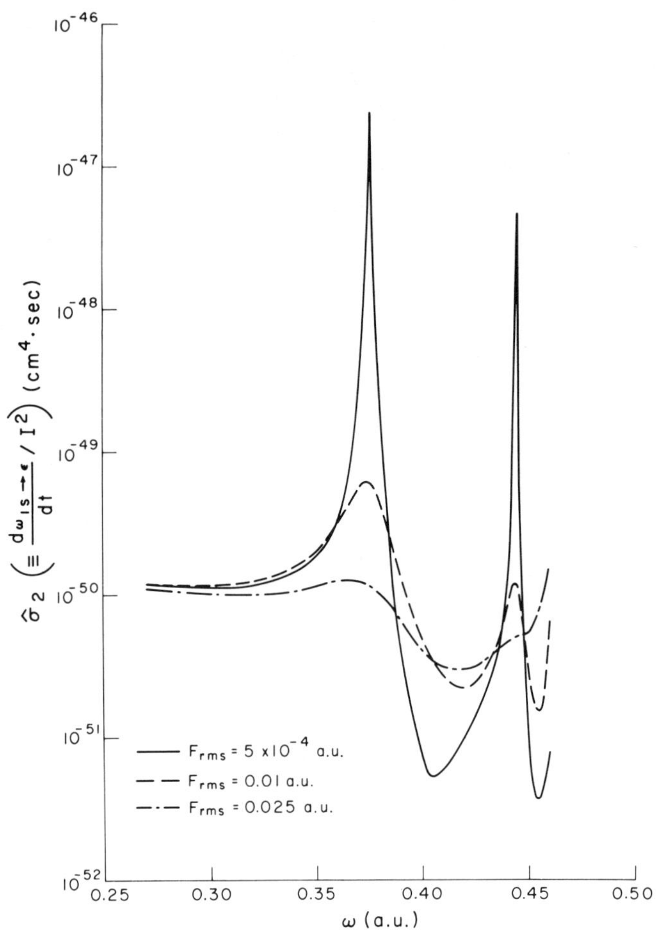

FIGURE 2.   Intensity dependent generalized cross section, $\hat{\sigma}_2$, for ionization of the bare 1s state of H in the frequency region dominated by resonant two-photon ionization. The intensity dependence of the resonant line shapes is qualitatively in agreement with that of the two-state plus continuum model of Ref. 12.

TABLE III.   Generalized cross sections for low field multiphoton ionization of atomic hydrogen.[a] In this
low field limit two-photon processes totally
dominate in the wavelength region of interest
and we see that the L^2-Floquet cross section, $\hat{\sigma}_2$
(Floquet), is essentially identical to the perturbation cross section of Chan and Tang (Ref. 2),
providing a check on the numerical methods, and
basis completeness.

$\lambda(\text{Å})$	$\hat{\sigma}_2(\text{Floquet})\left[\dfrac{cm^4}{w}\right]$	$\hat{\sigma}_2(\text{P.T.})\left[\dfrac{cm^4}{w}\right]$
1700	1.026×10^{-32}	1.025×10^{-32}
1400	8.457×10^{-33}	8.450×10^{-33}
1100	4.013×10^{-34}	4.013×10^{-34}
1000	1.927×10^{-34}	1.892×10^{-34}
980	1.148×10^{-33}	1.149×10^{-33}

[a]Floquet calculations carried out at 5×10^{-4} a.u. assuming
linear polarization. Angular momentum algebra is in agreement with that of Y. Heno, A. Maquet and R. Schwarcz, Phys.
Rev. A14, 1931 (1976).

In conclusion, we have shown that the Floquet-dressed atom
picture may be extended to treat continuum as well as bound processes, and that the resulting generalization, when combined
with analytic continuation and L^2 discretization, provides a
powerful non-perturbative technique for the treatment of intense
field processes.

The authors acknowledge the support of the National Science
Foundation and ERDA, stimulating conversations with Drs. P. L.
Knight and A. Szöke, and the computational acumen and assistance
of Chela V. Kunasz and Linda Kaufman.

REFERENCES

1.   For example, M. Lambropoulos, S. Moody, S. J. Smith and
W. C. Lineberger, Phys. Rev. Lett. 35, 159 (1975);

J. Morellec, D. Normand and G. Petite, Phys. Rev. A14, 300 (1976); P. Lambropoulos, E. A. Power and T. Thirunamachandran, Phys. Rev. A14, 1900 (1976); see also the review: J. S.Bakos, Adv. Electron, Electron Phys. 36, 57 (1974).

2.   Y. Gontier and M. Trahin, Phys. Rev. A4, 1896 (1971); F.T. Chan and C. L. Tang, Phys. Rev. 185, 42 (1969); see also the recent review: P. Lambropoulos, Adv. At. Mol. Phys. 12, 87 (1976).

3.   P. Lambropoulos and M. R. Teague, J. Phys. B9, 587 (1976); M. R. Teague, P. Lambropoulos, D. Goddmanson and D. W. Norcross, Phys. Rev. A14, 1057 (1976).

4.   Y. Gontier, N. K. Rahman and M. Trahin, Phys. Rev. Lett. 34, 779 (1975); Phys. Rev. A14, 2109 (1976).

5.   J. Gersten and M. H. Mittleman, Phys. Rev. A11, 1103 (1975); C. K. Choi, W. C. Henneberger and F. O. Sanders, Phys. Rev. A9, 1895 (1974); S. Geltman and M. R. Teague, J. Phys. B7, L22 (1974).

6.   a) J. H. Shirley, Phys. Rev. B138, 979 (1965); b) C. Cohen-Tannoudji and S. Haroche, J. Physiq. 30, 125, 153 (1969);   c) S. Autler and C. H. Townes, Phys. Rev. 100, 703 (1955).

7.   R. H. Young, W. J. Deal and N. R. Kestner, Mol. Phys. 17, 369 (1966); Ya. Zel'dovich, Sov. Phys. Usp. 16, 427 (1973).

8.   See, for example: M. L. Goldberger and K. M. Watson, Collision Theory (John Wiley, N.Y., 1964), Chap. 8, pp. 445–453.

9.   A mathematical review designed for physicists has been given by B. Simon, Ann. Math. 97, 247 (1973). See T. N. Rescigno and V. McKoy, Phys. Rev. A12, 522 (1975) for application to the photoeffect in H.

10.   H. A. Yamani and W. P. Reinhardt, Phys. Rev. A11, 1144 (1975).

11.   W. P. Reinhardt, Int. J. Quantum. Chem. S10, 359 (1976).

12.   S. Feneuille and L. Armstrong, J. Physiq. (Letters) 36, L235 (1975).

Multiphoton Ionization of Alkali Atoms

N. R. ISENOR
Department of Physics
University of Waterloo
Waterloo, Ontario, Canada

I. INTRODUCTION

The alkali atoms are particularly attractive for both experimental and theoretical study of multiphoton processes. Their ionization potentials are such that only a few photons are required at visible wavelengths to achieve ionization. Thus, in the case of nonresonant ionization, the slope of the log-log plot of ion yield versus laser intensity, is easy to determine accurately. Deviations from the expected slope are easy to investigate. The low photon number required also gives large ionization probabilities.

The simplicity of the alkali energy level scheme also is an advantage in interpreting results and making theoretical calculations. Experiments using linearly and circularly polarized light have been done and the results have been readily interpreted in relation to these energy levels and angular momentum selection rules.

Atomic beam and cell methods have been used to obtain samples of alkali atoms at suitable densities. The beam method, although more complicated, has the advantage of making the avoidance of contamination and spurious signals relatively easy. In both methods, suitable temperature distributions must be provided to avoid the formation of dimers of the alkali atoms. The ionization probabilities of the dimers are typically many orders of magnitude greater than those of the atoms and thus produce large signals of both monomer and dimer ions. The former can of course not be distinguished from those originating from atoms.

Recently very accurate absolute "cross-sections" for the three photon ionization of the alkali atoms at the ruby laser wavelength have been determined experimentally. The use of a single mode (longitudinal and transverse) laser and the saturation of ionization made this accuracy possible. This latter phenomenon, geometrical in nature, enables one to determine clearly the laser intensity at which the ionization probability at the focal point during the pulse becomes essentially unity. Thus no absolute measurement of the number of ions is needed to

determine absolute cross-sections. The effects of multimode beams
have been investigated by Mainfray's group [1].

In non-resonant multiphoton ionization involving k photons,
the rate of the process is given by

$$r = W_k F^k ,$$

(1)

where F is the photon flux and W_k a constant called "cross-section" for the purposes of this article. The number of ions
yielded during a single transverse mode laser pulse is given
by [2]

$$N = \rho \sum_{n=1}^{\infty} \frac{(-1)^{n-1}}{n!} \left(\tau_k W_k F_o^k\right)^n V_{nk} ,$$

(2)

where ρ is the atom density, τ_k is the effective pulse duration
($\equiv \int g^k dt$) for a kth order process and V_{nk} is the effective volume
($\equiv \int D^{nk} dv$) for an (nk)th order process. The single mode pulse
flux is given by

$$F(\underline{r},t) = F_o D(\underline{r})g(t).$$

(3)

This separation of variables is possible for a single transverse
mode laser pulse regardless of the temporal structure. Of course
a single longitudinal mode simplifies matters in practice. The
quantity $\tau_k W_k F_o^k$ (called P_f) is the parameter that determines the
probability of an atom being ionized at the focal point. When
this is $\ll 1$, we have $N \propto F_o^k$, for larger values, $N \propto F_o^{3/2}$ for spher-
ical lens focussing, and $N \propto F_o^2$ for cylindrical lens focussing.
These relationships are manifestations of the double cone and
double wedge geometry for the two types of focussing. The two
portions of the log N vs log F_o curve join smoothly. By fitting
the ion quantity and laser flux data to a normalized curve
($\log(N/\rho V_k)$ plotted against ($\log P_f$) we can determine a value of
W_k. The quantities that need to be measured absolutely are g(t),
the pulse energy, the focal length of the lens and the laser beam
waist size (which determine V_k and F_o).

II. SOME EXPERIMENTAL RESULTS

Absolute cross-sections for the three-photon ionization of
Na, K, Rb and Cs have been determined by Cervenan et al. [3]
using the data treatment method described in the preceding section. An atomic beam method with a double-chamber oven source
was used. The experimental setup is shown in Fig. 1. The top

chamber of the oven, at a higher temperature than the bottom, ensured that dimers were not present in the beam. The temperatures were carefully controlled so as to maintain a constant atom density in the focal region. The ions produced were detected by an electron multiplier and identified by their time-of-flight.

The laser used produced consistent pulses gaussian in spatial dependence and in time. Durations of ~ 12 ns and peak power values of up to ~ 0.5 MW were used. Neutral density filters and the use of lenses of different focal lengths enables one to extend the data from very low ion yields to well into the saturation regime.

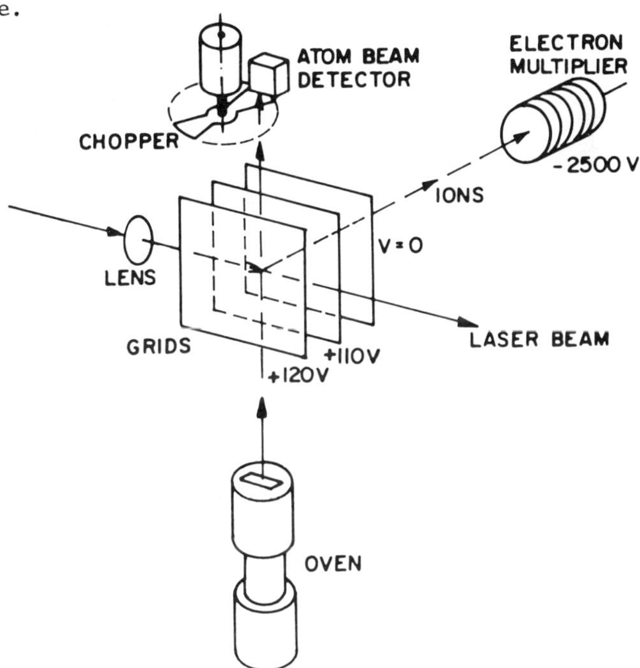

FIGURE 1.   Arrangement for the observation of laser-produced alkali ions using an atomic beam source. (Reproduced with permission from ref. 3.)

Sets of data collected for Na and Cs are shown in Figs. 2 and 3. The displacement of the curves for circular and linear polarization is for clarity on these normalized plots. The adjustment of the values of W_k to obtain these fits yielded the values of Table 1. The assigned errors result from the combination of the scatter in the results for several runs such as shown together with the errors in laser pulse energy calibration and beam parameters. The relationship between these results and the various theoretical values shown will be discussed later.

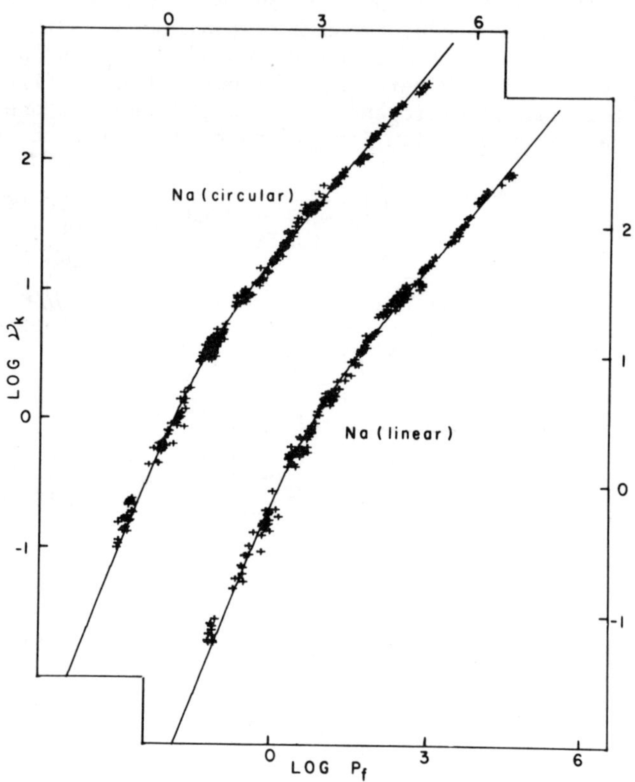

FIGURE 2.   Data points fitted to the normalized ion yield curve
for the three-photon ionization of Na at 694 nm.
(Reproduced with permission from ref. 3.)

Agostini and Bensoussan [4] have shown clearly the consequences of angular momentum selection rules in the three photon
ionization of K. They used a ruby laser pumped dye laser to tune
through the $6S_{1/2}$ and $4D_{3/2,5/2}$ two-photon resonances. The atoms
are ionized from these states by the further absorption of a
single photon. The relevant energy levels are shown in Fig. 4.
For circularly polarized light only the D state resonance occurs,
while for linearly polarized light the S state resonance

predominates. For circular polarization an atom must ionize by
acquiring k (in this case 3) units of angular momentum and therefore only the D state resonance can be involved. For linear
polarization both the S state and D state resonances contribute.
These resonances occur at different frequencies and thus are
readily distinguishable.

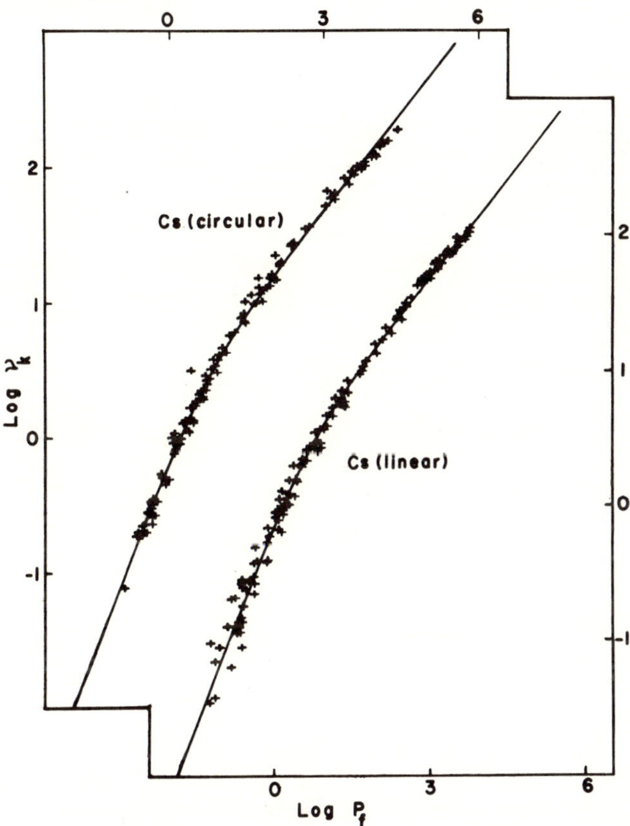

FIGURE 3.   Data points fitted to the normalized ion yield curve
            for the three-photon ionization of Cs at 694 nm.
            (Reproduced with permission from ref. 3.)

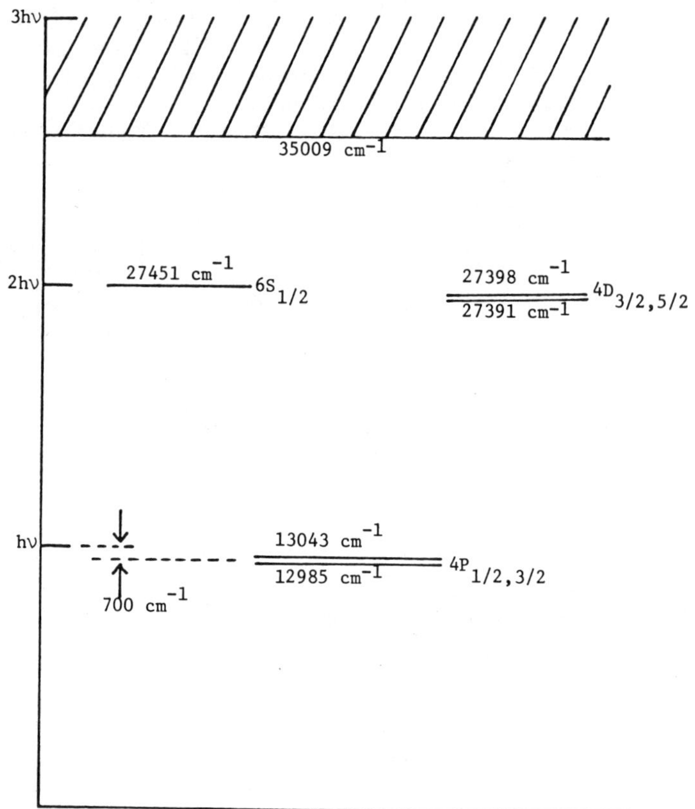

FIGURE 4.   Energy levels of K relevant to the experiment of Agostini and Bensoussan.

Morellec et al [5] have studied resonance shifts in the multiphoton ionization of Cs. In this work a single transverse and longitudinal mode Nd:glass laser was used to tune through the 6S-6F transition. The curves of intensity versus detuning from resonance, for constant ion yield, beautifully show the negative level shift with laser intensity. The level shifts are also demonstrated by their effect on the experimental order,

$$k_{exp} = \frac{d(\log N)}{d(\log I)} \, , \tag{4}$$

of the interaction. These authors also conclude that the shift of resonance observed should not be interpreted as a shift of the 6F level alone and that a shift of the 6S level must also be taken into account in any calculation of the effect. Of course in any experiment of this type where intensity dependent shifts occur, both temporal and spatial variations of the rate of ion production are of utmost importance. The level shifts would have different values at different points in the focal region at any given time.

The problem of dimers in the alkali atom sample has already been mentioned. Cervenan [6] observed that the yield of K^+ and K_2^+ ions was very large even with unfocussed laser beam (ruby) and observed that the production rate was roughly proportional to F^2, indicating a predominately 2-photon process. Chan [7] observed a similar effect with Na^+ and Na_2^+ from Na_2 in a sodium beam. The relevant energy levels of Na_2 are shown in Fig. 5. The $A^1\Sigma_u^+$ level is near the one photon level. From it the dimer could readily ionize by two-photon absorption. Only a small amount of additional energy is required to dissociate the dimer ion. Similar effects in Cs_2 were observed by Held et al [8,9] and by Collins et al [10]. These effects, although interesting, present a problem in studying processes in atoms. They were obviated in the atomic beam experiments by use of a two tempera-ture atom beam source [3]. In experiments using a cell, the temperature of evaporating metal was kept low to attain the same results [5].

III. DISCUSSION OF THE THREE-PHOTON CROSS-SECTION RESULTS

Table 1 shows the experimental results for the three-photon ionization of the alkalis at the ruby laser wavelength together with the theoretical results available. The cross-sections are quoted for linear and circular polarization along their ratio $(R \equiv W_3^c/W_3^\ell)$. The relative errors quoted for the ratio are con-siderably less than those for the corresponding cross-sections. The reason is that laser power calibration and focal volume er-rors do not contribute to the ratio error. Also, a greater con-sistence is found for this ratio from run to run than for the cross-sections.

The results of Table 1 are probably of sufficient accuracy to serve as indicators of applicability of the different theoret-ical calculations. The early calculations of Bebb [11] and Morton [12] as well as the recent calculations of Manakov et al. [13], Flank and Rachman [14] and Teague and Lambropoulos [15] are included. The exact agreement between the experimental re-sults for W_3^ℓ for Rb and Cs and the calculated results of Bebb is only apparent. The calculated numbers were taken from published

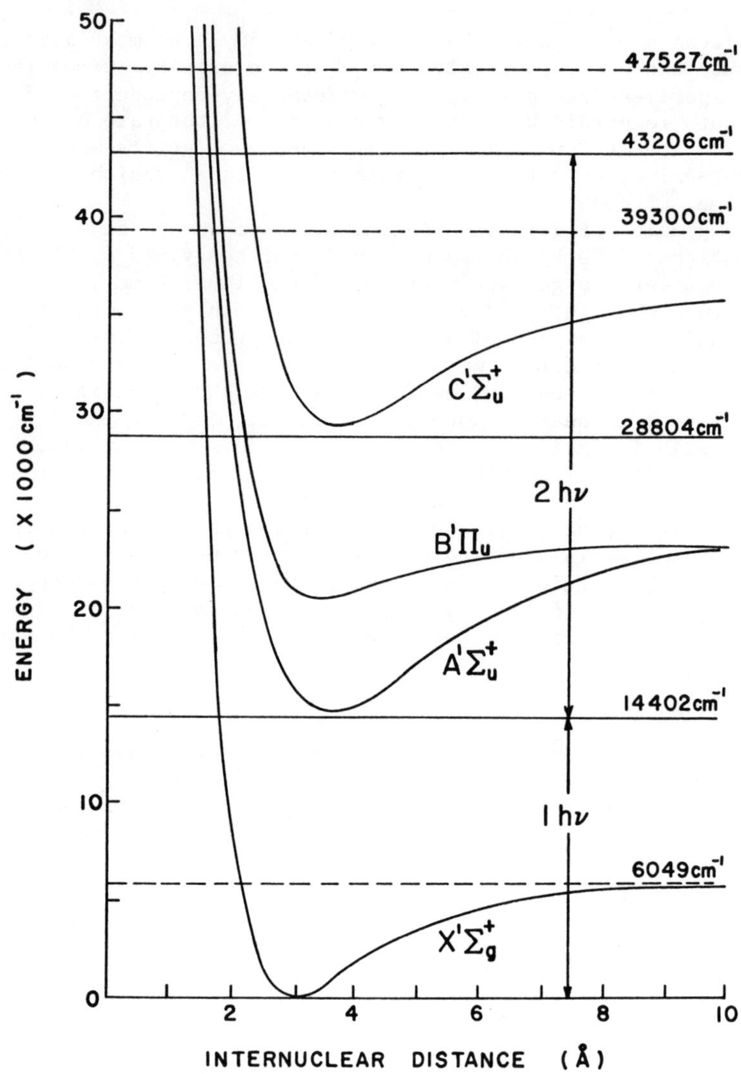

FIGURE 5.   Energy levels of Na_2 relevant to ionization at 694 nm.

curves and may be in error by a factor of ~ 5.

The value of R had been the subject of several theoretical papers. For non-resonant ionization it can have a maximum value of $(2k-1)!!/k!$ [16,17,18] for low values of k, but for higher values (> 4) it can have much lower values [19,20]. We see that R is close to the theoretical limit for the experimental results given here. The calculated values for R by Manakov et al. [13] agree quite well except for Cs. The values of R for Cs calculated by Flank and Rachman [14] and by Teague and Lambropoulos [15] are in much better agreement with the experimental result. The only other experimental result relevant to these is the R for Cs measured by Fox et al, a value of $2.15 \pm .40$. The latter measurement was to a great extent originally responsible for the theoretical interest in this quantity.

It had been reported by Evans and Thoneman [21] that a large dip existed in the ion yield curve for Cs in the flux regime of 0.5 to 1×10^{29} photons $cm^{-2}s^{-1}$. No such result was observed by Cervenan et al [3] (see Fig. 3). An attempt was made to explain their result in terms of the $9D_{3/2}$ and $9D_{5/2}$ levels. More recently, Georges and Lambropoulos [22] have attempted to justify the Evans and Thoneman dip and have come to the conclusion that it could be the result of spatial beam effects.

IV. SUMMARY

Some of the main experimental results for multiphoton ionization of alkali atoms have been presented. While accurate absolute "cross-sections" exist at the ruby wavelength, most of the experimental results involve the variation of ionization with wavelength and polarization and interesting resonance effects. While the calculations agree quite well among themselves and with experiment for the polarization dependence, the results for absolute cross-sections show considerable variation. There is no question that resonance and level shift effects are sufficiently interesting to warrant further experimental and theoretical study.

TABLE 1.   The cross-section for 3-photon ionization of Na, K, Rb and Cs for linearly (W_3^ℓ) and circularly (W_3^c) polarized single mode ruby laser light ($\lambda = 694$ nm). The units are $10^{-78}\text{cm}^6\text{s}^2$. The calculated values of Bebb [11], Morton [12], Manakov et al [13], Flank and Rachman [14] and Teague and Lambropoulos [15] are included. ($R = W_3^c/W_3^\ell$).

Element		Cervenan et al.		Bebb	Morton	Manakov et al.	Flank & Rachman	Teague & Lambropoulos
Na	W_3^ℓ	56	±13	80	9.4	9.33	--	--
	W_3^c	131	±30	--	--	23.2	--	--
	R	2.35	± 0.10	--	--	2.49	--	--
K	W_3^ℓ	0.35	± 0.09	0.05	0.89	0.187	--	--
	W_3^c	0.93	± 0.25	--	--	0.46	--	.0568
	R	2.66	± 0.11	--	--	2.46	--	2.18
Rb	W_3^ℓ	14	± 4	14	72.5	1.14	--	--
	W_3^c	30	± 9	--	--	2.68	--	--
	R	2.16	± 0.13	--	--	2.35	--	--
Cs	W_3^ℓ	18	±10	18	108	957	9.6	--
	W_3^c	41	±20	--	--	1360	23.5	11.7
	R	2.24	± 0.11	--	--	1.42	2.45	2.22

REFERENCES

1. C. Lecompte, G. Mainfray, C. Manus, and F. Sanchez, Phys. Rev. Letts. $\underline{32}$, 265 (1974).

2. M. R. Cervenan and N. R. Isenor, Opt. Commun. $\underline{13}$, 175 (1975).

3. M. R. Cervenan, R. H. C. Chan and N. R. Isenor, Can. Jour. Phys. $\underline{53}$, 1573 (1975).

4. P. Agostini and P. Bensoussan, Appl. Phys. Letts. $\underline{24}$, 216 (1974).

5. J. Morellec, D. Normand and G. Petite, Phys. Rev. A$\underline{14}$, 300 (1976).

6. M. R. Cervenan, Ph.D. Thesis, University of Waterloo, Waterloo, Canada (1975).

7. R. H. C. Chan, M.Sc. Thesis, University of Waterloo, Waterloo, Canada (1975).

8. B. Held, G. Mainfray, C. Manus and J. Morellec, Phys. Rev. Letts. $\underline{28}$, 130 (1972).

9. B. Held, G. Mainfray, C. Manus, J. Morellec and F. Sanchez, Phys. Rev. Letts. $\underline{30}$, 423 (1973).

10. C. B. Collins, B. W. Johnson, M. Y. Mirza, D. Popescu and I. Popescu, Phys. Rev. A$\underline{10}$, 813 (1974).

11. H. B. Bebb, Phys. Rev. $\underline{149}$, 25 (1966).

12. V. M. Morton, Proc. Phys. Soc. $\underline{92}$, 30 (1967).

13. N. L. Manakov, M. A. Preobragansky and L. P. Rapoport, Proc. XIth Intl. Conf. Phen. Ionized Gases, 23. Prague (1973).

14. Y. Flank and A. Rachman, Phys. Letts. A$\underline{53}$, 247 (1975).

15. M. Teague and P. Lambropoulos, J. Phys. B$\underline{9}$, 1 (1976).

16. P. Lambropoulos, Phys. Rev. Letts. $\underline{28}$, 585 (1972).

17. S. Klarsfeld and A. Maquet, Phys. Rev. Letts. $\underline{29}$, 79 (1972).

18. F. H. M. Faisal, J. Phys. B$\underline{5}$, L233 (1972).

19. H. R. Reiss, Phys. Rev. Letts. $\underline{29}$, 1129 (1972).

20. Y. Gontier and M. Trahin, Phys. Rev. A$\underline{7}$, 2069 (1973).

21. R. G. Evans and P. C. Thoneman, Phys. Letts. A$\underline{39}$,133 (1972).

22. A. T. Georges and P. Lambropoulos, Phys. Rev. A$\underline{15}$,727 (1977).

Experiments on Weakly Bound Electrons in Electromagnetic Fields

JAMES E. BAYFIELD
Department of Physics and Astronomy
University of Pittsburgh
Pittsburgh, Pennsylvania

Recent advances in experimental techniques have made possible a number of studies of highly excited states in static electric fields. In this zero frequency limit the Stark shifts and splittings of atomic energy levels are generally well understood, and a developed quantal tunneling theory of field-induced ionization of the atom exists. On the other hand, the tunneling mechanism is unimportant in laser induced multiphoton ionization of ground state atoms. It is expected on theoretical grounds [1] that situations may exist where true multiphoton absorption and tunneling are in some sense both important. The principal objective of the research of my students and myself has been to experimentally ascertain whether the microwave ionization of highly excited hydrogen atoms is such an intermediate case. Let the atom's parabolic quantum numbers be n, n_1, n_2, m.

Figure 1 shows schematically the difference between the two mechanisms. For present purposes we apply the term multiphoton ionization (MPI) to those processes involving a net absorption of a nonzero number of photons from an external electromagnetic field. Within perturbation theory, nonresonant MPI has a power law dependence upon the field strength F_o:

$$W_{MPI} \sim F_o^{2k_o}$$

where $88 < k_o < 173$ for $45 \lesssim n \lesssim 57$ and microwave frequencies between 9.4 and 11.6 GHz. The theories of tunneling predict a different but also strong dependence:

$$W_{Tunnel} \sim \alpha(F_o) e^{-2/(3n^3 F_o)}$$

The prefactor $\alpha(F_o)$ often has a power law form that appears to depend upon the details of the approximations used in the quantum calculations. If the field is made to oscillate slowly at

192   J. E. Bayfield

frequency ω, a semiclassical ionization calculation predicts a
correction to the tunneling exponent as well as corrections to
the prefactor that depend upon ω [2]. This would reduce the experimental critical field for ionization to values lower than the
static field value. Such a reduction of a factor of two has been
observed for values of n near 50 and values of F_o near 50 volts/cm
[3].

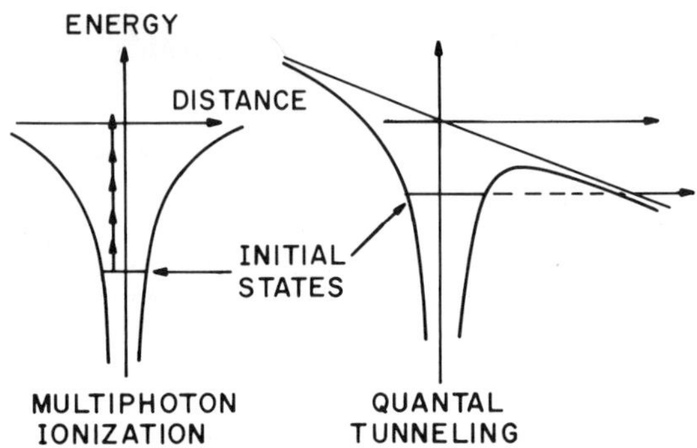

FIGURE 1. Multiphoton ionization and quantal tunneling
           processes.

We can identify three aspects of the interaction of the atom
with the field, although they may be strongly intermixed. The
strong field coupling problem of a single hydrogenic bound state
with the continuum, already discussed, is characterized by the
Keldysh parameter $\gamma = \omega/nF_o$ a.u., whose square is essentially the
ratio of electronic energies for the initial bound state to the
final state electron oscillating in the field. Our experiments
span the expected transition region near $\gamma = 1$.

The other two aspects involve the coupling of different bound
states. The nearly degenerate states of given n but different ℓ
and m are strongly coupled in a first order AC Stark interaction
as discussed for two states by Townes [4] and developed further
for any n by Kovarskii and Perel'man [5]. Expected features of
this interaction are sizeable energy level shifts and high populations of photon replica or satellite states in the dressed atom's
spectrum. These occur when the peak electric dipole coupling
strength $\mu F_o = 3/2\ n(n_1-n_2)F_o$ becomes comparable to the photon
energy $\hbar\omega$. Since $\mu F_o/\omega = 3/2\ (n_1-n_2)(1/\gamma)$, our experiments are
in this sense also in a strong field regime.

Finally there is the coupling between states with different
principal quantum number. If an atom is initially prepared with

the value $n = n_0$ near 50, then $\Delta n = 1$ transitions at frequencies near 10 GHz require the absorption of a few photons. Locally the atom's energy levels are almost equally spaced, having an effective local anharmonicity of $\Delta E_n/E_n)\big|_{n_0} = 4/n_0$. In a multiphoton picture the higher lying levels can resonantly enhance the ionization process at certain microwave frequencies. In a modified tunneling picture, transitions at crossings of the upper Stark sublevels for $n = n_0$ with the lower sublevels for $n = n_0 + 1$ might also enhance ionization, but not in a strongly frequency-dependent manner.

So much for the possibilities. What are the experimental facts of the matter? Figure 2 shows the type of apparatus used to study multiphoton excitation and ionization of highly excited hydrogen atoms [6]. Basically, a fast atomic beam at energies near 10 KeV is formed by charge exchange collisional production of a low excited state (n = 10) followed by Doppler-tuned laser excitation (using specific CO_2 laser frequencies) of a selected level with $n = n_0$. Such beams have been passed through a number of microwave field structures, primarily a TE_{10} waveguide containing a variable frequency field and a long TM_{0M0} cavity operating at its fixed frequencies. The atoms passing through the waveguide were exposed to a pulse containing about 100 field cycles and having a Gaussian-like envelope. The field within the cavity was different in that a long nearly rectangular pulse was produced, with all atoms in the beam exposed to the same peak field strength to within four percent. Ionization in either field structure was detected by observation of the resultant fast protons in the beam. Microwave induced transitions from the state n_0 to higher bound levels were detected by ionization in a subsequent second weaker microwave field. This ionization was separated from ionization within the first field by the spatial resolution provided by a voltage-labeling technique that alters the kinetic energy of protons formed in the field structure of interest. Energy analysis of all the protons in the beam then selects the field ionization of interest.

A number of experiments investigating the dependence of the field ionization on field strength have been completed. The best of these using the long cavity still gave results consistent with a power law dependence much weaker than that for nonresonant MPI [7]. At n = 40 and a frequency of 9.917 GHz, the experimental slope of the traditional log-log plot of ionization probability vs field strength was $k_{eff} = 12 \pm 3$, too weak to be due primarily to field nonuniformity. Such reduced values of $k_{eff} \ll k_0$ have been predicted for nonresonant MPI in strong fields [2] and can be viewed as arising from the essentially complete mixing of the upper bound levels with the continuum, as is the case for static fields; also F_c varies only a factor of two for ω between essentially zero and 10 GHz. Thus this data does little to distinguish between the two mechanisms of "strong-field nonresonant MPI" and "low frequency tunneling", and it would even seem given just this information that these might be the one and the same process.

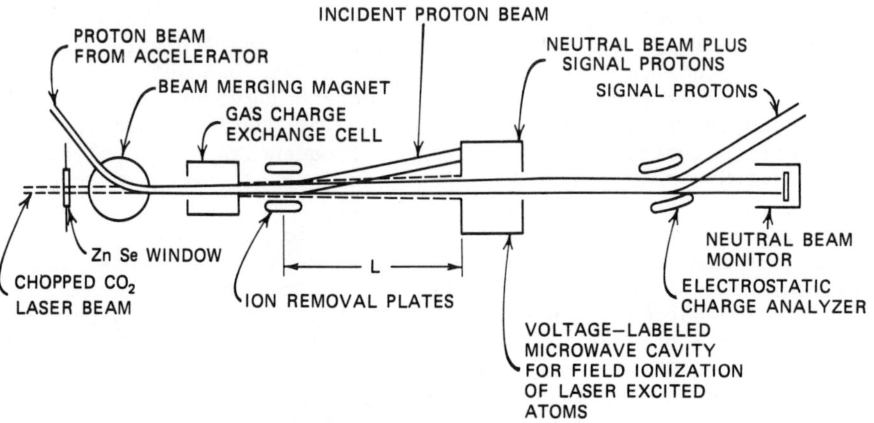

FIGURE 2.   Schematic diagram of the apparatus used in experiments on microwave interactions with highly excited hydrogen atoms.

Recent experiments that continuously vary the frequency of the field in the TE_{10} waveguide have considerably increased our knowledge beyond that just described [8]. Resonant structure was observed in the field ionization for $45 \leqslant n \leqslant 57$ and frequencies $9.4 \leqslant \omega/(2\pi) \leqslant 11.6$ GHz. The experiments were difficult, for in spite of efforts to hold the microwave power level fixed, it did vary by a few percent at certain frequencies. The observed frequency dependence was a convolution of the true field ionization spectrum with an instrumental spectrum enhanced by the high degree of nonlinearity of the field strength dependence of the ionization. Thus a critical experimental test was to observe changes in the spectrum obtained by Doppler tuning to a different value of n_0 while leaving the microwave system unaltered. Such

spectral changes did dominate the overall observed spectra.
Figure 3 shows an ionization spectrum for $n_o = 48$ as well as an
excitation spectrum for transitions from $n_o = 48$ to higher bound
states. The high point with the large error bar located at 11.1
GHz is indicative of the largest instrumental distortions that
were ever observed. The excitation spectrum shown was taken under identical waveguide microwave system conditions and contains
effects due to the same frequency dependent microwave power fluctuations. Thus the largest peaks in each spectrum in Fig. 3 must
be due to resonant frequency dependences in the atomic processes
involved, for the resonances in excitation and ionization occur
at different frequencies. The same cannot be said for the smaller
peaks at 9.3 GHz. One might wonder if the large peaks were due
to resonant-like variations in k_{eff} which would enhance the
atomic response to the instrumental field variations. Effects of
this type may be present, as resonance effects in k_{eff} are known
to accompany amplitude resonances in laser multiphoton ionization
experiments involving an important intermediate state [9]. The
conclusion that resonant effects exist in the ionization process
is, however, unaltered by this further possibility.

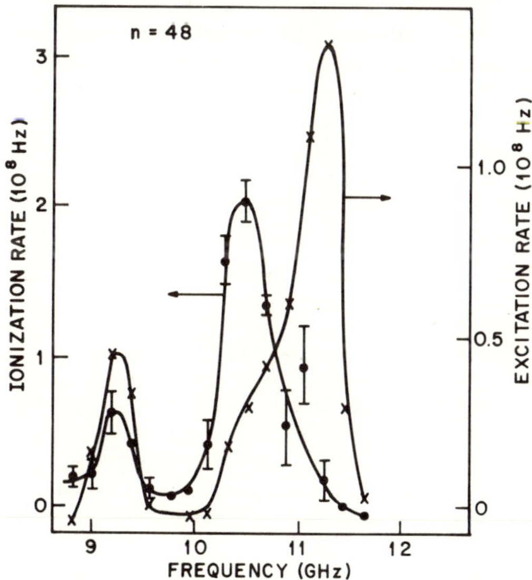

FIGURE 3. A comparison of the frequency dependences for the
           ionization (left scale) and excitation (right scale)
           of hydrogen atoms with n = 48 in a microwave field.

In summary, the microwave ionization of highly excited hydrogen must be viewed as some sort of strong field coupling to the continuum that is resonantly enhanced by the presence of the higher lying states. Resonant transitions to these states do occur at rates that compete with ionization. The frequencies of the excitation maxima are close to that expected for $\Delta n = 1$ multiphoton transitions within the field-free atom, with Stark shifts that range from 250 MHz for a five photon transition ($k' = 5$) in $n_o = 49$ to 350 MHz for a seven photon transition at $n_o = 45$. The ionization resonances are however red-shifted from the $\Delta n = 1$ field-free atom frequencies by about 10% (1 GHz) consistent with the theory of multilevel resonant enhancement [10]. A simple view of the ionization process is schematically indicated in Fig. 4, where the interaction with the continuum is represented by the static Stark potential at a field strength near the peak value F_o, while the field-free bound states enhance vertical transitions from the state n_o up to levels near the top of the barrier. While this picture is consistent with all our data, it needs theoretical justification.

A remaining interesting feature of the excitation resonances concerns their widths. The three important contributions to the width are the ionization width, the "field width" or power-broadening component and uncertainty principle broadening. For the $n_o = 48$ resonance at 11.4 GHz, the rates (see Fig. 3) give about 50 MHz and 100 MHz for the first two components respectively. If one estimates the third contribution using the geometrical length of the waveguide field, one obtains 100 MHz. However, for a Gaussian-like pulse envelope $f(t)$, the effective interaction time T_{eff} is less because high-order transitions occur mainly near the field maximum. Thus T_{eff} is determined by an integral

$\int f^{2k'}(t)dt$ over the envelope and has a value reduced by about

a factor k', which for the $k' = 5$, $n_o = 48$ resonance gives an effective uncertainty principle width of 500 MHz, not far from the observed value. Under these short pulse time conditions, a t^3 variation has been theoretically predicted for the time dependence of the ionization probability [11]. Further experiments to verify this would be interesting, as would experiments with long, tunable field structures where uncertainty principle broadening becomes unimportant.

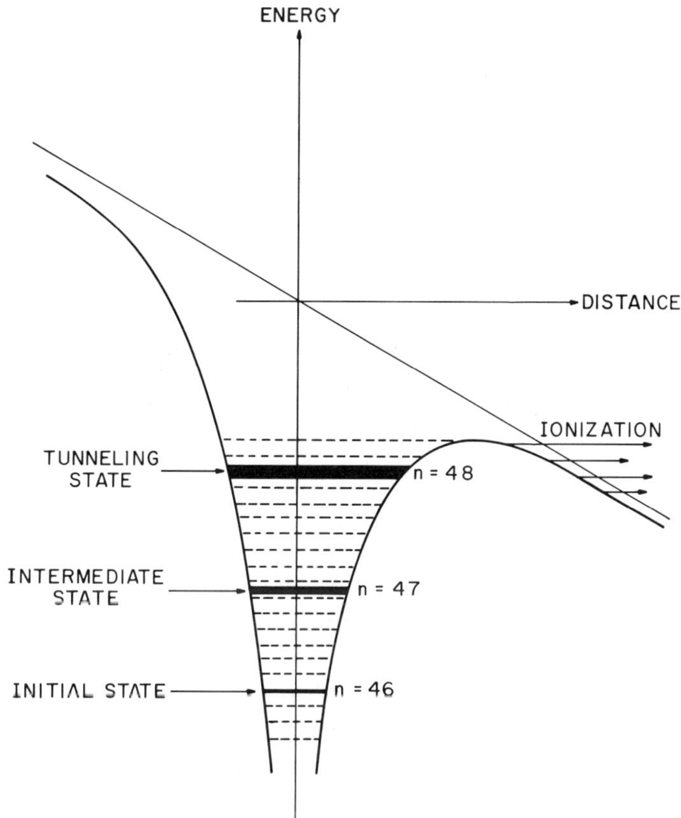

FIGURE 4.   A simplified picture of a combined multiphoton-quantal tunneling process. The absorption of a number of photons is postulated to enhance field ionization. The dressed-atom states are schematically indicated as dashed lines, and are likely different than that shown.

REFERENCES

1.   V. A. Kovarskii and N. F. Perel'man, Sov. Phys.-JETP $\underline{34}$, 738 (1972).

2.   V. S. Popov, V. P. Kuznetsov and A. M. Perelomov, Sov. Phys. JETP $\underline{26}$, 222 (1968).

3.   P. M. Koch and J. E. Bayfield, Contributed Papers of the Fourth International Conference on Atomic Physics, Heidelberg, Germany, 1974, J. Kowalski and H. G. Weber, editors (University of Heidelberg, 1974), p. 336.

4.   C. H. Townes and A. L. Schawlow, Microwave Spectroscopy (McGraw-Hill, New York, 1955), p. 280, case 2.

5.   V. A. Kovarskii, N. F. Perel'man and S. S. Todirashku, Sov. J. Quant. Electron. $\underline{4}$, 1344 (1975), and references therein.

6.   J. E. Bayfield, Rev. Sci. Instr. $\underline{47}$, 1450 (1976).

7.   P. M. Koch, L. D. Gardner and J. E. Bayfield, in Beam-Foil Spectroscopy, Vol. 2, D. J. Pegg and I. A. Sellin, editors, (Plenum Press, New York 1976).

8.   J. E. Bayfield, L. D. Gardner and P. M. Koch, Phys. Rev. Lett. $\underline{39}$, 76 (1977).

9.   D. T. Alimov and N. B. Delone, Sov. Phys. JETP $\underline{43}$, 15 (1976).

10.   D. M. Larsen and N. Bloembergen, Opt. Commun. $\underline{17}$, 254 (1976).

11.   M. V. Fedorov and A. E. Kazakov, Opt. Commun. $\underline{22}$, 42 (1977).

Polarization Effects in Two-Photon Ionization

M. J. VAN DER WIEL AND E. H. A. GRANNEMAN
F.O.M.-Institute for Atomic and Molecular Physics
Amsterdam, The Netherlands

I. INTRODUCTION

This article discusses a narrow selection of topical problems
in the field of two-photon absorption processes; it by no means
purports to be a survey of this very large field. The effects
mentioned in the title refer to the influence of the incident
light polarization - linear or circular - on the rate of the two-
photon ionization process (in alkali atoms), and thereby on associated quantities like the photoelectron spin polarization and
the angular distribution.

Investigation of these effects is of interest from a fundamental point of view, as it leads to knowledge of the matrix elements involved in the transitions, in particular the bound-free
elements. Usually the bound-bound elements connecting the ground-
and intermediate states are already known to a fair degree of
accuracy. The essence of measuring polarization effects is that
one is dealing with *ratios* of intensities, which can be obtained
with much greater precision than absolute intensities. From such
intensity ratios one derives ratios of matrix elements, which
normally provide a sufficiently sensitive test of the theory.

Besides the fundamental aspect, it should be mentioned that
both resonant [1,2] and near-resonant [3] two-photon processes
have been put forward as candidates for use in a source of spin
polarized electrons, for which a wide range of uses exist (see
e.g. [4]).

Let us consider how the polarization effects come about, and
for simplicity, let us take the resonant process. Quite in general, the excitation leads to orientation or alignment of the
excited state, depending on whether circular or linear polarization of the light is used. Such orientation or alignment is
always retained at least partially. This holds even if, in the
subsequent time evolution of the excited state, coupling occurs
of the total electronic angular momentum J with the nuclear spin
I, which leads to a reduction of the J-orientation.

It is this orientation which causes polarization effects in
the two-step processes: even for unpolarized light in the ionizing
step,a non-zero spin polarization will be found and a non-unity
ratio of ion rates for circular and linear light polarization
$(\hat{\sigma}_{circ}/\hat{\sigma}_{lin})$.

For non-resonant processes, the situation is more complex.
In the domain of validity of second-order perturbation theory,
i.e. for field strengths small compared to electron binding
energies, a virtual intermediate state is described in terms of
a linear combination of the complete set of eigenstates. Now
the orientation of each of these states contributes to the polarization effects. On the other hand, problems of coupling of I
and J do not occur due to the short lifetimes involved; the uncertainty principle requires the product of lifetime and energy
mismatch $(E_n - E)$ between photon (E) and one of the eigenstates n
to be at least equal to h. For example, a mismatch of 1 eV leads
to a lifetime of 10^{-15} sec.

II. RESONANT TWO-STEP IONIZATION

Although the polarization effects generally contain information on ratios of radial overlaps, the actual value of the effect
depends on other factors as well, which we wish to call "angular
momentum" factors. By this expression we refer to all factors in
the calculation apart from the radial overlaps, i.e. direction(s)
and polarizations(s) of the light beam(s), pulse lengths and possible time difference between the two light pulses.

In most of the analyses so far, the angular-momentum factors
have been dealt with on the basis of the fine-structure scheme
for ground- and excited state. This has given rise to rather
puzzling discrepancies with experimental observations. Pioneer-
ing work on polarization effects in two-step ionization in Cs
was reported by Lambropoulos and Lambropoulos [1] and by Zeman
[2]. In both instances, the use of circular polarization led to
a sizeable spin polarization of the photoelectrons. However, the
expected values were not attained, neither at the power levels
used nor when extrapolating to zero light intensity. In a set of
more recent experiments [5-7], it was recognized however, that
the time-evolution of the excited state orientation (which is
the basis of quantum-beat spectroscopy), had generally been over-
looked in work on ionization processes. Due to (I,J) coupling,
a precession sets in at the moment of excitation, i.e., the initial J orientation is shared periodically between I and J.
Alternatively, one may describe the phenomenon as coherent excitation of a set of hyperfine levels. The evident result of the

hyperfine coupling is an average orientation markedly lower than that created initially.

It now depends on the time interval between excitation and ionization, what the outcome of the ionizing event will be. This means that in general no simple ionization rate can be formulated. Note that for the moments of excitation and ionization to be well-defined, we require bandwidths of the incident radiation larger than the natural or induced widths. It is also understood that we are dealing with excited states for which the hyperfine splitting is large compared to the natural linewidths (low-n levels of the alkalis). Moreover, optical pumping is ignored throughout this discussion.

Only in two limiting cases does the time-interval-dependent ionization probability reduce to an ordinary rate. One is the case in which the time interval is short compared to the precession periods τ_F of all (I,J) combinations. This condition is fulfilled, if use is made of

i)   one (or two coincident) light pulse(s), whose width is small compared to τ_F;

ii)   ionizing light of such intensity that the lifetime of the intermediate state against ionization is lowered to less than τ_F.

For those conditions it suffices to use the fine-structure scheme*.

The second case requires time intervals between excitation and ionization, which on the average are long compared to τ_F, while in addition time integration should be made over many periods of the precession. Such conditions obtain, e.g., in low-intensity CW excitation and ionization; they necessitate a description which takes into account the hyperfine coupling.

Another interesting situation is that of two light pulses with a variable time interval. For that situation quantities like $\hat{\sigma}_{circ}/\hat{\sigma}_{lin}$ or the spin polarization, when recorded as a function of time difference between two short pulses ($\tau_p \ll \tau_F$), directly reflect the oscillatory behavior of the excited-state orientation (Fig. 1). Such "polarization beats" might even be used for a determination of FS or HFS splittings for cases, which for some reason are not readily accessible to the ordinary quantum-beam spectroscopy.

---

*Obviously the choice of basis set is free, as long as appropriate angular momentum algebra is applied. Here we refer to the most convenient scheme, i.e. that in which the excited-state density matrix contains only diagonal elements.

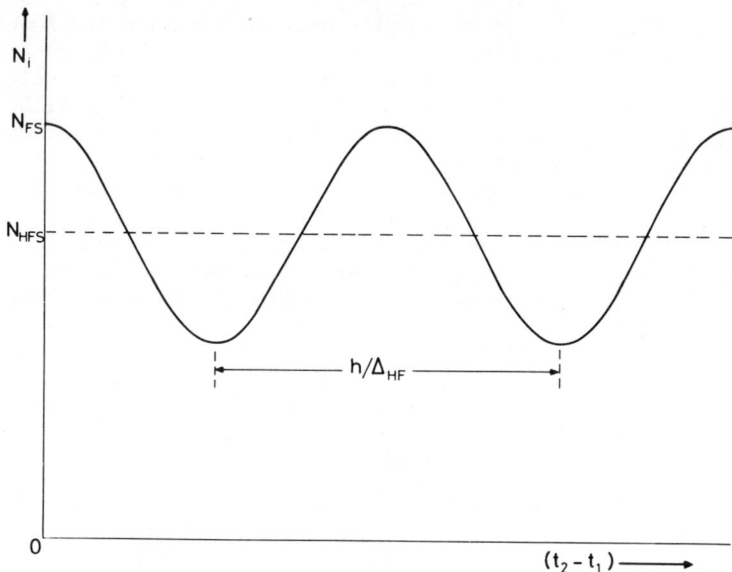

FIGURE 1.   Two-photon ion production rate via an intermediate
            state with two hyperfine sublevels separated by Δ_{HF}.
            The abscissa gives the time difference between ex-
            citing (t_1) and ionizing (t_2) event. Bandwidths of
            exciting and ionizing radiation are assumed suffi-
            ciently large for these moments to be defined to an
            accuracy better than the magnitudes of (t_2-t_1)
            considered here. N_{FS} is the fine-structure rate,
            for time differences small compared to h/Δ_{HF}; N_{HFS}
            is the average of the ion rate over many periods
            of the precession. The photoelectron spin polariza-
            tion is defined as a ratio of two N_i with different
            modulation depths, i.e. it will also exhibit this
            quantum-beat structure. The same holds for the ratio
            of ion rates for linear and circular polarization.

   Apart from the pulse duration, we should also consider the
effect of pulse intensity. Again the time the atom spends in
the excited state is the only criterion, and induced emission
tends to reduce this time. If the lifetime is reduced to less
than τ_F by saturating CW resonant light, we are dealing with a
saturated population of fine-structure sublevels from which ionization occurs. A similar saturated population can also be
created during a short exciting pulse, but this will be followed
by a redistribution over hyperfine sublevels due to (I,J)-coupl-
ing, as long as the ionization probability is small.

We shall now illustrate some of the points in the above discussion. Experimental work on the resonant process in Cs via
the $6P_{3/2}$, $7P_{1/2}$ and $7P_{3/2}$ states has been performed [5-7] in an
arrangement as shown in Fig. 2.

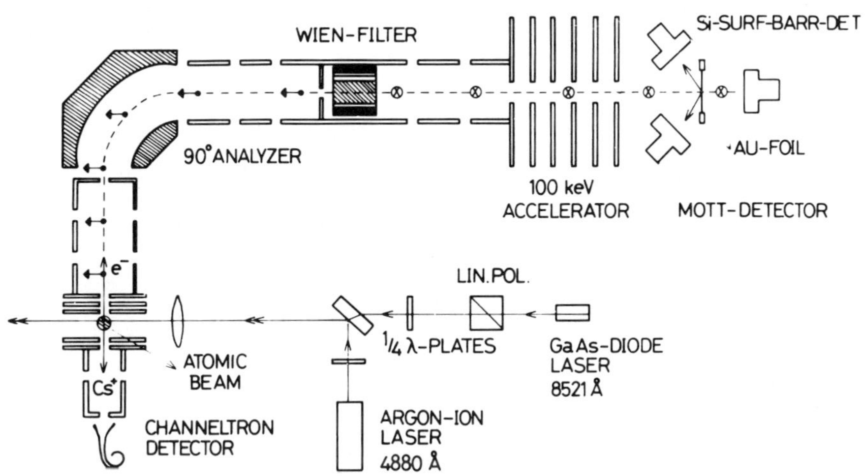

FIGURE 2.   Experimental arrangement for measurements of the ion
count rate (Cs^+) and electron spin polarization (e^-)
in two-step photoionization via the $6^2P_{3/2}$ state.
The electron spin is analyzed in a Mott-detector [6].
The Wien filter, a device with crossed electric and
magnetic fields, is used to convert the incoming
longitudinal polarization into a transverse polarization, necessary for the spin analysis ($\leftarrow$ electron spin
in the plane of the drawing; $\otimes$ electron spin perpendicular to the plane of the drawing). In the two-
step photoionization via the 7P states, the Argon-ion
and GaAs lasers are replaced by a nitrogen-laser
pumped dye-laser.

For the $6P_{3/2}$ case, the resonant step was made with a pulsed
(and therefore fast wavelength-scanning), narrow-band IR diode
laser, which was temperature tuned to scan over the resonance.
This laser provided IR absorption to two separate groups of hyperfine transitions $F_g = 3$ and $4 \rightarrow F_e = 2,3,4$ and $3,4,5$. Ionization
was produced by a ⁓50 nsec light pulse from an Ar-ion laser,
which could be made to coincide in time with either the first or
the second group of HF transitions. The ionizing intensity was

quite low enough for the average time spent in the excited state
to be long compared with the $6P_{3/2}$ precession periods (2 - 7 nsec).
In the ionizing step, interference occurs from three con-
tinuum states: $6P_{3/2} \rightarrow \varepsilon S_{1/2}$, $\varepsilon D_{3/2}$ and $\varepsilon D_{5/2}$. Usually it is
assumed that spin-orbit coupling in the D-continuum can be neg-
lected. However, a measurement of more than one polarization
effect permits this assumption to be tested. To that end, we
introduce two ratios, X and Y, of the radial parts of bound-free
matrix elements, where

$$X = \frac{R(6P_{3/2} \rightarrow \varepsilon D_{5/2})}{R(6P_{3/2} \rightarrow \varepsilon D_{3/2})} \quad \text{and} \quad Y = \frac{R(6P_{3/2} \rightarrow \varepsilon S_{1/2})}{R(6P_{3/2} \rightarrow \varepsilon D_{3/2})} \quad (1)$$

We can then work through the angular momentum algebra in
the HF scheme [6] and arrive at expressions for $A = \hat{\sigma}_{circ}/\hat{\sigma}_{lin}$
and spin polarization P of the form:

$$A = A(|X|,|Y|) \quad \text{and} \quad P = P(X,|Y|).$$

A, the most accurate quantity, provides us with a relation be-
tween $|X|$ and $|Y|$. It is now possible to determine X and $|Y|$
from the measured value of P. It appears that no significant
spin-orbit coupling occurs in the D-continuum: $X = 1.07 \pm 0.05$.
The corresponding result for Y is $|Y| = 0.55 \pm 0.02$, to be com-
pared with a calculated value of 0.44 [8].
Figure 3 demonstrates what happens when the exciting-light
intensity is increased into the saturated regime. Apart from
the expected non-linearity of the total ion rate, we observe a
decrease of P to a value of 0.38. This value represents saturation in the FS scheme, followed by relaxation over the various
HF levels before ionization occurs. Note that the alternative
of saturation in the HFS scheme would lead to the much lower
value of P = 0.26.
In our second example, an experiment on the 7P state as intermediate, we used a pulse from an N_2-laser pumped dye-laser of
1.4 nsec pulse width for excitation and ionization [7]. The precession periods connected with the 7P HF coupling are: 5 - 20 nsec
for the $7P_{3/2}$ and 2.7 nsec for the $7P_{1/2}$. This implies that for
the $7P_{3/2}$ state results of a FS description can be employed,
while for the $7P_{1/2}$ the result will be intermediate between FS
and HFS.

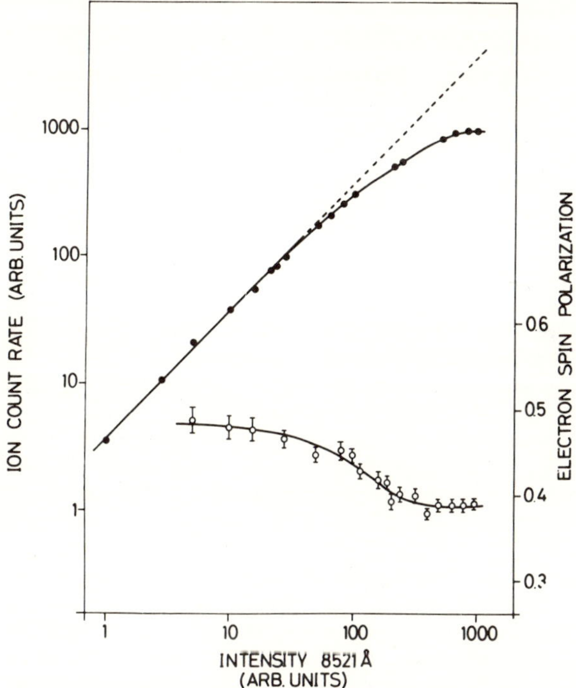

FIGURE 3.   The ion countrate and the electron spin polarization
via the transitions $6^2S_{1/2}(F=4) \rightarrow 6^2P_{3/2}(F=3,4,5)$
$\rightarrow \varepsilon^2S_{1/2}, \varepsilon^2D_{3/2,5/2}$ as a function of the relative
intensity of the exciting radiation (8521 Å).   Both
exciting and ionizing radiation (4880 Å) are right-
hand circularly polarized.   The excitation of ground
state atoms from only the F=4 hyperfine substate is
possible by accurate timing of the exciting GaAs-
laser and ionizing Argon-ion laser pulses (see text).

Indeed, it appears that the observed $7P_{3/2}$ data are in
agreement with a FS calculation [9]:

$$A_{3/2} = 1.22 \pm 0.01 \qquad (1.23; \text{ calcul.})$$
$$P_{3/2} = +0.80 \pm 0.05 \qquad (0.82; \text{ calcul.})$$

In the analysis we have assumed that $X = 1$ (see definition, Eq.(1));
we then find $|Y| = 0.50 \pm 0.01$, in close agreement with the theo-
retical value of 0.49.   For the $7P_{1/2}$ state, however, a direct
comparison is not possible.   In order to still extract information on the bound-free matrix elements, we express the ion production rate as a function of the time interval $(t_1 - t_2)$

between excitation and ionization:

$$N(t_1 - t_2) = N_{HFS} + (N_{FS} - N_{HFS}) \cos 2\pi \left(\frac{t_2 - t_1}{\tau_F}\right),$$

i.e., a sinusoidal variation around the average HFS value (see
Fig. 1). An appropriate integration over all time intervals contained in the pulsewidth leads -- for triangular pulse shapes --
to a simple analytical dependence of N_{AV} on τ_p/τ_F. Using this
dependence, we can correct the measured countrates for linear and
circular polarization to yield a value of $A = 1.32$. This quantity
for the $7P_{1/2}$ state depends on only one ratio of radial overlaps,
$Z = R(7P_{1/2} \rightarrow \epsilon S_{1/2})/R(7P_{1/2} \rightarrow \epsilon D_{3/2})$ which turns out to be
$|0.55 \pm 0.05|$, again in satisfactory agreement with the calculation:
0.51.

   With regard to saturation phenomena, the above values were
all measured under conditions of ion intensity quadratically dependent on light power. In order to achieve this, it was necessary to detune the laser peak wavelength considerably away from
resonance. Figure 4 shows the result for P when scanning the
laser stepwise over a broad range around the resonances. When on
resonance, there was a strong deviation from quadratic power dependence in the ion count rates. The corresponding polarizations
are close to those expected for full saturation: +0.60 for $7P_{3/2}$
and -0.60 for $7P_{1/2}$. At the extreme ends of the wavelength
scale we are still dealing with resonant excitation, since the
probability for the off-resonance two-photon process drops much
faster with wavelength than does the light power still available
at the resonance with the peak detuned.

III.   NON-RESONANT TWO-PHOTON IONIZATION

   Much of the work on non-resonant multiphoton transitions
has been concerned with checking the limits of the validity of
the nth-order perturbation theory. Many examples have been found
of high-order processes, which behave perfectly as I^n over a large
range of light power I, and which also exhibit the expected strong
dependence on the photon statistics. Very impressive work on
these high intensity and temporal coherence effects is due to the
group at Saclay [10].
   For a precise, quantitative application of the theory, the
main problem evidently is the choice of wave functions and the
method of performing the infinite summation in the ionization as
given by second-order perturbation theory:

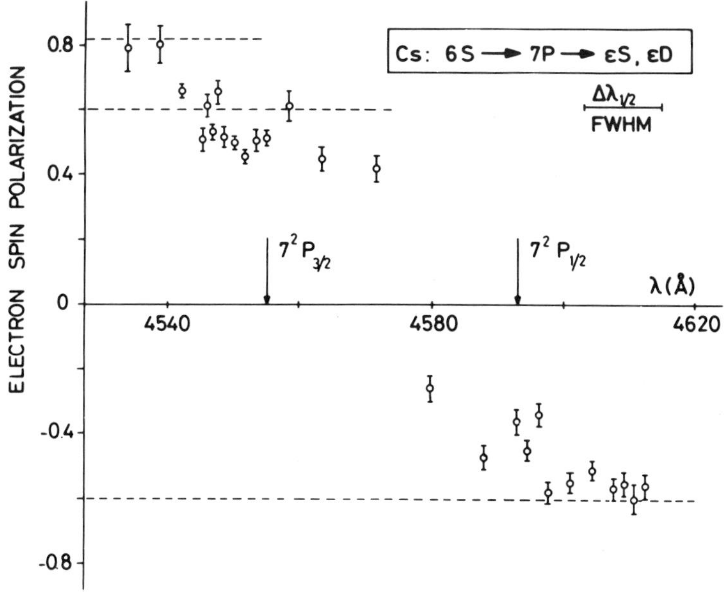

FIGURE 4.   Spin-polarization of the photoelectrons obtained in
resonant two-photon ionization of caesium via the
$7^2P_{1/2,3/2}$ intermediate states. The polarization
is measured as a function of wavelength of the laser
light; excitation as well as ionization is provided
by the same laser pulse. The bandwidth of the light
is 12 Å (FWHM). By varying the peak wavelength the
intensity of the exciting light is varied while the
ionizing light intensity remains the same. The
broken horizontal lines indicate values for the
electron polarization as calculated in the fine-
structure scheme: +0.82 and +0.60, two-photon ionization via the $7^2P_{3/2}$ intermediate state for unsaturated
and saturated excitation, respectively; -0.60, two-
photon ionization via the $7^2P_{1/2}$ intermediate state
for unsaturated as well as saturated excitation. In
all experimental cases the ionization process was not
saturated, i.e., the number of excited atoms ionized
was a small fraction of the total number of excited
atoms.

$$\hat{\sigma} \simeq \left| \sum_n \frac{<f|\vec{r}|n> <n|\vec{r}|g>}{\omega_n - \omega} \right|^2 \qquad (2)$$

In this expression $\hat{\sigma}$ is the probability per (photon flux)^2, f and g are the final- and ground-state wavefunctions, the summation is over all bound and continuum eigenstates n, and ω_n and ω are the energies of eigenstate n and the incident photon.

With regard to the experimental test of the theory, the situation for non-resonant cases is in marked contrast with that for the resonant process. Examples of a comparison of experiment [11-14] and theory [15,16] for $\hat{\sigma}_{circ}/\hat{\sigma}_{lin}$ are given in Table 1, as taken from references [15] and [16]. Although the comparison does not provide a severe test of the theory, the agreement was considered to be satisfactory.

TABLE 1. Ratios of multiphoton ionization rates for circular and linear polarization

atom	n-photon n =	laser wavel. (Å)	$\hat{\sigma}_{circ}/\hat{\sigma}_{lin}$ experiment	theory	
Cs	2	3471	1.28 ± 0.2 [11]	1.14	[15]
	2	4545 – 5017	see Fig. 5 [17]	see Fig. 5	[9, 20]
Cs	3	6943	2.24 ± 0.11[12]	2.45	[16]
			2.15 ± 0.4 [13]	2.22	[15]
Rb	3	6943	2.16 ± 0.13[12]	2.31	[16]
K	3	6943	2.66 ± 0.10[12]	2.18	[15]
			2.34 ± 0.22[14]	2.46	[16]
Na	3	6943	2.35 ± 0.10[12]	2.49	[16]

However, a really detailed theoretical study of such effects in two-photon ionization of Cs over a rather wide range of wavelengths, presents a rather different picture. From the comprehensive analysis of Teague et al. [9] it appears that $\hat{\sigma}_{circ}/\hat{\sigma}_{lin}$ is very sensitive to the choice of wavefunction and the method of summation. In fact, at certain wavelengths different choices lead to changes in the value of $\hat{\sigma}_{circ}/\hat{\sigma}_{lin}$ over the entire allowed range (0 to 1.5). A definite conclusion is therefore impossible.

This situation is all the more disturbing, since it implies that a measurement of such ratios (Fig. 5 and ref. [17]) does not help solve another serious problem with regard to the absolute rates. In the wavelength region of Fig. 5, i.e. near the 7P resonances, the experimental $\hat{\sigma}$ [18] differs drastically from the theoretical curve; the deep minimum in the calculated curve is entirely missing in the experimental result (Fig. 6). This dip is a very fundamental consequence of the cancellation of the dominant 6P and 7P contributions to the summation in Eq.(2); all signs are equal except those of the energy denominators. Variation of the wavefunctions used leads to only insignificant shifts in the position of the minimum and at most one order of magnitude change in the value at the minimum.

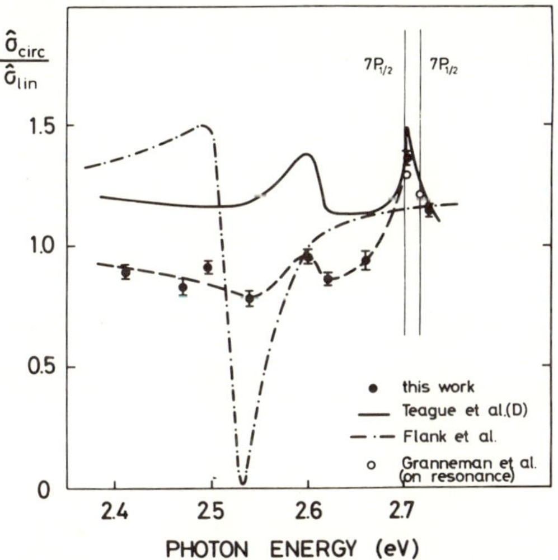

FIGURE 5. Non-resonant two-photon ionization of caesium:measurement of the ratio of the ion countrates for circularly and linearly polarized light for the wavelengths of the argon-ion laser. The experimental points ($\bullet$,Klewer et al. [17]) are compared with theoretical results as obtained by Teague et al.(— [9]). In the latter paper, four different sets of wave functions were used in the calculation of the ratio $\hat{\sigma}_{circ}/\hat{\sigma}_{lin}$. Only the results obtained with set "D",which Teague et al. consider to be the most accurate, were plotted in this figure.The open circles (o) represent the measured values on the $7^2P_{1/2,3/2}$ resonances (Granneman et al. [7]).

It is tempting to interpret the absence of the dip in the experimental result as an indication of incomplete suppression of molecular effects, i.e. resonant processes in Cs_2 leading to $Cs^+ + Cs$. However, this possibility was given ample attention and, to our knowledge, any molecular effect, or effects from surfaces in the ionization region, have been completely ruled out.

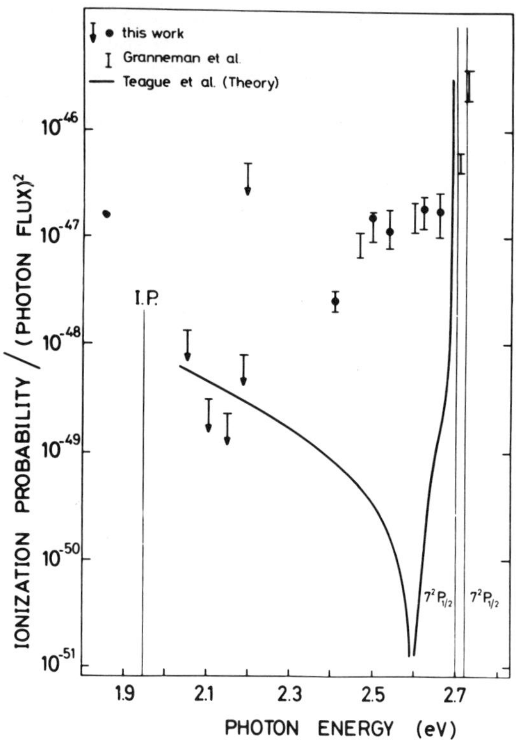

FIGURE 6.   Non-resonant two-photon ionization probability of Cs as a function of photon energy in the energy region between the ionization potential (I.P.) and the 7P resonances. The nine experimental points in the argon-ion laser wavelength region (I,2.4-2.7 eV) are those measured by Granneman and Van der Wiel [18]. The points (•) and the arrows (↓), which indicate upper limits for the ionization probability, are from Klewer et al.[17]. The upper limits are determined in the wavelength region of the argon-ion laser pumped Rhodamine -6 G dye laser (2.0-2.2 eV). The solid line represents the theoretical calculation of Teague et al. [9] (set "D"). See also ref. [21].

The experiment discussed here should have been quite well suited for a test of the perturbation theory, in particular from the point of view of the laser used. The Ar-ion laser delivers a power density in the ionization region of only $<10^5$ W/cm^2, but at a high enough repetition rate to retain a detectable signal. The uncertainty due to photon statistics cannot exceed a factor of two, since we are dealing with a two-photon process. The laser may be operated in the CW mode, such that the radial density distribution is accurately known. Finally, frequency doubling of one of the lines permits an accurate calibration of the apparatus on a known UV cross section for one-photon ionization.

From the Ar-ion laser data points (Fig. 6) it might be concluded that the dip could be shifted to energies below 2.4 eV. Such a conclusion is not inconsistent with a more recent attempt towards a measurement, in which a dye laser was used in the region of 1.9–2.2 eV [17]. The power of this laser was insufficient, however, to arrive at more than a number of upper limits for $\hat{\sigma}$ in this region.

In this rather puzzling situation, what is clearly needed is an independent check of the experiment near the 7P resonance, preferably down to the ionization threshold at 1.95 eV. In order to have sufficient intensity, it seems advisable to use a flash-lamp or N$_2$-laser pumped dye laser rather than an Ar-ion pumped CW dye laser. With the enormous discrepancy mentioned above, it is evident that so far the non-resonant work has not contributed to the derivation of bound-free matrix elements.

IV. CONCLUSION

In summary, we have discussed a few selected problems in the area of two-photon processes. In the resonant case, interesting physics is encountered in the behaviour of polarization effects with (I,J) coupling, and a sensitive check of calculated bound-free matrix elements appears possible. Moreover, the polarized-electron currents obtained in the 6P$_{3/2}$ case during the IR laser pulse were of the order of 10^{-7} A. The definitive use of the process in a continuous source of polarized electrons requires CW operation of the diode lasers. That this is indeed possible, was demonstrated recently [19].

In the non-resonant case, the problem is still wide open, even under the most well-defined conditions. Clearly other, independent, experiments of polarization effects and absolute rates as a function of wavelength are needed.

This work is part of the research program of the Stichting FOM (Foundation for Fundamental Research on Matter) and was made possible by financial support from the Nederlandse Organisatie voor Z.W.O.

REFERENCES

1. P. Lambropoulos and M. Lambropoulos, Proc. Int. Symp. on Electron and Photon Interactions with Atoms, Ed. by H. Kleinpoppen and M.R.C. McDowell, Plenum Press (New York, London 1976) p. 525.

2. H. D. Zeman, Proc. Int. Symp. on Electron and Photon Interactions with Atoms, Ed. by H. Kleinpoppen and M.R.C.McDowell, Plenum Press (New York, London 1976) p. 581.

3. P. Lambropoulos, Phys. Rev. $\underline{A9}$, 1992 (1974).

4. J. Kessler, Polarized Electrons, Springer Verlag (Berlin, Heidelberg, New York 1976).

5. E.H.A. Granneman, M. Klewer, K. Nygaard and M. J. Van der Wiel, J. Phys. $\underline{B9}$, L87 (1976).

6. E.H.A. Granneman, M. Klewer and M.J. Van der Wiel, J. Phys. $\underline{B9}$, 2819 (1976).

7. E.H.A. Granneman, M. Klewer, G. Nienhuis and M.J. Van der Wiel, J. Phys. B (1977) to be published.

8. P. Lambropoulos, private communication.

9. M. R. Teague, P. Lambropoulos, D. Goddmanson and D. W. Norcross, Phys. Rev. $\underline{A14}$, 1057 (1976).

10. See review by: G. Mainfray, this volume.

11. R. M. Kogan, R. A. Fox, G. T. Burnham and E. J. Robinson, Bull. Am. Phys. Soc. $\underline{16}$, 1411 (1971).

12. M. R. Cervenan, R.H.C. Chan and N. R. Isenor, Can. J. of Physics $\underline{53}$, 1573 (1975).

13. R. A. Fox, R. M. Kogan and E. J. Robinson, Phys. Rev. Lett. $\underline{26}$, 1416 (1971).

14. M. R. Cervenan and N. R. Isenor, Opt. Communic. $\underline{10}$, 280 (1974).

15. M. R. Teague and P. Lambropoulos, to be published.

16. G. Laplanche, Y. Flank, M. Yaouen and A. Rachman, Phys. Lett. A, to be published.

17. M. Klewer, M.J.M. Beerlage, E.H.A. Granneman and M.J. Van der Wiel, J. Phys. $\underline{B10}$, L243 (1977).

18. E.H.A. Granneman and M.J. Van der Wiel, J. Phys. $\underline{B8}$, 1617
     (1975).

19. J.L. Picqué, S. Roizen, H. H. Stroke and O. Testard, Appl.
     Phys. $\underline{6}$, 373 (1975).

20. Y. Flank, G. Laplanche, M. Jaouen and A. Rachman, J. Phys.
     $\underline{B9}$, L409 (1976).

21. T. B. Cook, F. B. Dunning, G. W. Foltz and R. F. Stebbings,
     Phys. Rev. $\underline{A15}$, 1526 (1977); this new determination of the
     Cs ground-state photoionization cross section affects the
     calibration of the data of ref. [*18*] and brings the absolute
     rates down by a factor of two.

Multiphoton Spectroscopy
of the Alkaline Earths

J. J. WYNNE, J. A. ARMSTRONG AND P. ESHERICK
IBM Thomas J. Watson Research Center
P. O. Box 218
Yorktown Heights, New York

I. INTRODUCTION

Optical spectroscopy is replete with a long and continuing
history of achievements which have been fundamental in the development of the understanding of atomic structure. However,
many aspects of the electronic structure of many-electron atoms
are still not well understood. In many-electron atoms, the
Coulomb repulsion and spin-orbit coupling effects between electrons strongly influence the nature of the electronic states.
The one-electron picture, where each electron is viewed as moving
in a well-defined orbit, independent of the motion of the other
electrons, breaks down when these effects are strong enough. This
has hindered efforts to identify and classify the states of many-
electron atoms with one-electron labels. We have shown how multiphoton ionization spectroscopy provides us with new spectroscopic
data allowing us to identify previously unknown states of alkaline
earth atoms. These atoms may be thought of as two-electron atoms,
and we have shown how the Coulomb repulsion of the two outer
electrons may be taken into account via a model based on multi-
channel quantum defect theory (MQDT).

Many laser spectroscopists have been studying the alkali
metal atoms, where the energies of the bound states are known to
better than 1 cm^{-1} through the use of conventional, pre-laser absorption and emission spectroscopy. The laser spectroscopists
have been looking at these atoms with high resolution, multiphoton techniques to study fine and hyperfine structure. Conventional survey spectroscopy of the alkali atoms has been successful in identifying the states because these are essentially one-
electron atoms. They behave as one-electron atoms because they
have an outer electron "orbiting around" a closed electronic
shell. The bound states of the one-electron atoms fall into
series where the energy of each state is given by

215

$$E = I - R/(n^*)^2 \tag{1}$$

where R is the Rydberg constant and I is the ionization limit. For hydrogen, n* is an integer, n, and for the alkali metal atoms

$$n^* = n - \delta \tag{2}$$

where δ, the so-called quantum defect, is found to be nearly constant for states of the same orbital angular momentum, l.

The contrast between the alkali metal atoms and the alkaline earth atoms Ca, Sr and Ba is brought into focus by considering the series of even-parity, $J = 2$ states of Ca, converging on the first ionization limit, that have been labeled 4snd $^1D_2$. Of these, only the 4s3d, 4s4d, 4s5d, and 4s6d states were correctly identified in the literature before our multiphoton ionization spectroscopic studies. The 4s7d $^1D_2$ had been incorrectly identified by emission spectroscopy [1], and we found the correct position more than 100 cm^{-1} away (Fig. 1) [2]. Furthermore, this state is more than 1200 cm^{-1} higher than the 4s6d $^1D_2$. Were we able to look for this state using high resolution lasers by scanning and recording on chart paper at the rate of 10 MHz/cm starting from the 4s6d state, we would have had to use 30 km of chart paper before we reached the 4s7d state! It would still be 3 km away from the expected position based on the results of emission spectroscopy! This points to the need for doing survey spectroscopy on atoms other than the alkali metals.

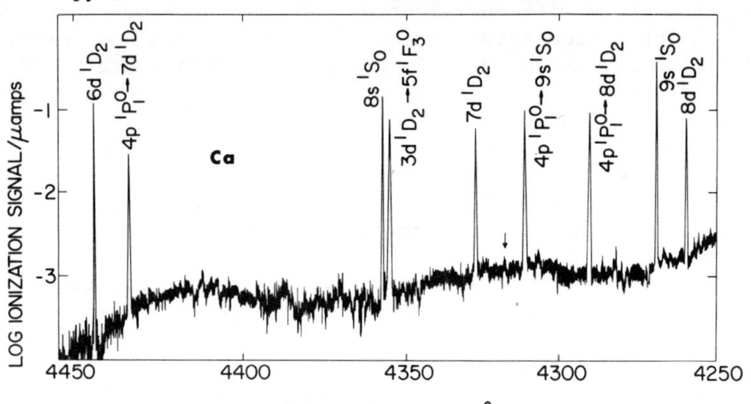

FIGURE 1.   Multiphoton ionization spectrum showing the 4s7d $^1D_2$ state. The arrow marks the position where this state would have appeared according to Ref. 1.

In this paper, we shall describe the experimental technique of multiphoton ionization spectroscopy and show how we have used it, supplemented with absorption spectroscopy where needed, to study series of even- and odd-parity states in Ca, Sr and Ba. We shall indicate how to use MQDT to analyze the data, describe some interesting trends in the results and point out some future directions for studies in these atoms.

II. EXPERIMENTAL TECHNIQUES

In the technique of multiphoton ionization spectroscopy, two or more photons excite atoms from the ground state to an excited state. If the excited state is bound, there are several processes which may ionize it, including photo-ionization, collisional ionization, or chemi-ionization. Whatever the mechanism, strong ionization signals occur when an atomic vapor is irradiated by lasers tuned to multiple photon resonances. The simple set-up shown in Fig. 2 may be used to detect ionization. An ionization probe consisting of a thin wire is inserted into a pipe containing the atomic vapor. For Ca, Sr and Ba, the pipe is heated to ~700 °C. If the probe is negatively biased relative to the walls of the pipe, thermionic emission from the probe will lead to space-charge limited current. Ions produced by the laser excitation partially neutralize the space charge, thereby allowing an increased electron current to flow [3]. One may then detect ionization by observing the pulsed current, synchronous with the pulsed lasers, which flows through the external load resistor. Figure 3 is a typical spectrum obtained when Ca vapor is irradiated with a single, repetitively pulsed nitrogen-laser-pumped dye laser. Bias voltages of only ~1V are needed to produce this high signal-to-noise ratio. Although the simple ionization probe shown in Fig. 2 serves as an excellent ion detector, the electric field near the probe is not uniform. To remove the uncertainty as to the magnitude of the applied field, we have adopted a parallel plate configuration. An alternative is to irradiate the atoms in a field free region and let them drift out of this region into the region where they can be collected by the electric field around the probes [4].

As we shall see, the spectrum of Fig. 3 is due to two-photon excitation of $^1S_0$ and $^1D_2$ bound states of Ca. By using more than one laser, variations of this basic experiment may be performed. For example, we have used three dye lasers, simultaneously pumped by the same nitrogen laser, to stepwise excite Ca, Sr and Ba, from the $^1S_0$ ground state to high-lying odd parity states which are primarily triplet in character. These states are not observed in absorption spectra from the ground state because of the spin-forbidden nature of the transitions. The specific three laser technique we have used, which will be described below, overcomes this barrier.

For calibration purposes, and for verifying data available in the literature, we have also used standard absorption spectroscopy to supplement our multiphoton ionization data. However, we use the dye lasers to supply us with the light which is subsequently absorbed by one-photon transitions.

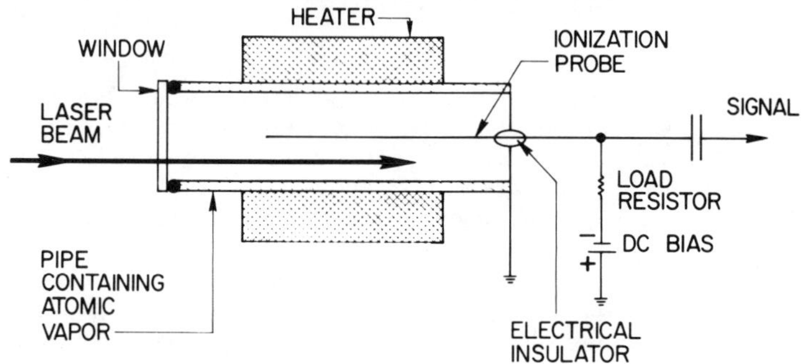

FIGURE 2. Experimental setup for detecting multiphoton ionization in a hot atomic vapor.

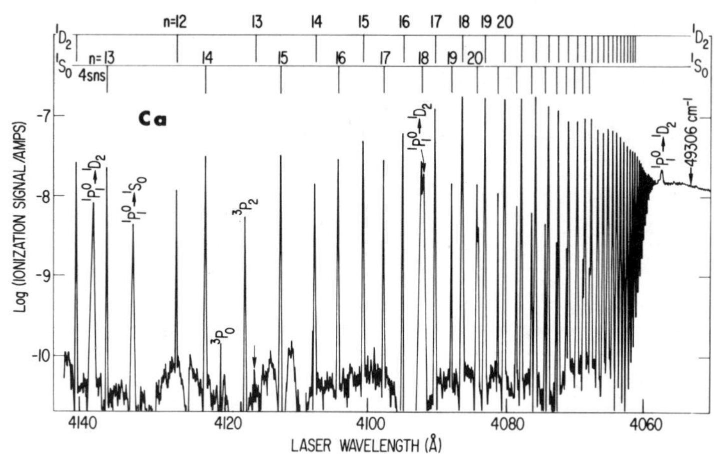

FIGURE 3. Multiphoton ionization spectrum of even-parity states of Ca converging on the ionization limit at 49306 cm^{-1}. Experimental conditions: laser polarization, linear; laser intensity, 10^7 W/cm^2; Ca pressure, 0.1 Torr; buffer gas pressure, 10 Torr; voltage on probe, 1 V.

III. EXPERIMENTAL RESULTS: EVEN-PARITY STATES

The spectrum shown in Fig. 3 shows strong peaks at many places. By tuning the laser frequency ν to lower energies than Fig. 3, we verified that a peak was observed when 2ν was tuned to the energies of the highest previously known $^1S_0$ and $^1D_2$ states of Ca, namely the 4s12s $^1S_0$ and the 4s6d $^1D_2$ states. We also observed that the signal strengths of the peaks varied as the square of the input laser power. We concluded that we were seeing two photon excitation of series of 4sns $^1S_0$ and 4snd $^1D_2$ states converging on the 4s $^2S_{\frac{1}{2}}$ ionization limit of Ca.

To verify the values of J and separate the two spectra from one another, we used the laser polarization. Both $J = 0$ and $J = 2$ states can be observed with linearly polarized light, but only $J = 2$ states can be observed with circularly polarized light. This is easily understood with reference to Fig. 4. It is important to recognize that these selection rules apply only if the intermediate state is polarized, i.e., when it is a *virtual* state. As such, it is excited only when the laser field is present. In contrast, the intermediate state can be a *real* state when the laser is tuned directly on resonance (Fig. 4c). In the presence of collisions, this populated *real* state dephases. In this case, the $J = 1$ intermediate state is unpolarized with the substates $M = +1, 0$ and -1 all equally populated and mutually dephased. The peaks in Fig. 3 are labeled $^1S_0$ and $^1D_2$ on the basis of the determination of J. This method of J selection is vital if one is to sort out a spectrum of states which do not follow the simple Rydberg formula (Eq. 1).

We have applied these techniques to the bound, even-parity $^1S_0$ and $^1D_2$ states of Ca and Sr. In order to get accurate energy values for each new state, we need a calibration procedure. Accurate calibration of the laser wavelength was obtained by recording, simultaneously with the ionization signal, the absorption of the second harmonic of the laser from the ground state to the well-known $^1P_1^{\circ}$ states of Ca and Sr. The ionization signals and the calibrating absorption signals were recorded on a single strip of chart paper using the experimental setup shown in Fig. 5. Using data from several scans and averaging, state energies were determined to an accuracy of better than 0.2 cm^{-1} in almost every case. Detailed results and descriptions of the experimental procedure are presented in Refs. 2 and 5.

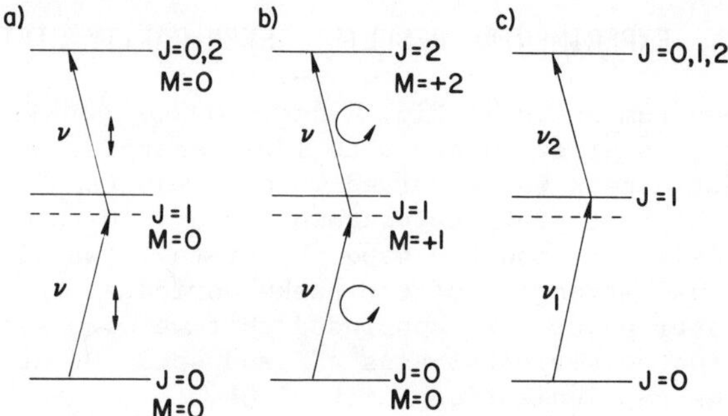

FIGURE 4.   Allowed two-photon transitions from a J=0 initial
state. For a direct two-photon transition to a state
at energy 2ν, the final state may have a) J=0 or 2
when the laser beam at frequency ν is linearly polarized, but b) only J=2 when the laser is circularly
polarized. In case c) one laser at ν_1 excites the intermediate J=1 state on resonance, collisions dephase
this state, and the second laser at ν_2 excites final
states with J=0,1 or 2 at the energy $\nu_2 + \nu_2$.

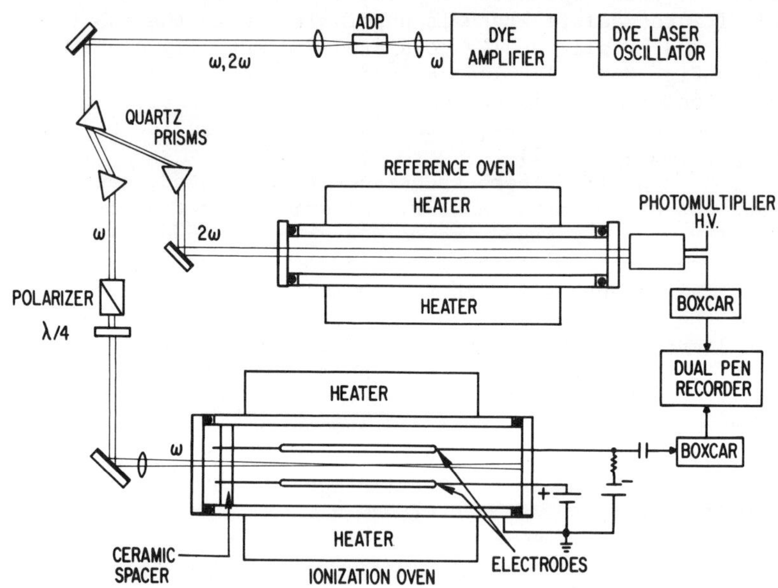

FIGURE 5.   Experimental configuration for measuring ionization
spectrum and calibrating it with an absorption spectrum.

IV.   EXPERIMENTAL RESULTS: ODD-PARITY STATES

The literature shows that the msnp $^3P^\circ$ states of Ca, Sr, and Ba (m = 4,5, and 6 respectively) were correctly known only to n = 7. We have observed three-step transitions from the ms^2 $^1S_0$ ground state to these states by using a three laser excitation technique. One laser excites atoms from the ground state to the lowest energy ms(m + 1)p $^3P^\circ_1$ state (the intercombination line transition). A second laser then excites these atoms to the ms(m + 1)s $^3S_1$ state. To obtain a spectrum, a third laser is tuned and one measures the ionization signal that occurs when the atoms make transitions from the populated $^3S_1$ level to higher energy msnp $^3P^\circ$ states. In this way, we have observed the $^3P^\circ$ states of Ca and Sr to n > 70 [6]. Additional signals were obtained from transitions from the populated $^3S_1$ level to the msnp $^1P^\circ_1$ levels whose energies are known accurately from absorption spectroscopy. Signals were also obtained from known two-photon transitions in the same atomic species (i.e., Ca or Sr) or in impurity atoms (e.g., Na, Ba). Furthermore the third laser was monitored by splitting off part of the beam and measuring its transmission through an air-spaced Fabry-Perot interferometer. The interferometer was calibrated from the ionization signals obtained on known transitions in Ca, Sr, Ba and Na. The calibrated interferometer then provided a means of calibrating the third laser frequency.

The data for Ba are complicated by both configuration mixing and strong spin-orbit effects. Whereas the $^3P^\circ$ states of Ca and Sr are only perturbed by one interloper (the 3d4p and 4d5p respectively), in Ba the 5d6p, 5d7p, 5d8p and 5d4f configurations all affect the spectrum. Furthermore, the high n $^1P^\circ_1$ states produce stronger ionization signals than the high n $^3P^\circ$ states when excited from the 6s7s $^3S_1$ levels. These problems have been resolved by complementing the excitation spectrum from the 6s7s $^3S_1$ with the excitation spectrum from the 6s7s $^1S_0$ intermediate state, and by taking one photon absorption spectra of Ba from the 6s^2 $^1S_0$ ground state. The result is that the J = 1 spectrum of Ba is now well understood and the 6snp $^3P^\circ_2$ states have been observed and identified to n = 50. The detailed results will be published in the near future.

V.   ANALYSIS OF RESULTS

Once the energies of the states, as well as their parity and J are determined, one calculates the effective quantum number from Eq. 1 and looks for trends toward constant quantum defect, δ. If δ is reasonably constant, one says that the states form a relatively unperturbed Rydberg series. On the other hand, if δ is changing from one state to the next, there is strong evidence for configurational mixing and spin-orbit effects. For the $^3P^\circ$ states of Ca and Sr, the quantum defects are relatively constant. Each spectrum shows evidence for a single low-lying perturber, the 3d4p and 4d5p configuration in Ca and Sr respectively, which mixes into the 4snp and 5snp series, respectively, over a limited number of states. Conventional second-order stationary state perturbation theory is perfectly adequate to describe this situation. On the other hand, in Ba, as already mentioned, there are several low-lying perturbers. While the states we label 6snp $^3P^\circ$ have relatively constant quantum defect, the perturbers cause the $^1P^\circ_1$ spectrum to have a continuously changing δ. Perturbation theory is not adequate for explaining this situation.

For the even-parity $^1S_0$ states of Ca and Sr, the data shows evidence for a single configuration, $4p^2$ or $5p^2$, perturbing the 4sns or 5sns series, respectively. But the $^1D_2$ spectra of Ca and Sr show continuously changing δ.

These spectra can be successfully analyzed by MQDT. This theory allows one to treat series of states as channels instead of as individual states. A channel describes a set of states, both bound and unbound, without regard to the specific energy of the highly excited electron. Specification of the angular momentum of the outer electron and the core, along with a description of their coupling, completes the description of the channel. For example, the designation 4snd $^1D_2$ describes a channel which contains bound states of even-parity and $J = 2$, where the core is in the 4s state, the outer electron has $l = 2$, and the electrons are coupled to form a $^1D_2$ type state. This channel also contains the continuum states designated by $4s\varepsilon d$ $^1D_2$, where ε designates the kinetic energy of the unbound outer electron. Within its range of applicability, MQDT determines the energy levels and wavefunctions of the interacting Rydberg series in terms of interactions between a small number of channels. The interaction is expressed in terms of a small number of physically meaningful parameters.

This is not the place for an exposition of MQDT in detail. A more complete description can be found in Refs. 2 and 7. But a graphical presentation of MQDT in terms of Lu-Fano plots [8] will help the reader to better understand what can be learned from our data. In such a plot, the effective quantum number, $\nu_i = n_i^*$ (mod 1) of a state, calculated from Eq. 1 as if it were part of a series converging on an ionization limit I_i, is plotted against $\nu_j = n_j^*$ (mod 1) as if the series converged on a different limit I_j. The limits are chosen as those to which the principal and perturbing series converge. As an example, the 4snd $^1D_2$

series of Ca is perturbed by 3d5s $^1D_2$. Thus an appropriate Lu-Fano diagram plots ν_{4s} vs. ν_{3d}. Because unperturbed Rydberg series are periodic in the ν_i, it turns out to be necessary to plot only the non-integral parts of the ν_i.

Lu-Fano plots of our $^1D_2$ data in Ca, $^3P^\circ$ data in Sr, and $^1P^\circ_1$ data in Sr, are shown in Figs. 6, 7 and 8 respectively. The data is plotted as points. In Fig. 6, we see that the value for ν_s (the effective quantum number assuming the series converges on the 4s $^2S_{\frac{1}{2}}$ limit) is continuously changing. This reflects strong configuration interactions of the 4snd $^1D_2$ series with the $4p^2$, $3d^2$ and 3d5s $^1D_2$ configurations. The continuously changing quantum defect of Fig. 6 cannot be dealt with by perturbation theory. But MQDT allows us to understand the data in terms of an inter-action between channels. The solid line in Fig. 6 is an MQDT fit to the data [2]. It is seen to be very good in passing through the positions of the data points. Note, in particular, the positions of the open square and the closed square. The open square corresponds to the incorrectly assigned 4s7d $^1D_2$ state, while the closed square corresponds to the state we found and have labeled as 4s7d. It is seen that a smooth curve drawn through the data points gives the experimentalist a guide to identifying states from his data even if he has no detailed understanding of MQDT. Having found the values for the MQDT parameters which give the solid curve in Fig. 6, one may calculate the admixture of each contributing configuration into the real states. The results show that whereas the "$4p^2$" configuration is almost entirely in the state labeled $4p^2$ in Fig. 6, the "$3d^2$" and "3d5s" configura-tions are spread out over dozens of states. (The labels are en-closed in quotation marks because the perturbers are not neces-sarily pure configurations but may be mixed amongst themselves.) In fact, no individual bound $^1D_2$ state contains more than 3% of the "$3d^2$" configuration, nor more than 13% of the "3d5s" con-figuration. A more detailed discussion of these results may be found in Ref. 2.

By contrast, the Lu-Fano plot of Fig. 7 shows that there is a single perturbing configuration, 4d5p, on the 5snp $^3P^\circ$ channel of Sr. This configuration perturbs the 5s6p and 5s7p states and has negligible influence on the other states. It is interesting to note that the singlet analogs of these states shown a much stronger configuration interaction with the influence of the 4d5p $^1P^\circ_1$ state felt over many of the 5snp $^1P^\circ_1$ states (Fig. 8) [5]. More details on these results are found in Refs. 5 and 6.

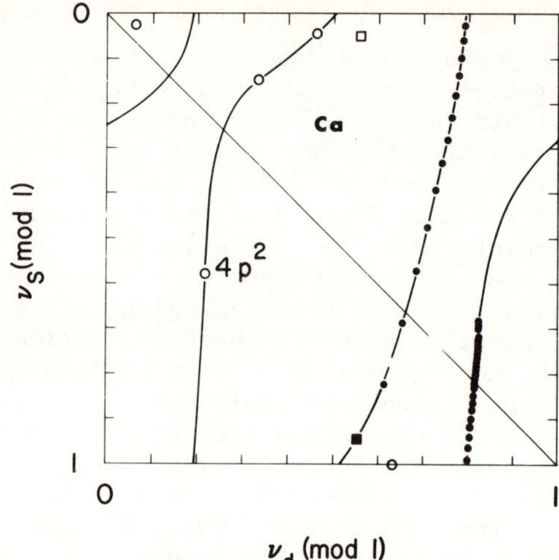

FIGURE 6.   Lu-Fano plot of the $^1D_2$ states of Ca. The open circles
correspond to previously observed states (Ref. 1) and
the solid circles correspond to our new data (Ref.2),
with the exceptions that the open square is the state
labeled by 4s7d in Ref. 1 and the solid square is the
state we identify as 4s7d. The solid curves are MQDT
fits (Ref. 2).

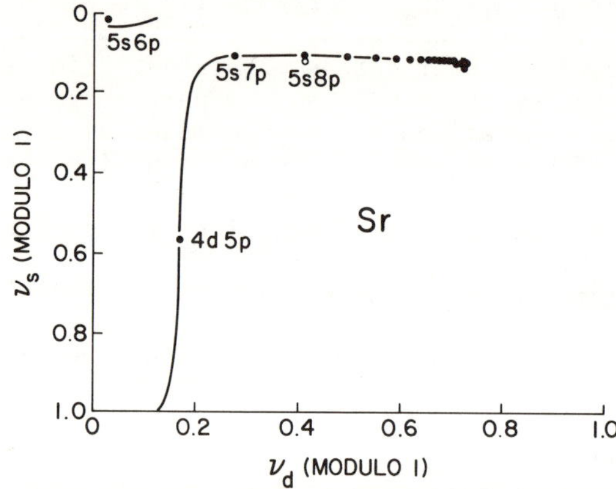

FIGURE 7.   Lu-Fano plot of the $^3P°$ states of Sr. The solid circles
are data points. The open circle corresponds to an in-
correct assignment of the 4s8p $^3P°_1$ in the literature.
The solid line is the result of a two channel MQDT fit
to the data (Ref. 6).

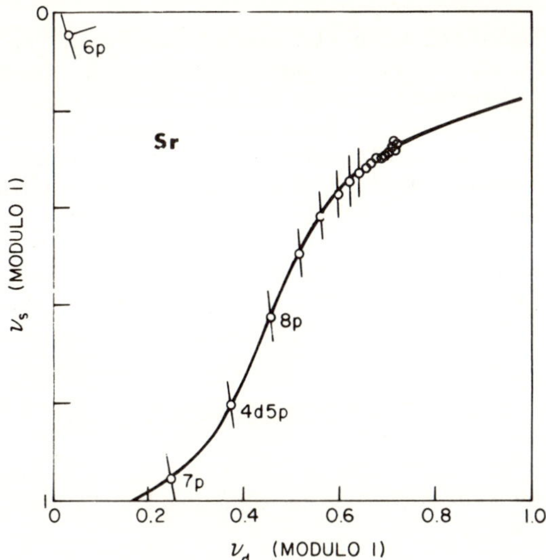

FIGURE 8. Lu-Fano plot of the $^1P_1^o$ states of Sr (Ref. 5).

VI. DISCUSSION

The ease with which new spectroscopic data can be acquired
using multiphoton techniques opens up numerous possibilities for
study. In the alkaline earths, there are some important unanswered
questions about the $^1S_0$ series. In Ca and Sr, the $4p^2$ and $5p^2$
configurations, respectively, make their presence known in the
bound state region. But one would also expect to find evidence
of the $3d^2$ and $4d^2$ configurations. No such evidence has been
found in our studies in the bound state region. These techniques
should be extended to the autoionization region to study even-
parity states and look for these missing configurations. A preliminary study of the even-parity (J = 0,1 and 2) autoionization
spectrum of Ca has already been reported [9].

Our studies in Ca, Sr and Ba show that the configuration interactions are much weaker in the triplet spectra than in the
singlet spectra. This is consistent with the tendency of the
two electrons to avoid one another in triplet wave functions.

We have seen enormous Stark shifts on the high-lying Rydberg
states, but a systematic study remains to be done. The strong
Stark effects are accompanied by easily observed DC electric
field induced second harmonic generation (SHG). This SHG is

reasonantly enhanced when the sum of two input photons is tuned to an even-parity or an odd-parity state. A careful investigation of this phenomenon is an attractive project to undertake. Further studies should be done on the Zeeman effect, two-photon pumped optical parametric oscillation, the ionization mechanism, resonantly enhanced optical sum mixing into the vacuum ultra-violet and the autoionization spectrum in these atoms.

Finally, the multiphoton ionization studies can be extended to other atomic and molecular systems [10]. Such studies should be very helpful in unraveling the mysteries of the electronic structure of multi-electron systems.

ACKNOWLEDGMENTS

We thank Mr. L. H. Manganaro for his technical assistance. The U. S. Army Research Office supported this work.

REFERENCES

1.   G. Risberg, Ark. Fys. $\underline{37}$, 231 (1968).

2.   J. A. Armstrong, P. Esherick and J. J. Wynne, Phys. Rev. A$\underline{15}$, 180 (1977).

3.   D. Popescu, M. L. Pascu, C. B. Collins, B. W. Johnson and I. Popescu, Phys. Rev. A$\underline{8}$, 1666 (1973), and references cited therein.

4.   K. C. Harvey and B. P. Stoicheff, Phys. Rev. Letts. $\underline{38}$, 537 (1977).

5.   P. Esherick, Phys. Rev. A$\underline{15}$, 1920 (1977).

6.   P. Esherick, J. J. Wynne and J. A. Armstrong, Optics Letters $\underline{1}$, (July, 1977).

7.   U. Fano, J. Opt. Soc. Am. $\underline{65}$, 979 (1975), and references cited therein.

8.   K. T. Lu and U. Fano, Phys. Rev. A$\underline{2}$, 81 (1970).

9.   J. J. Wynne, J. A. Armstrong and P. Esherick, Bull. Am. Phys. Soc. $\underline{22}$, 64 (1977).

10. P. M. Johnson, J. Chem. Phys. $\underline{64}$, 4143 and 4638 (1976).

Part 4

COHERENCE AND RESONANCE EFFECTS

Multiphoton Processes in Intense Radiation Fields

Y. GONTIER AND M. TRAHIN
Service de Physique Atomique
Centre d'Etudes Nucleaires de Saclay
B.P. N° 2
91190 - Gif-Sur-Yvette, France

The interaction of atomic systems with intense radiation
fields has stimulated many theoretical and experimental works,
principally during the past decade [1].

This talk does not contain any exhaustive analysis of all
the phenomena which the interaction of radiation with matter can
originate. It will be limited to the discussion of multiphoton
ionization of atoms by intense radiation fields within the framework of time independent perturbation theory [2]. At moderate
intensity one usually considers the lowest order nonvanishing
contribution provided by the perturbation series. This approximation is not sufficient when very intense sources are involved or
in resonant processes. In both cases one deals with a more realistic model in considering the contributions of the higher-order
terms [3]. For N-photon ionization these terms account for arbitrary number of emission and absorption of photons in such a way
that the net number of absorbed photons remains equal to N.

The general formulas we obtain this way are consistent with
respect to electrodynamics since each diagram is counted once.
Therefore the analytic function representing the sum of the
series is to be considered as an analytic continuation of this
series when its convergence is not satisfactory. Thus it is of
central interest in the two circumstances mentioned above.

We begin now with the derivation of the expressions for the
transition amplitudes. Then they will be used for a detailed
discussion of multiphoton resonant process occurring at moderate
intensity.

The time evolution operator is calculated from the integral
[2,3]:

$$U(t) = \frac{1}{2\pi i} \oint G(E) e^{-iEt} \, dE \, , \qquad (1)$$

where $G(E)$, the resolvent operator of the total Hamiltonian H in the Schrödinger picture, can be written in terms of the interaction V as

$$G(E) = G^{o}(E) + G^{o}(E)VG(E). \qquad (2)$$

$G^{o} = (E-H_{o})^{-1}$ is the resolvent of the free Hamiltonian H_{o}. In taking the inner product of Eq. (2) with two definite states of the system "atom plus field" and iterating one obtains the whole series of higher order contributions to a definite process. For example, in the case of one-photon ionization, the only (non-vanishing) contributions to $G(E)$ accounting for the depletion of a single photon are given by the (infinite) series

$$G^{(1)}(E) = G^{o}V^{-}G^{o} + G^{o}V^{+}G^{o}V^{-}G^{o}V^{-}G^{+} + G^{o}V^{-}G^{o}V^{+}G^{o}V^{-}G^{o}$$

$$+ G^{o}V^{-}G^{o}V^{-}G^{o}V^{+}G^{o} + \ldots \qquad (3)$$

where V^{+} and V^{-} are to be used when a photon is created and annihilated respectively.

More generally, the series form of $G(E)$ corresponding to the absorption of N photons can be found from the same argument. The sum of this series has been performed after a suitable and systematic association of operators $A \equiv G^{o}V^{-}$ and $B \equiv G^{o}V^{+}$ [3]. As a result of the summation of the whole perturbation series one finds that the expression of $G(E)$ to be used for N-photon absorption is

$$G^{(N)}(E) = \tau_{\square} \left[A\tau_{o} \right]^{N}, \qquad (4a)$$

where $\tau_{\square}$ is defined in terms of the continued fractions

$$\tau_{o} = \frac{1}{1 - A\tau_{o}B}, \qquad (4b)$$

and

$$\tau_{\phi} = \frac{1}{1 - B\tau_{\phi}A}, \qquad (4c)$$

by

$$\tau_{\square} = \frac{1}{1 - A\tau_{o}B - B\tau_{\phi}A}. \qquad (4d)$$

It is to be noted that the operator $G(E)$ corresponding to N-photon emission is given by

$$G^{(N)}(E) \equiv \tau_{\square}\left[B\tau_\phi\right]^N , \tag{5}$$

and that the following useful relations hold between the continued fractions

$$\tau_\square A\tau_o \equiv \tau_\phi A\tau_\square \tag{6a}$$

$$\tau_\square B\tau_\phi \equiv \tau_o B\tau_\square \tag{6b}$$

In contrast to their apparent simplicity, the expressions shown in Eqs. (4) give rise to serious computational difficulties. The reason is that in atoms the operators A and B are represented by matrices whose sizes depend on the number of states considered. In addition, for resonant processes, one must isolate the resonances appearing in the continued fractions. To this end, the projection operator technique is used.

Let ε be a subspace spanned by some particular eigenstates of H_o and P the projection operator onto ε, one finds [4]:

$$[E - H_o - \tilde{R}(E)]\tilde{G}(E) = 1, \tag{7a}$$

where

$$\tilde{G}(E) = PG(E)P, \tag{7b}$$

and

$$\tilde{R}(E) = PR(E)P. \tag{7c}$$

The operator R(E) is defined as:

$$R(E) = V + VQG_o V + VQG_o VQG_o V + \ldots \tag{8}$$

where

$$V = V^+ + V^- , \tag{9}$$

and Q is the projection operator outside ε, i.e., $P + Q = 1$. Denoting

$$\bar{A} = QG_o V^- , \tag{10a}$$

and

$$\bar{B} = QG_o V^+ , \tag{10b}$$

the diagonal matrix element of R(E) with respect to the photon states are to be calculated from

$$R^D(E) = V^+ \bar{\tau}_\square \bar{A}\tau_o + V^- \bar{\tau}_\square \bar{B}\tau_\phi \tag{11a}$$

where as the nondiagonal matrix elements accounting for the depletion of N photons is obtained from [3]

$$R^{(ND)}(E) = V^+ \bar{\tau}_\Box (\bar{A}\bar{\tau}_o)^{N+1} + V^- \bar{\tau}_\Box (\bar{A}\bar{\tau}_o)^{N-1},\tag{11b}$$

where

$$\tau_o = \frac{1}{1 - \bar{A}\tau_o \bar{B}},\tag{12a}$$

$$\tau_\Box = \frac{1}{1 - \bar{A}\tau_o\bar{B} - \bar{B}\tau_\phi\bar{A}}.\tag{12b}$$

In combining Eqs. (7), (11) and (12), one obtains G(E) through the inversion of the matrix $E - H_o - \tilde{R}(E)$. Thus the matrix element of $\tilde{G}(E)$ describing the resonant transition which takes place between the states $|a\rangle$ and $|b\rangle$ is given by

$$G_{ba}(E) = \frac{\mathbb{R}_{ba}(E)}{(E-E_a-\mathbb{R}_{aa}(E))(E-E_B-\mathbb{R}_{bb}(E)) - |\mathbb{R}_{ba}(E)|^2},\tag{13a}$$

where the operator $\mathbb{R}$ is defined in terms of the resonant state $|c\rangle$ as

$$\mathbb{R}(E) = R(E) + \frac{R(E)|c\rangle\langle c|R(E)}{E-E_c - R_{cc}(E)}\tag{13b}$$

As a result of the projection technique, the matrix elements of the operator R(E) are slowly varying functions of E. Therefore the poles of $G_{ba}(E)$ in the resonance region are approximately given by

$$E^\pm \simeq \frac{\bar{E}_a - \bar{E}_c}{2} \pm \sqrt{\left(\frac{\bar{E}_a - \bar{E}_c}{2}\right)^2 + |R_{ac}|^2}\tag{14a}$$

and

$$E_B \simeq \tilde{E}_b + \frac{|R_{ab}|^2}{\tilde{E}_b - \tilde{E}_a} + \frac{|R_{bc}|^2}{\tilde{E}_b - \tilde{E}_c}.\tag{14b}$$

where

$$\bar{E}_a = \tilde{E}_a + \frac{|R_{ab}|^2}{\tilde{E}_a - \tilde{E}_b} , \tag{14c}$$

$$\bar{E}_c = \tilde{E}_c + \frac{|R_{cb}|^2}{\tilde{E}_c - \tilde{E}_b} , \tag{15}$$

and

$$\tilde{E}_i = E_i + R_{ii}(E_i) .$$

In Eqs. (14) the matrix elements of $R(E)$ are calculated at $E \simeq E^{\pm}$ or $E \simeq E_B$.

The probability is determined from Eqs. (1), (13) and (14) by using the residue theorem.

In considering the only resonant terms one finds

$$|U_{ba}(t)|^2 \simeq \frac{\alpha - \beta \cos \Omega t}{\Omega^2} |R_{bc}R_{ca}|^2 , \tag{16a}$$

where

$$\alpha = \frac{1}{(E^+ - E_B)^2} + \frac{1}{(E^- - E_B)^2} , \tag{16b}$$

$$\beta = \frac{2}{(E^+ - E_B)(E^- - E_B)} , \tag{16c}$$

and

$$\Omega = \sqrt{\left(\frac{\bar{E}_a - \bar{E}_c}{2}\right)^2 + |R_{ac}|^2} . \tag{16d}$$

One observes that the probability is a periodic function of time and is inversely proportional to the square of the distance between the resonant poles. Due to the term $|R_{ac}|^2$, Ω never vanishes. Thus, if we plot the value of the poles $E^{\pm}$ against the intensity, the curves never cross each other. They give rise to anticrossing [4]. The smaller the distance between the poles, the greater the enhancement of the resonance. It is to be noted that anticrossing gives rise to a damping of the probability which flattens more and more the resonance peak as the intensity is increased. It may happen that the resonance disappears completely. This situation is encountered when anticrossing takes place at high intensity, i.e., when the shift of the ground state is not too large. Thus, such a formalism will

prove to be of central interest in multiphoton ionization of rare
gases. In every case it provides even through the approximate
formulas of Eqs. (13), (14) and (16), a simple but consistent interpretation of any resonant process.

In ionization experiments, one does not measure the number
of ions produced at a definite time but the ones created during
a time interval. Therefore the probability is to be integrated
over the temporal pulse shape and the number of ions calculated
can be directly compared to the result of measurements. We note
in passing that the standard procedure [2] used to determine the
probability per unit time is of no use. The first reason is that
in most cases of interest Ω^{-1} can be comparable to the interaction time. Such a situation makes irrelevant any time limiting
procedure. The last reason is that the function Ω does not show
any root. Thus, energy conservation cannot be expressed in the
form of a delta function. Only the usual integration of the
probability over the continuum states is to be performed.

Let $\rho(E_f)$ be the density of continuum states, one defines
the transition probability induced by a single mode radiation
field of well defined occupation number by

$$P(t) = \int \rho(E_f) \left| U_{fg}^{n(t)}(t) \right|^2 dE_f \, , \qquad (17)$$

where $n(t)$ is the number of photons which are in the mode at
time t. It is determined from the temporal pulse shape and
chosen to have the following simple form

$$n(t) = n_o, \quad 0 \leqslant t \leqslant t_o \quad , \quad n(t) = 0, \ t > t_o \qquad (18)$$

where t_o is the pulse duration.

In combining Eqs. (17) and (18) the number of ions $N^+(t)$ is
to be calculated from

$$\frac{dN^+(t)}{dt} = N_o(t) \frac{dP(t)}{dt}, \qquad (19)$$

where $N_o(t)$ is the number of neutral atoms at time t. By express-
ing the results obtained from the exact summation of the perturbation series in the formalism of the dressed atom [4] we have
elaborated a consistent and accurate method which allows a simple
interpretation of any resonant multiphoton process. It has per-
mitted, from approximate formulas, to put forward some fundamental features of resonant ionization which have been confirmed by
accurate computational analysis.

REFERENCES

1. See for example, the following papers and the references
    quoted therein; P. Lambropoulos, Phys. Rev. $\underline{A9}$, 1992 (1974);
    Y. Gontier and M. Trahin, Phys. Rev. $\underline{172}$, 83 (1968); $\underline{A4}$,
    1896 (1971); J. Morellec et al., Phys. Rev. $\underline{A14}$, 300 (1976).

2. M. L. Goldberger and K. M. Watson, Collision Theory (John
    Wiley and Sons, Inc., New York, 1974), Chap. 8.

3. Y. Gontier, N. K. Rahman and M. Trahin, Phys. Lett. $\underline{A54}$,
    341 (1975); Phys. Rev. $\underline{A14}$, 2109 (1976).

4. C. Cohen-Tannoudji, Cargese Lectures in Physics, edited by
    M. Levy (Gordon and Breach, New York, 1967) Vol. 2, pp. 347-
    393; C. Cohen-Tannoudji and S. Haroche, J. Phys. (Paris)
    $\underline{30}$, 125 (1969); $\underline{30}$, 153 (1969).

Theory of Resonant Multiphoton Ionization in Atoms

LLOYD ARMSTRONG*
Physics Department
The Johns Hopkins University
Baltimore, Maryland

and

SERGE FENEUILLE
Laboratoire Aime Cotton
Orsay, France

I. INTRODUCTION

Resonance occurs in the multiphoton ionization of an atom
when the energy of an excited state of the atom is equal to the
energy of the ground state of the atom plus some number of photons. The presence of this resonance is usually reflected by an
increase in the probability of ionization of the atom. We shall
discuss the structure of this resonance as a function of such
parameters as detuning of the photon frequency from the exact
resonance, the relative sizes of various probabilities which appear in the description of the process, and the correlation properties and pulse length of the photon field. We shall show that
in the general case transition rates cannot be defined for resonant multiphoton ionization and will define the particular circumstances under which such rates can be defined.

There are a number of excellent reviews [1] of theoretical
formalisms which can be used to study resonant multiphoton ionization. Our emphasis here shall not be on the formalism, however, but on the general aspects of the phenomena which may lead
to interesting and observable results. In order to keep our discussion as clear and succinct as possible, we will describe this
process using a simplified model [2]. Obviously not all real
situations can be fit quantitatively by a single model, but the
general qualitative aspects of the process should be well described by our model. We will also indicate as we go along how

*Work supported in part by the United States National Science
Foundation.

more realistic descriptions of certain aspects of the process will affect our results.

We will consider the case in which m photons of energy ω ($\hbar = 1$) are needed to ionize the atom, and there is an n ($n < m$) photon resonance between the ground state $|g)$ and an excited state $|a)$ of the atom. We assume that there are no other atomic states which can be placed in resonance or near resonance with either $|g)$ or $|a)$ by any number $q \leqslant m$ of photons. We shall also neglect spontaneous emission, assuming the ionization lifetimes are much shorter than spontaneous lifetimes.

As has been demonstrated by several calculations, perturbation theory works very well in the calculation of atom-field interactions in the off-resonance case. Therefore, we can to a very good approximation treat all off-resonant processes in the lowest nonvanishing order of perturbation theory. This enables us to introduce an effective Hamiltonian operator [3] which describes the ionization through all intermediate states except $|a)$:

$$H^{(m)} = |g,p) \sum_{i,j\ldots\neq a} \frac{(g,p|H_{AF}|i,p-1)\ldots}{(E_g-E_i+\omega)\ldots}$$

$$\times\ (\ell,p-m+1|H_{AF}|E,p-m)(E,p-m|$$

$$+\ H.C. \tag{1}$$

where $|g,p)$ describes an atom in the ground state $|g)$ and p photons of energy ω, $|E)$ is a state of the ionized atom plus free electron with total energy E, etc; H_{AF} is the atom-field interaction. Likewise we can define

$$H^{(n)} = |g,p) \sum_{i\ldots} \frac{(g,p|H_{AF}|i,p-1)\ldots}{(E_g-E_i+\omega)\ldots}$$

$$\times\ (\ell,p-n+1|H_{AF}|a,p-n)(a,p-n|$$

$$+\ H.C. \tag{2}$$

which describes the n photon interaction between $|g)$ and $|a)$; and ($r = m-n$)

$$H^{(r)} = |a,p-n) \sum_{i\ldots} \frac{(a,p-n|H_{AF}|i,p-n-1)\ldots}{(E_a-E_i+\omega)\ldots}$$

$$\times\ (\ell,p-m+1|H_{AF}|E,p-m)(E,p-m|$$

$$+\ H.C. \tag{3}$$

which describes the r photon interaction between $|a)$ and $|E)$. The resulting model is shown in Fig. 1.

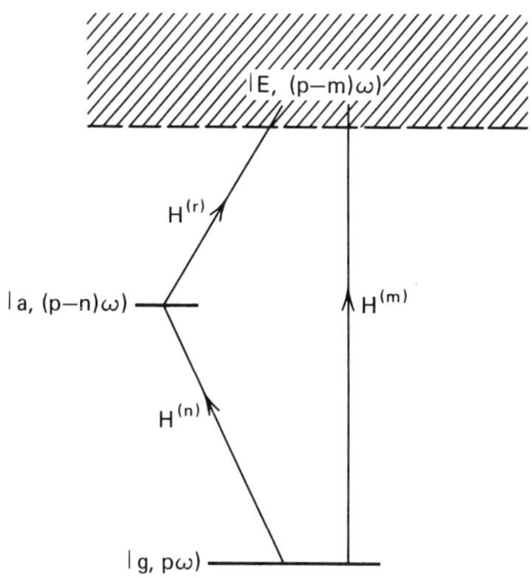

FIGURE 1. Model of two level atom.

We shall assume that the atom-field state is given by $|g,p)$ at $t = 0$, and follow the evolution of the system in time. The model is sufficiently simplified that this time dependence can be obtained analytically; the resulting expression [2] for the probability of ionization is, however, very complicated and will not be repeated here. The general aspects of the solution can, nevertheless, be easily discussed in terms of a few parameters:

$$\Gamma_g = 2\pi \left| (g,p|H^{(m)}|E,p-m) \right|^2 \equiv 2\pi |H_{gE}|^2$$

$$\Gamma_a = 2\pi \left| (a,p-n|H^{(r)}|E,p-m) \right|^2 \equiv 2\pi |H_{aE}|^2$$

$$H_{ga} = (g,p|H^{(n)}|a,p-n)$$

$$q = 2H_{ga}/(\Gamma_a \Gamma_g)^{1/2} \tag{4}$$

Although all of these parameters are, strictly speaking, functions of ω, for simplicity we shall take them to be constant. The relative magnitudes of these three parameters determine the characteristics of the solution of this model problem. It is therefore of use ot have some idea how Γ_g, Γ_a and H_{ga} depend on m, n, and r. In general, if $m = 2$ or 3, or if $m >> 3$ and $r << n$, then $|H_{ga}| >> \Gamma_a$. If $m >> 3$, and $r << n$, then $\Gamma_a >> |H_{ga}|$. It is usually the case that $\Gamma_g << \Gamma_a$, H_{ga}; however, if $m >> 3$, $n = 1,2$, one may have $\Gamma_g \simeq \Gamma_a$. To give a specific example in two photon ionization of Cs (7p state near resonant) $H_{ga} \simeq 10^{-19}$ $I^{\frac{1}{2}}$ erg, $\Gamma_a \simeq 10^{-27}$ I erg, $\Gamma_g \simeq 10^{-39}$ I^2 erg, where I is in watts/cm^2. In time units, rather than energy units, $H_{ga} \simeq 10^8$ $I^{\frac{1}{2}}$ sec^{-1}, $\Gamma_a \simeq I$ sec^{-1}, $\Gamma_g \simeq 10^{-12}$ I^2 sec^{-1}.

II. TIME EVOLUTION OF THE RESONANCE PROFILE

The quantum field model we have introduced corresponds to a step function in the photon field occurring at $t = 0$. Such a sharp rise is, of course, unphysical, but demonstrates clearly some of the effects produced when an atom suddenly enters a photon field, or when a laser is pulsed. We will describe in Section III some (usually relatively minor) modifications to our results when more realistic pulse shapes are used.

An abrupt increase in the photon field has the effect for times shortly after the increase of changing the single frequency field into a field with a very broad spectral width (roughly 1/T where T is the time of the pulse). As a result, for short times ("short" will be dependent on the circumstances, as will be described below), details of the resonance structure will be washed out since the applied field is much broader than the structure itself. That is to say, the observed width of the resonance is really that of the field. Figure 2 shows how the resonance structure develops in time for a particular choice of parameters. The quantity $\delta = E_g + n\omega - E_a + \Delta_s$ is the detuning from resonance, where Δ_s is the difference in Stark shifts of $|g)$ and $|a)$. Obviously an experiment involving very short pulses will not detect a distinct resonance structure, a phenomena which has been experimentally observed by the Saclay group [4] using picosecond pulses.

The complete solution to this model problem shows that the lack of resonance structure is caused (mathematically) by damped oscillating terms of the type $(\sin Wt)e^{-\Gamma t}$ and $(\cos Wt)e^{-\Gamma t}$, where $\Gamma = (\Gamma_a + \Gamma_g)/2$ and $W = [4|H_{ga}|^2 + \delta^2]^{\frac{1}{2}}$. These terms will average to zero for times $t_1 >> 1/W$ if $|H_{ga}| >> \Gamma_a, \Gamma_g$, or will decay to zero for times $t_2 > 1/\Gamma_a$ when $\Gamma_a >> |H_{ga}|$, Γ_g. These two times, t_1 and t_2, define "short" insofar as supression of resonance structure is concerned.

The times t_1 and t_2 also define very roughly the onset of what might be called the "rate region", in which a probability per unit time of ionization can be defined in the region near the resonance. If we neglect for the moment the effects of all

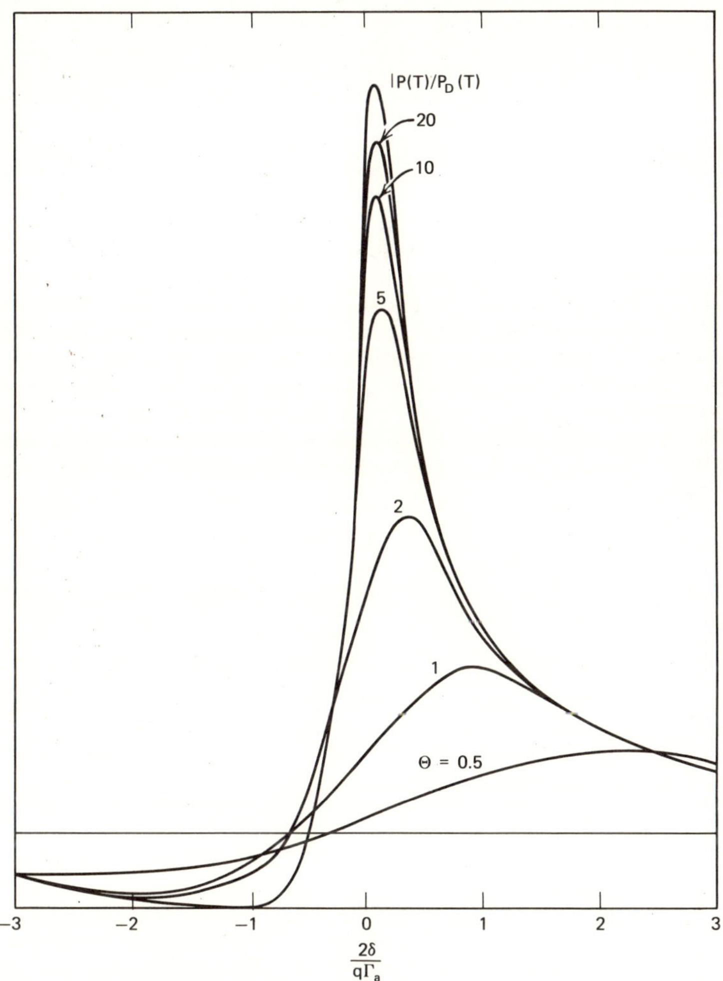

$$\frac{2\delta}{q\Gamma_a}$$

FIGURE 2. Typical resonance profile for the two level model. $\theta = \Gamma_a T$, where T is the pulse length. $P_D(T)$ is the probability of ionization through all nonresonant states: $P_D(T) = 1 = e^{-\Gamma_g T}$ (ref. 7).

states other than $|a)$, this transition rate is given by

$$w\alpha \left| \frac{H_{ga} \, H_{aE}}{\delta + i\gamma} \right|^2 \tag{5}$$

where γ is the larger of $2 H_{ga}$ and $\Gamma_a/2$. (See also Sec. IV.)

The region in which $|H_{ga}| \simeq \Gamma_a$ is particularly interesting in that no probability per unit time of ionization can be defined in any time interval. If one follows the time evolution of $P(t)$, the probability of ionization as a function of time, one finds a curve filled with "kinks". These kinks are produced because the ionization time in this case is of the same order as the time for a Rabi oscillation of the population between $|g)$ and $|a)$.

One cannot define a probability per unit time of ionization near $\delta = 0$ when $\Gamma_g \simeq \Gamma_a$, either. In this region, the ionization profile will be almost flat for times $t_3 < 1/\Gamma_g$. Since the probability of ionization far from the resonance goes roughly as $1 - e^{-\Gamma_g t}$, one sees that t_3 corresponds to a time fairly near saturation. When structure does appear in the ionization profile its most dominant feature will be a minimum near $\delta = 0$.

It is well known from perturbation formulations of multi-photon ionization cross sections [5] that there are minima in the cross section produced by interference between contributions from different intermediate states. For example, in two-photon ionization, if there are only two possible intermediate states, and they have identical transition matrix elements linking them to both the ground state and the continuum, there will be a complete cancellation in the ionization cross section when the photon frequency is exactly intermediate between the energies of the two states.

In the simple model we have been considering, there is also a minimum in the probability of ionization, produced in this case by the interference between the process being considered exactly $(|g) \rightarrow |a) \rightarrow |E))$ and all other processes which have been absorbed into Γ_g. One of the most interesting predictions of this simple model is that the position and shape of this minimum changes greatly in time (Fig. 2). For short times $t < 1/\Gamma_a$ the minimum is seen to be quite shallow and broad. This is not surprising since, as discussed above, for short times the excitation has a very broad spectral range, washing out details of the resonance. For longer times, the minimum becomes sharper and much deeper; somewhat surprisingly, it also moves nearer to the resonant state as it deepens. For long times, $t \gg 1/\Gamma_a$, the position of the minimum δ_{long} is that which would be predicted from the usual perturbation expansion; for short times, the minimum is at $\delta_{short} = 2\delta_{long}$. In order to get a physical feeling for the time

regimes involved we can use the parameters given above for two
photon ionization of Cs; "short" times will be those for which
TI < 1, where T is the pulse length in seconds, I, the intensity
in watts/cm^2.

The physical basis of this "moving minimum" is apparently
again the large spectral width of the exciting photons at small
t. This spectral width does not affect processes which are far
off resonance (i.e. those described by Γ_g). By "broadening" the
resonant processes, however, it makes their effects felt over a
wider frequency range than would normally be the case. This has
the effect of pushing the minimum away from the "near resonant"
intermediate state.

In order to investigate whether this predicted behavior of
the minimum is due to an oversimplfied model, we have carried out
calculations [6] using a model of two photon ionization in which
three excited states of the atom are treated explicitly, with all
other states again being absorbed into an effective Γ_g. This
model is indicated schematically in Fig. 3. Matrix elements were
chosen so that a minimum would appear for a photon tuned to the
region between states $|a)$ and $|b)$. Typical results are shown in
Fig. 4. Again, the minimum is seen to deepen and shift toward
the nearest resonant state as time increases. Thus it seems
likely that this behavior of the minimum is a real effect which
should be observable.

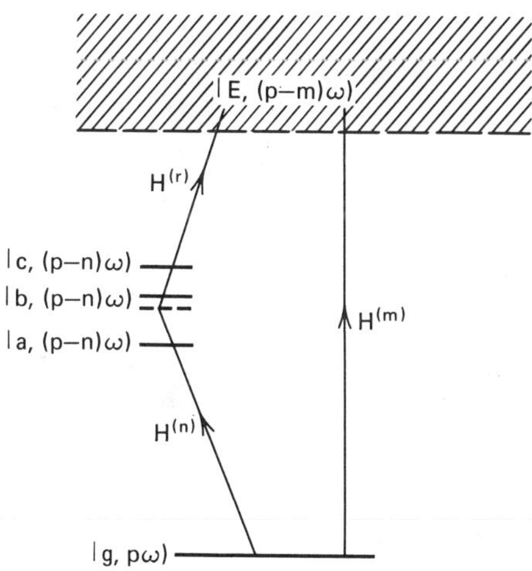

FIGURE 3. Model of four level atom.

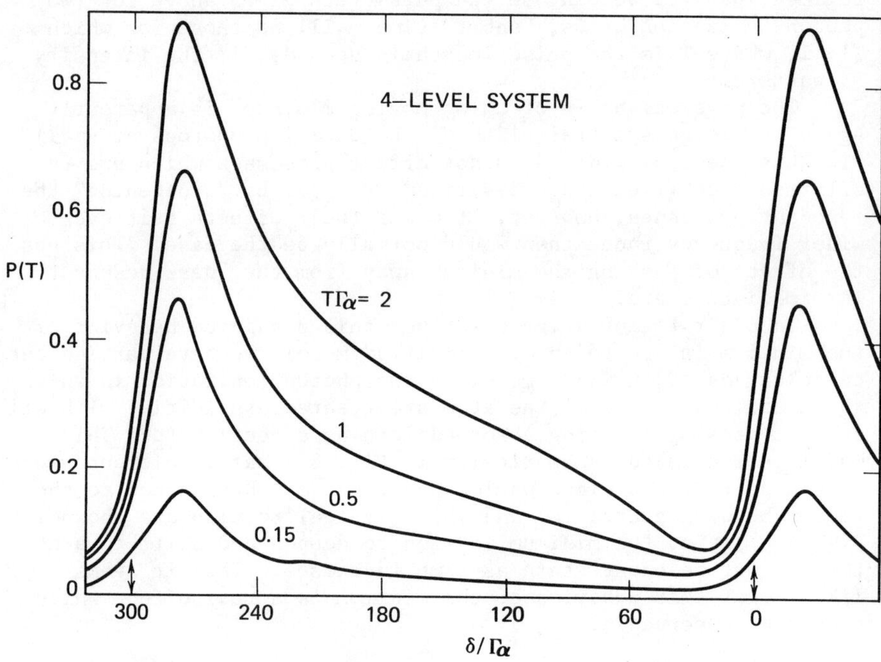

FIGURE 4.   (a) Typical resonance profile for the four level model. δ is the detuning from level $|b)$; level $|a)$ lies at $\delta = -300\Gamma_b$.

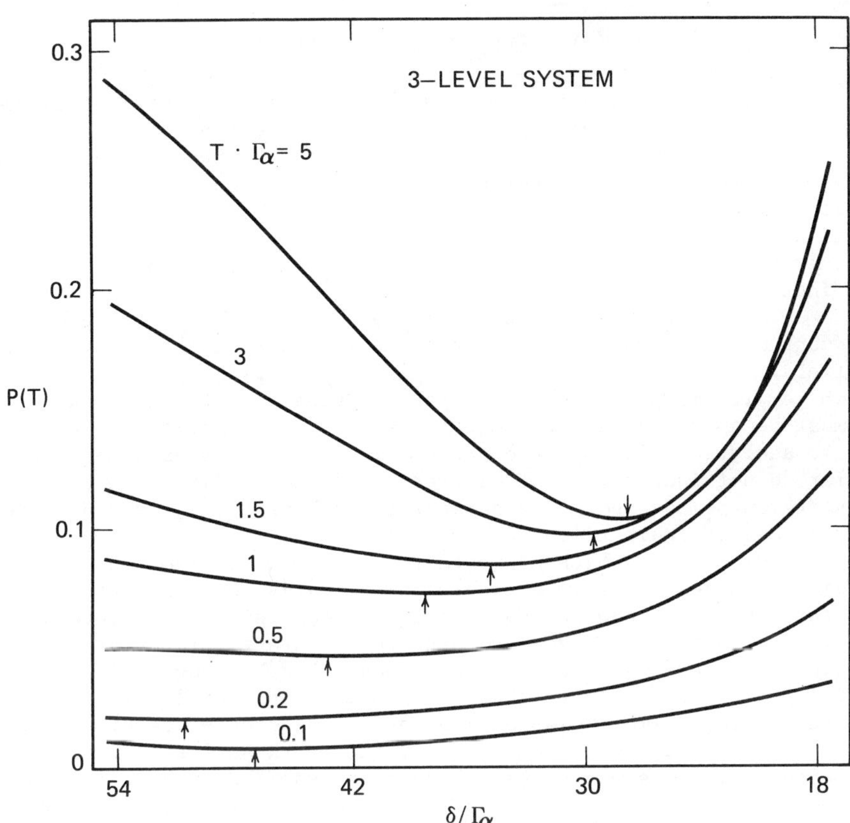

FIGURE 4. (b) Enlargement of the region of the minimum from 4(a).

III. EFFECT OF PULSE SHAPE

All of the results described above have been obtained assuming a square pulse of light. A real laser does not, of course, produce a square pulse. How realistic are the results described above, then? We shall not disucss this question in detail, since it is the subject of a paper by M. Crance in a contributed session* later in this meeting. However, we can report that the answer seems generally to be that pulse shape plays a rather minor role.

The question has been studied [7-10] by transforming the quantum field problem above into one involving a classical driving field. This classical field is given by $E = g(t)[E_0\sin\omega t + E_0'\cos\omega t]$ where $g(t)$ is the pulse envelope. In general, it seems that $g(t)$ must be specified and the equations resolved numerically; that is, analytic solutions are not possible. Results obtained thus far do not indicate great divergences in most cases from the statements made above, if parameters such as Γ_a, Γ_g, and H_{ga} are replaced by their time-averaged values.

One case in which a smoothly varying pulse leads to quite different results from a square pulse [8] is when the strong Stark shifts of the resonant intermediate state due to interactions with the continuum and other bound states of the atom. (In the equations given above, we have absorbed such shifts into δ.) For a square pulse those shifts are constant in time, and thus cause a simple displacement of the resonance position. For a smooth pulse, on the other hand, the Stark shifts will be time dependent. This may cause a line to be shifted into or out of resonance during some portion of the pulse. As a result, the line shape becomes asymmetric, as shown in Fig. 5. [8].

Finally, it should be noted that by using the classical field equations, one can also consider the effects of mode structure on the ionization probability. It is likely that presence of laser mode structure will produce some effects, but that seems at the moment to be "unknown territory".

---

*See ICOMP Abstracts (Department of Physics and Astronomy, Univ. of Rochester, Rochester, NY 14627), p.

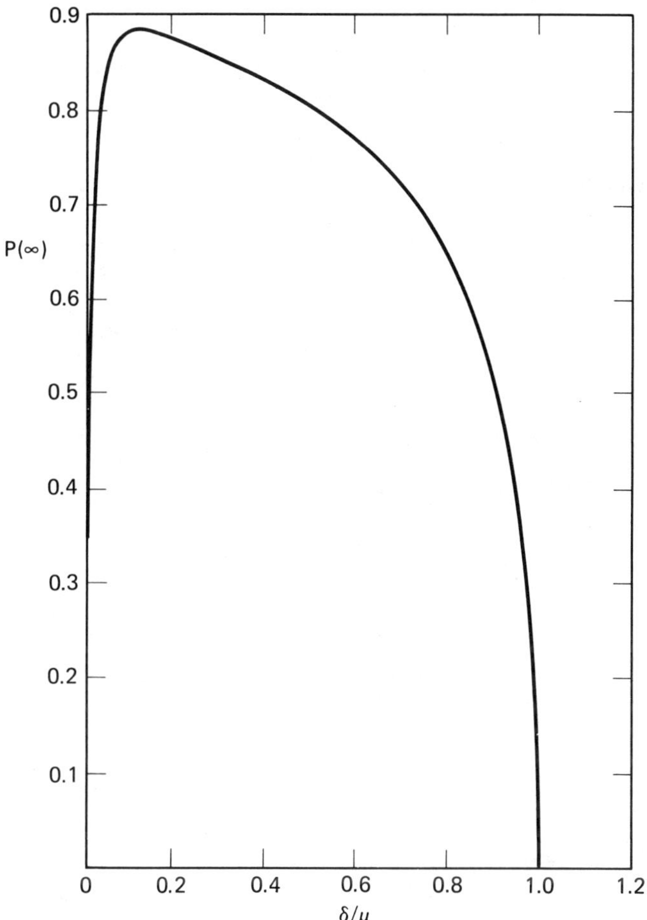

FIGURE 5.   Asymmetric line profile resulting from a pulse shape
for which $[g(t)]^2 = [1 + (t/\tau)^2]^{-3}$; $\tau = 1/\Gamma_a$;
$\mu = \Delta_s/[g(t)]^2$.

IV.   EFFECTIVE ORDER OF NONLINEARITY;
EFFECTS OF PHOTON CORRELATIONS

The transition rate for nonresonant m-photon ionization is proportional to I^m where I is the intensity of the photon beam ($I = p\omega$). As a resonance is approached, however, the probability of ionization no longer has such a simple dependence on I. In this section we will consider the I dependence of two types of transition rates which have been obtained using various perturbation approaches. The first of these [2,11]

$$w_1 = \frac{\Gamma_g}{2}\left[1 + \frac{(\delta + q\Gamma_a)^2}{\delta^2 + 4|H_{ga}|^2}\right] \tag{6}$$

should be approximately correct when $|H_{ga}| \gg \Gamma_g$, Γ_a, and $t < 1/\Gamma_a$; the second [2,3]

$$w_2 = \Gamma_g\left[\frac{(\delta + q\Gamma_a/2)^2}{\delta^2 + \Gamma_a^2/4}\right] \tag{7}$$

should be approximately correct when $\Gamma_a \gg |H_{ga}|$, Γ_g. The rather complicated intensity dependence of w_1 and w_2 can be conveniently analyzed by introducing the effective order of nonlinearity k defined by

$$k = d(\log w)\,/d(\log I) . \tag{8}$$

A rough idea of the dependence of w on I and k can then be obtained by setting $w = \text{const } I^k$; this is exact, of course, only if $k \neq k(I)$.

Using the rate of Eq. (6) in Eq. (8), one finds [11]

$$k_1 = d(\log w_1)/d(\log I)$$

$$= m - \frac{n}{(y^2 + A^2)}\left(\frac{A^2(y+1)^2}{y^2 + A^2 + (y+1)^2}\right) \tag{9}$$

where

$$y = \delta/q\Gamma_a \quad , \quad A^2 = \Gamma_g/\Gamma_a .$$

In deriving Eq. (9), we have set the Stark shifts of $|a)$ and $|g)$ equal to zero; their presence would not significantly affect the form of (9). As $\delta \to \infty$, $k_1 \to m$; that is, the transition rate approaches the perturbation limit $w \propto I^m$. This limit is also reached when $y = -1$, i.e. at the position of the minimum in w_1. As $\delta \to 0$

$$k_1 \to m - n[1/(1 + A^2)]$$

$$= r + (nA^2)/(1 + A^2) . \tag{10}$$

On resonance (i.e. when $\delta = 0$), one speaks of the transition $|g) \to |a)$ as having "saturated", such that the I dependence of w is approximately given by the intensity dependence of the last step, $|a) \to |E)$; i.e., I^r. The degree of saturation possible can be seen to depend on the magnitude of the quantity A^2, which is a function of the decay constants of the ground and resonant states.

Repeating the calculation for w_2, but keeping this time the Stark shifts which we take to be of the form aI^b, one finds

$$k_2 = m - 2r \frac{\Gamma_a^2/4}{\delta^2 + \Gamma_a^2/4} + 2abI^b\left(\frac{1}{\delta + q\Gamma_a/2} - \frac{\delta}{\delta^2 + \Gamma_a^2/4}\right). \tag{11}$$

As $\delta \to \infty$, $k_2 \to m$ as before; however, as $\delta \to 0$, $k \to m - 2r + 4abI^b/q\Gamma_a$. In addition, we see that because of the Stark shift term aI^b, $|k_2| \to \infty$ as $\delta \to -q\Gamma_a/2$. That is, $|k_2| \to \infty$ at the minimum in the cross section w_2. Obviously, such extreme behavior will not be observed in a real atom where effects not included in our model will almost certainly prevent such infinities; however, it is not unreasonable that values of $|k_2| >> m$ may in fact occur near minima in the ionization cross section.

We have thus far considered the case in which the number of photons present in the beam at $t = 0$ is p. A more realistic description of an actual laser beam can be obtained by describing the original photon field by a probability distribution ρ_p. Specification of ρ_p is equivalent to specification of the coherence (correlation) properties of the light. Then, if w_p is one of the transition rates obtained above for an incident field of p photons, the transition rate in the more realistic field will be

$$w = \sum_p w_p \rho_p . \tag{12}$$

In terms of the effective order of nonlinearity, this becomes

$$w \cong \text{const} \sum_p (p)^k \rho_p . \tag{13}$$

Since k is a function of, among other things, the detuning from
resonance, the coherence properties of the light will not simply
have a constant effect over the resonance profile, but rather can
be expected to make discernible changes in the profiles given by
w_1 and w_2.

Equation (12) has been evaluated [12] for two photon ioniza-
tion using the rate of Eq. (5) with $|H_{ga}| \gg \Gamma_a$ for the case of
coherent light (Glauber state) [13]:

$$\rho_p = e^{-<p>} <p>^p / p!$$

and for the case of chaotic light [13]:

$$\rho_p = <p>^p / (1 + <p>)^{p+1}$$

where $<p>$ is the average number of photons in the field:

$$<p> = \sum \rho_p p .$$

The results obtained [12] showed that chaotic light leads to a
resonance linewidth about 20% larger than that obtained with coherent light. For large δ, one obtains the well-known off-resonant result that chaotic light is twice as efficient as coherent
light in inducing two photon ionization. Exactly on resonance,
however, the magnitude of the two photon cross section is, in
fact, independent of the photon statistics, depending only on
$<p>$.

In the more general case of m-photon ionization, coherent
and chaotic light will not be equally effective in producing
ionization right on the resonance. As shown above, for situa-
tions in which $|H_{ga}| \gg \Gamma_a, \Gamma_g$, k at resonance is very close to r,
i.e., $w \propto I^r$. As is well known from calculations describing off-
resonance multiphoton ionization [14], the cross section for
chaotic light will in this case be r! times the cross section for
coherent light. When $\Gamma_a \gg |H_{ga}|$, Γ_g, k at resonance is $m - 2r$
$+ 4 abI^b / q\Gamma_a = f$. In this case, $w_{chaotic}/w_{coherent} \cong f!$ at
resonance.

One can also use w_1 or w_2 in Eq. (12) rather than the w of
Eq. (5); this will enable one to evaluate the "effectiveness" of
chaotic and coherent light in producing ionization in the region
of the interference minimum. As indicated above in the discussion of k, it is likely that in the region of the minimum
$w_{chaotic}/w_{coherent} > m!$. Such an effect has recently been predicted by Dixit and Lambropoulos [15].

V. CONCLUSIONS

It is obvious from these brief considerations that resonant multiphoton ionization is a complex phenomenon which can lead to resonance profiles of widely varying types. In contrast to "non-resonant" multiphoton ionization which almost always can be described by a transition rate obtainable using perturbation theory, resonant multiphoton ionization exhibits profiles whose shapes may change according to the length and intensity of the laser pulse. Because of this, the concept of ionization rate must be used very cautiously for resonance phenomena. In particular for "short" laser pulses - which need not really be very short in some cases (e.g. Cs as described above) - the observed profile will be quite different from that predicted by lowest order perturbation theory.

The effective order of nonlinearity can also vary greatly from point to point in the resonance profile. This in turn leads to large variations in the effectiveness of chaotic and coherent light in causing ionization; this will be reflected in quite different resonance profiles corresponding to laser light of differing coherence properties.

There are a number of problems which remain to be resolved in this area. The influence of laser pulse shape and mode structure has not yet been fully explored, even using semiclassical field equations. This problem has scarcely been touched using quantum fields except through use of perturbation theory; obviously this is not always justified. The question of photon statistics has only been studied thus far in cases in which ionization rates exist.

We have not even mentioned many aspects of the resonant problem, such as the difficulty in obtaining atomic wavefunctions for evaluation of matrix elements, or the question of the effectiveness of one polarization of light compared to another. The omission of these aspects should not be taken as a comment on their importance, but rather as simply an indication that they fell outside the scope of this work.

REFERENCES

1.   See e.g. P. Lambropoulos, <u>Advances in Atomic and Molecular Physics</u>, <u>12</u>, 87 (1976); Phys. Rev. A<u>9</u>, 1992 (1974); M. L. Goldberger and K. M. Watson, <u>Collision Theory</u>, (Wiley, New York, 1964); Y Gontier and M. Trahin, Phys. Rev. A<u>7</u>, 2069 (1973).

2.   B. L. Beers and L. Armstrong, Jr., Phys. Rev. A$\underline{12}$, 2447
     (1975). Similar results to those obtained in this work
     were found by A. E. Kazkov, V. P. Makarov, and M. V.
     Federov, Zh. Eksp. Theor. Fiz $\underline{70}$, 38 (1976); (Soviet
     Physics, JETP $\underline{43}$, 20 (1976)).

3.   L. Armstrong, Jr., B. L. Beers, and S. Feneuille,
     Phys. Rev. A$\underline{12}$, 1903 (1975).

4.   Private communication.

5.   See e.g., H. B. Bebb and A. Gold, Phys. Rev. $\underline{143}$, 1 (1966).

6.   C. Theodosiou and L. Armstrong, Jr. to be published.

7.   M. Crance and S. Feneuille, Phys. Rev. A, to be published.

8.   C. W. Choi and M. G. Payne, Oak Ridge National Laboratory
     Report ORNL/TM - 5754 (1977).

9.   J. R. Ackerhalt, to be published.

10. M. V. Federov, P. N. Lebedev Physical Institute Preprint
     144 (1976).

11. S. Feneuille and L. Armstrong, J. de Physique $\underline{36}$, L235
     (1975).

12. L. Armstrong, P. Lambropoulos, and N. K. Rahman, Phys.
     Rev. Lett. $\underline{36}$, 952 (1976).

13. R. Glauber, Phys. Rev. $\underline{130}$, 2529 (1963); $\underline{131}$, 2766 (1963).

14. G. S. Agarwal, Phys. Rev. A$\underline{1}$, 1445 (1970).

15. S. N. Dixit and P. Lambropoulos, to be published.

Resonance and Laser
Temporal Coherence Effects
in Multiphoton Ionization of Atoms

G. MAINFRAY
Service de Physique Atomique
Centre d'Etudes Nucléaires de Saclay
B.P. 2, 91190 Gif-Sur-Yvette, France

I.   LASER TEMPORAL COHERENCE EFFECTS

The theoretical calculations of the multiphoton ionization
rate generally assume a monochromatic light generated by a single-
mode laser. However, in the past few years, most of the multiphoton ionization experiments have been performed with multimode
pulsed lasers which had spectral bandwidths of a few cm^{-1}. As it
is well known, a multimode laser has a spectrum which consists of
a very large set of evenly, closely spaced frequencies. The beat-
ing between adjacent longitudinal modes modulates the temporal
laser pulse, and produces peak intensities. As far as high-power
Q-switched dye or Nd-glass lasers are concerned, these temporal
fluctuations could be removed only lately, by selecting a single
mode [1]. When the laser operates in a single mode no fluctuation occurs, as shown in Fig. 1a. The laser pulse has a regular
and well-defined temporal behavior; the photons arrive in file
during the laser pulse duration. When the laser oscillates in
two adjacent modes, the modulation of the pulse is sinusoidal,
with a period corresponding to a round trip time of the light in
the oscillator cavity, as shown in Fig. 1b. The modulation depth,
at the highest part of the pulse,varies from shot to shot, depending on the relative intensities of the two modes. A two-mode
pulse is a special case for which the phase of the two-modes does
not play any role. When the laser oscillates in several adjacent
modes, the laser pulse exhibits a quasiperiodical structure, with
a period corresponding to a round trip time of the light in the
laser cavity. Figure 1c shows the temporal behavior of a seven-
mode laser pulse, and Fig. 1d when the seven modes are phase-
locked. The photons arrive in groups during the pulse duration.
Roughly speaking, the duration of the peak intensities is re-
lated to the coherence time of the laser pulse τ_c, with $\tau_c=(\Delta\nu)^{-1}$,
where $\Delta\nu$ is the spectral bandwidth of the laser pulse. Figure 2

shows the overall laser spectral bandwidth $\Delta\nu$ and the corresponding coherence time τ_c as a function of the number of modes. τ_c
gets shorter and shorter as the number of modes increases. When
the number of modes is more than ten, the temporal fluctuations
are too short to be measured by the usual experimental detectors
with a resolution of about 150 ps. Whereas in a nonresonant
multiphoton ionization process, atoms play the role of very fast
(10^{-15} sec) and sensitive detectors concerning the temporal fluctuations and hence the statistical properties of the laser pulse.
It is expected that the multiphoton ionization rate will be higher if the photons arrive bunched, as with a multimode laser pulse,
than if they arrive in single file as with a single-mode laser
pulse. This is referred to as the effect of correlations or
photon statistics.

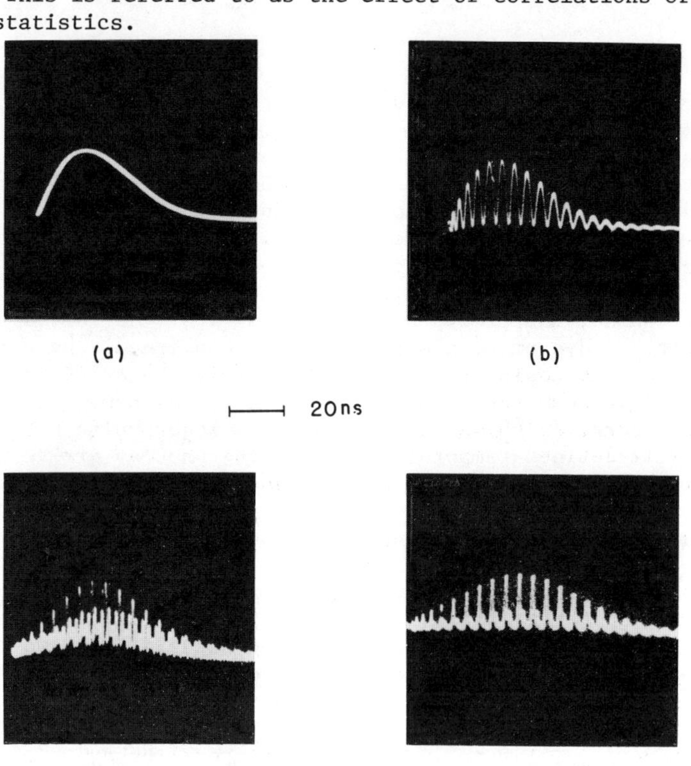

FIGURE 1.   Temporal distribution function of the laser intensity
            when the laser operates in (a) a single mode, (b) two
            modes with visibility $v = 0.8$, (c) seven modes, and
            (d) seven modes when these modes are phase-locked.

number of modes	overall bandwidth $\Delta \nu$	coherence time τ_c
1	11 MHz	40 ns
2	170 MHz	5.7 ns
7	1.1 GHz	0.93 ns
10	1.6 GHz	0.62 ns
30	26 GHz	38 ps
70	76 GHz	13 ps
100	113 GHz	8 ps

FIGURE 2. Spectral bandwidth and coherence time of a laser pulse when the laser operates in a number of modes from 1 to 100.

Let us see now how we can describe the temporal fluctuations of a typical laser pulse used for multiphoton ionization experiments. The instantaneous laser intensity seen by atoms can be expressed in the form:

$$I(t) = \overline{I_M} \cdot G(t)i(t)$$

where $\overline{I_M}$ is the maximum time-averaged intensity. $G(t)$ is the normalized temporal distribution function envelope of the laser intensity, its duration is about 10^{-8} sec for a Q-switched laser pulse. $i(t)$ is a periodic function, which will play a very fundamental role in this study. It has a stochastic pattern which depends on the number of modes and on both relative phases and amplitudes of modes. We generally measure only the time-averaged intensity

$$\overline{I}(t) = \overline{I_M} \cdot G(t)$$

without taking into account the peak intensity function $i(t)$. Whereas multiphoton ionization of atoms is a highly nonlinear process which is very sensitive to peak intensity function since the K-photon ionization probability is proportional to the Kth power of the instantaneous intensity. To insure real physical meaning it is essential to record $\overline{I}(t)$, which is easily measurable, and consider the ratio $i(t) = I(t)/\overline{I}(t)$, instead of $I(t)$ alone. This very simple remark has been ignored and has contributed to creating misunderstanding. Moments related either to i(t) or I(t) are very different as we shall see later on. i(t) shall be called the peak intensity function.

Let us define the Kth order time-independent peak intensity moment by,

$$f_K = \overline{\langle i^K \rangle}$$

where the bracket stands for the ensemble average.

We shall see later on the significance and importance of f_K to readily characterize the statistics of a laser pulse.

It is possible now to relate f_K with multiphoton ionization. The number of ions induced by a multimode laser pulse is:

$$N_m = \beta \int \overline{I^K(t)} \, dt = \beta \int \overline{I^K(t)} \cdot i^K(t) \, dt \, .$$

The number of ions produced by a single-mode laser pulse having the same average intensity is:

$$N_1 = \beta \int \overline{I^K(t)} \, dt = \beta \int \overline{I}^K(t) \, dt \, .$$

Hence,

$$\langle \frac{N_m}{N_1} \rangle = \langle \frac{\overline{I^K}}{\overline{I}^K} \rangle = f_K \, .$$

Thus f_K may be related to multiphoton ionization very simply. It is the enhancement of the ion signal due to the multimode operation of the laser. If we measure the number of multiphoton ions produced by an incoherent laser pulse of maximum average intensity $\overline{I}_M$, and divide it by the number of multiphoton ions produced by a monomode laser pulse of the same maximum average intensity $\overline{I}_M$, we obtain a direct measurement of f_K.

At this point one remark has to be made. For lasers operating with a sifficiently small number of modes, where i(t) may be directly registered by oscillographic techniques, we have two different methods for measuring f_K: direct registration and multiphoton ionization. For a higher number of modes where it is impossible to register fluctuations directly, we have just shown that f_K may still be measured by multiphoton ionization. For this, we need $\overline{I}(t)$, the average value of the intensity, and the value of the ionization for multimode and monomode operation at the same $\overline{I}(t)$. Now the way is clearly indicated. Observations should be made with a variable number of modes from $M = 1$ to the highest values. Let us mention that two other moments have been considered in the literature, g_K and b_K [2]:

$$g_K = \frac{\overline{<I^K>}}{\overline{<I>}^K} \quad , \qquad b_K = \frac{<\overline{I}^K>}{<\overline{I}>^K}$$

with

$$\overline{I} = \frac{1}{T} \int_T I(t)dt \quad , \qquad \overline{I^K} = \frac{1}{T} \int_T I^K(t)dt \ .$$

Many thoeretical papers dealt with g_K, which is difficult, if not impossible to measure with reasonable precision. Whereas f_K is a moment more directly connected with values experimentally available. Moreover f_K is the only moment which is directly connected with the phase correlation between modes.

To take into account the temporal coherence effects, theoreticians generally calculate the multiphoton ionization probability of an atom in two extreme cases: first with a laser light considered as a pure coherent field, and second with thermal light. From an experimental point of view, the situation is not so simple. The first question is whether a single-mode laser pulse can be considered as a pure coherent field. A single-mode laser pulse does not have a well stabilized amplitude; $\overline{I}_M$ is not exactly reproducible at each shot; however, the instantaneous intensity of a single-mode laser pulse can be measured at each shot. Second, most of the Nd-glass lasers which have been used in multiphoton ionization experiments in the past few years had spectral bandwidths of a few cm^{-1}. Such a laser light cannot be considered as being fully Gaussian as chaotic light is. In this paper, we investigate the laser temporal coherence effects on the eleven-photon ionization of xenon atoms, by varying the mode-structure of a Nd-glass laser pulse from a single-mode to about one hundred modes. Thus we do not deal with the absolute coherence properties of the laser field but only with relative coherence properties, since we shall measure the relative efficiency of the eleven-photon ionization rate of zenon atoms by varying the mode content of the laser pulse.

The laser operates in the lowest order transverse mode with Gaussian spatial intensity distribution. The laser radiation is linearly polarized, and is centered at 10643 Å. This wavelength has been selected to investigate a direct eleven-photon ionization of xenon atoms, without any resonant intermediate excited state. The experiment consisted of a measurement of the number of multiphoton ions as a function of the average laser intensity

for different values of the number of modes. Experimental results
are shown in Fig. 3. Curve (a) was obtained with single-mode
laser pulses. Curve (b) was obtained with two-mode pulses with
visibility 0.6. Curve (c) was obtained with seven-mode pulses.
The number of ions induced by a seven-mode pulse is enhanced by
400 over that induced by a single mode pulse of the same average
intensity. Experimental points obtained with seven-mode pulses
become scattered. This scattering is due to variation of phases
of modes from one laser shot to another one. The very significant influence of phase relationship appears by comparing the
two curves (c) and (d) obtained both with seven-mode pulses.
Curve (c) is obtained when the seven modes are assumed to have
random phases. For curve (d), the seven modes were phase locked,
by placing in the oscillator cavity a solvent cell which contained a dye to phase lock the seven modes. The enhancement due
to mode-locking for seven-mode pulses is 80, in good agreement
with theoretical considerations.

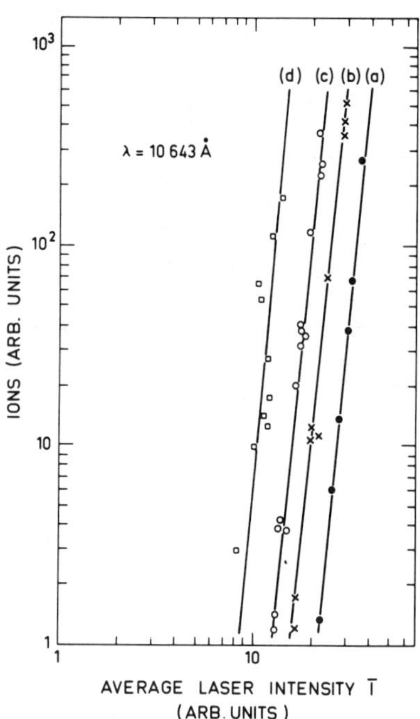

FIGURE 3. Log-log plot of the variation of the number of multiphoton xenon ions as a function of the average laser
intensity, when the laser operates in (a) a single mode,
(b) two modes with visibility v=0.6,(c) seven modes,
and (d) seven modes when these modes are phase-locked.

Figure 4 shows, in the solid line, the enhancement of the number of ions, that is the experimental value of the f_{11} moment, when the number of modes is increased from one to one hundred for the same average laser intensity. The number of ions is enhanced by nearly 10^7 when the number of modes is increased from one (coherence time 40 ns) to one hundred (coherence time 8 ps). This figure also shows, in the dashed line, the f_{11} moment calculated in assuming that phases of the modes are independent. These calculated f_{11} values are found to be greater than the experimental values corresponding to the enhancement in the number of ions due to the mode structure. This difference cannot be explained in terms of different intensities of the modes. The present experiment, as well as calculations, have shown that the enhancement factor in the number of ions is relatively insensitive to the distribution of the intensities of the modes. However, the phase distribution of the modes is the determining factor. The assumption of statistical independence of the phases of the modes in a Q-switched Nd-glass laser is not obvious. Several adjacent modes may have definite phase relationships that tend to decrease the f_K function, whereas distant modes can be considered as having independent phases. When the number of modes becomes large, the influence of such a phase relationship becomes less and less important. Experimental and calculated values are expected to tend towards each other and become identical at an asymptotic value for a very large number of modes. In the limit of an infinite number of independent modes, this asymptotic value has been calculated by several authors to be K! that is 4×10^7 for K = 11.

The off-resonant multiphoton ionization rate can be written as:

$$w = \alpha \, f_K \, \bar{I}^K$$

The Kth order peak intensity moment, or K^{th}-order autocorrelation function f_K is equal to unity for a single-mode laser pulse or a bandwidth-limited pulse, that is a pulse with a duration less or equal to the coherence time. In the limit of an infinite number of independent modes, f_K equals K! [3-6]. Thus a significant correction factor has to be applied for experimental results obtained with a multimode laser pulse. This correction factor depends both on the laser spectral linewidth and on the order K of the nonlinear process. This correction factor is small (≤ 2) for a two-photon ionization process, whereas it has a dramatic effect in high order nonlinear processes such as multiphoton ionization of rare gases. This effect gives, in many cases, the key to the explanation of discrepancies between theoretical and experimental values of multiphoton ionization rates. Conversely, multiphoton

ionization processes allow us to consider an atom irradiated by a
laser pulse, as an ideal photon detector concerning the statistical properties of the laser pulse. The determination of the Kth-
order autocorrelation function of the laser peak intensity is of
special interest to fully characterize a laser radiation.

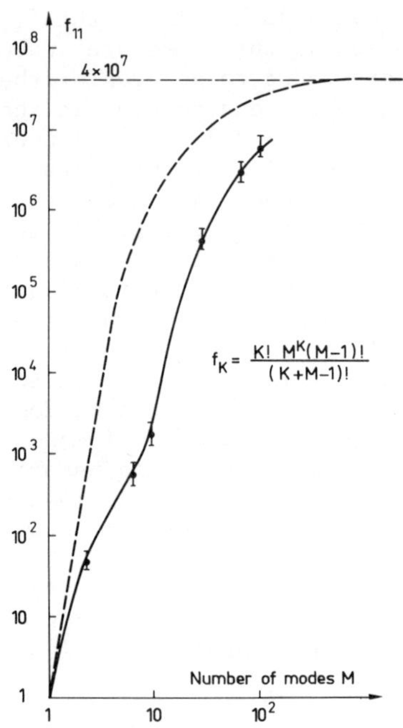

FIGURE 4.   Variation of the experimental value of the moment as
             a function of the number of modes, for the same aver-
             age laser intensity. This figure also shows (dashed
             line) the moment calculated in assuming that phases
             of the modes are independent.

II.   RESONANCE EFFECTS

The multiphoton ionization rate of an atom exhibits a typical resonant character when the energy of an integer number of
photons is very close to the energy of an atomic level satisfying
the selection rules. Figure 5 shows the calculated six-photon
ionization cross-section of atomic hydrogen in the ground state

as a function of the laser wavelength [7]. Peaks arise from the
resonant structure, that is when the energy of five photons equals
the energy of intermediate atomic levels $\dot{n} = 3,4$ and 5. The amplitudes of these peaks are indeterminate because damping terms have
been neglected.

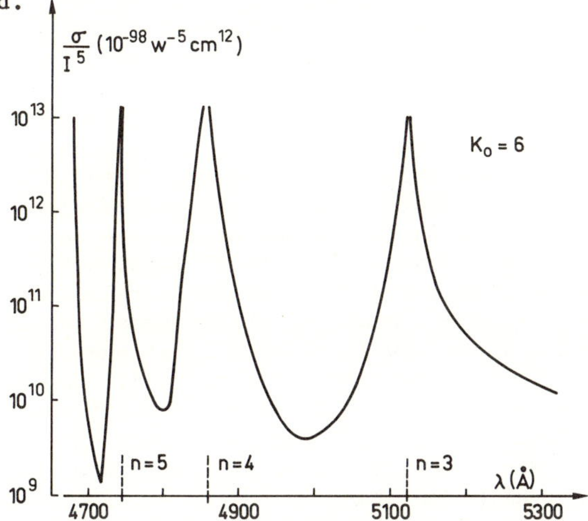

Y. GONTIER and M. TRAHIN
Phys. Rev. 172, 83 (1968)

FIGURE 5.   Variation of the calculated six-photon ionization
            cross-section of atomic hydrogen in the ground state
            as a function of the laser wavelength.

     The theoretical aspect of resonant multiphoton ionization
processes will be discussed in other invited papers. Here, we
shall pay particular attention to showing the outstanding properties exhibited experimentally in resonant multiphoton processes.
The investigation of resonant multiphoton ionization processes
requires a single-mode or a bandwidth-limited laser pulse in order
to avoid having to take into account the statistical properties
of the laser radiation. Such tunable-wavelength single-mode high
power lasers have become available only in the last three years.
We have conducted an investigation of a four-photon ionization of
cesium atoms using a single-mode Nd-glass Q-switched laser tunable
over 80 Å [8]. A typical example of a resonant multiphoton ionization process occurring here and which we shall discuss, is the
resonant three-photon exictation of the 6F levels. Figure 6 shows
the variation of the number of atomic cesium ions as a function
of the laser wavelength (for a constant laser intensity $I = 4 \times 10^8$
w/cm^2) in the neighborhood of the resonance with the 6F levels.

Several remarks have to be made. Firstly, the number of ions is enhanced by about five orders of magnitude due to the resonant process. It should be pointed out, however, that the ratio of the number of ions at the resonance maximum to the number of ions off-resonance is questionable. In off-resonant multiphoton ionization of cesium atoms, the collected atomic ions Cs^+ are formed both from a four photon ionization process of atomic cesium and from a dissociation of a part of the molecular ions by the laser radiation in a one-photon process [9]. This molecular contribution gives, in many cases, the key to the explanation of the discrepancies between theoretical and experimental values of the off-resonant multiphoton ionization rate of alkali atoms. Secondly, the maximum in the number of ions is shifted with respect to the wavelength position of the 6F level derived from spectroscopic tables. This resonance shift is related to the ground state 6S and resonant state 6F shifts under the influence of the laser field. These level shifts exhibit what is called the "ac Stark effect". The net effect is to modify the energy of an atomic state so that the multiphoton transitions take place between modified or "dressed" states as they were called by Cohen-Tannoudji. The energy shift of the resonance, ΔE, has been found to be linear with the laser intensity within the range 10^8-10^9 w/cm^2, $\Delta E = 2.2$ cm^{-1}/Gw.cm^{-2} [8,10]. These experimental values are in good agreement with shift values calculated by different authors. Thirdly, the width (FWHM) of the resonance profile is much larger (0.3 cm^{-1}) than the laser linewidth (10^{-3} cm^{-1}). The resonance width can be explained mainly in terms of the coupling of the resonant state with the continuum via the absorption of one additional photon. The resonance width was measured to increase with the laser intensity. This width will be compared, in the near future, with published theoretical calculations [11]. Moreover, since the resonant shift is proportional to the laser intensity, spatial nonuniformities of the laser intensity within the interaction volume induce a large distribution of resonance shifts. This effect causes the resonance profile to be broadened. Fourthly, Fig. 6 shows an asymmetry in the wings of the resonance profile. However, no significant asymmetry appears in the vicinity of the maximum of the, resonance profile.

Resonance effects are also characterized by a large variation of the slope $K = \partial Log\ N_i/\partial Log\ I$ as a function of the laser wavelength. Figure 7 shows the very significant variation of the slope K as a function of the resonance detuning $\delta = E_{6F}-E_{6S}-3E_P$, where E_P is the laser photon energy, for the three-photon resonance of the 6F levels. In this representation, a resonance appears as a very sharp phenomenon marked by a dramatic change of the slope from $K \simeq 30$ to $K = 2$. Thus in the vicinity of a resonance, the observed slope K no longer corresponds to the number

of photons absorbed by the atoms. However, when the resonance
detuning is large enough, we find again K values equal to four,
corresponding to a non-resonant four-photon ionization of cesium
atoms.

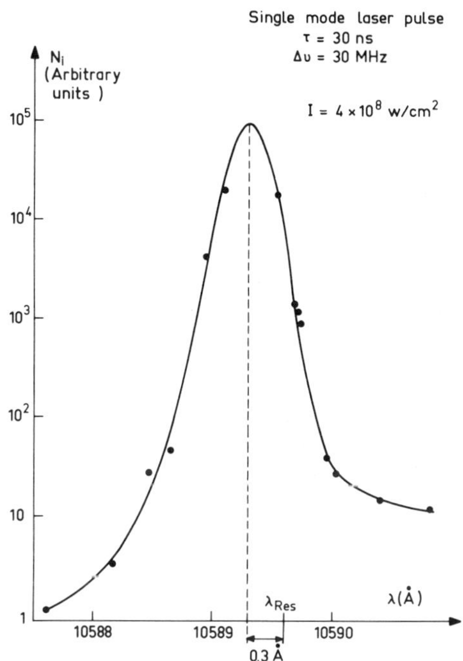

FIGURE 6.   Variation of the number of Cs ions as a function of
            the laser wavelength for laser intensity
            $I = 4 \times 10^8$ w/cm^2.

     To conclude this discussion on resonance effects, we shall
give some preliminary results on the temporal aspects of resonant
multiphoton ionization processes. Non-resonant multiphoton ionization of atoms occurs via virtual energy states within a time
on the order of 10^{-15} s. Resonant multiphoton ionization is
characterized by a time related mainly to the lifetime, in the
presence of the laser field, of the resonant atomic state involved. The characteristic time of the resonant process is always much longer than 10^{-15} s time scale of the non-resonant process. Up to now, resonance effects have been investigated with
laser pulse durations of about 10^{-8} s, a time longer than that
characteristic of the resonant process. This situation can be
considered as a quasi continuous excitation [12]. On the

contrary, when a very short laser pulse (10^{-12}, 10^{-11} s) is used, it can be expected that in many cases the resonant process does not have enough time to take place during the laser pulse duration. Thus the multiphoton ionization process would be governed only by the non-resonant process. This assumption was recently supported by an experimental investigation of multiphoton ionization of xenon, krypton and argon gases using a tunable-wavelength bandwidth-limited 30-psec laser pulse of 10^{13} w/cm^2 at 1.06 μm [13]. No resonant effects were observed although resonance conditions were satisfied.

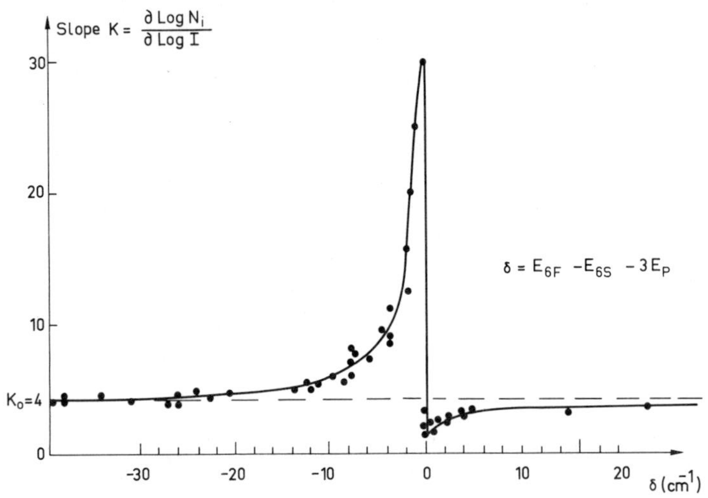

FIGURE 7.   Variation of the slope K as a function of the resonance detuning.

On the contrary, resonance effects in four-photon ionization of cesium atoms via the 6F levels were still observed recently using a tunable-wavelength bandwidth-limited 15-psec laser pulse of 10^9 to 10^{10} w/cm^2 at 1.06 μm, as shown in Fig. 8. The variation of the number of ions as a function of the laser wavelength exhibits a typical resonant character: enhancement in the number of ions and a resonance shift. The resonance shift for increasing values of the laser intensity is indicated by the dashed line. Moreover, the width (FWHM) of the resonance profile (1 Å) is essentially governed by the laser linewidth (1.5 Å).

The combination of photon statistics and resonance effects are beginning to be actively investigated [14-16]. However, this general problem is still far from being solved.

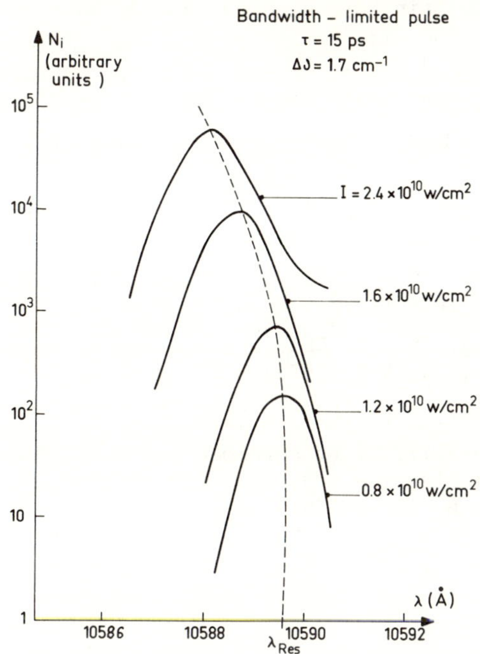

FIGURE 8.   Variation of the number of Cs ions as a function of
the laser wavelength for four different values of the
intensity of a 15 psec. laser pulse.

REFERENCES

1. C. Lecompte, G. Mainfray, C. Manus and F. Sanchez, Phys. Rev.
Lett., 32, 265 (1974).

2. F. Sanchez, Nuovo Cimento, B27, 305 (1975).

3. J. Ducuing and N. Bloembergen, Phys. Rev., A133, 1493 (1964).

4. P. Lambropoulos, C. Kikuchi and R. K. Osborn, Phys. Rev.,
A144, 1081 (1966).

5. G. S. Agarwal, Phys. Rev., A1, 1445 (1970).

6. J. L. Debethune, Nuovo Cimento, B12, 101 (1972).

7. Y. Gontier and M. Trahin, Phys. Rev., 172, 83 (1968).

8. J. Morellec, D. Normand and G. Petite, Phys. Rev., A14,
300 (1976).

9. B. Held, G. Mainfray, C. Manus and J. Morellec, Phys. Rev. Lett., $\underline{28}$, 130 (1972).

10. V. A. Grinchuk, G. A. Delone and K. B. Petrosyan, Fiz. Plasmy, $\underline{1}$, 320 (1975) [Sov. J. Plasma Phys., $\underline{1}$, 172 (1975)].

11. B. L. Beers and L. Amstrong, Jr., Phys. Rev., $\underline{A12}$, 2447 (1975).

12. M. Crance and S. Feneullie, J. Phys. Lett. (Paris) $\underline{37}$, L333 (1976).

13. L. A. Lompre, G. Mainfray, C. Manus and J. Thebault, Phys. Rev., $\underline{A15}$, 1604 (1977).

14. L. Amstrong, Jr., P. Lambropoulos and N. K. Rahman, Phys. Rev. Lett., $\underline{36}$, 952 (1976).

15. J. Mostowski, Phys. Lett., $\underline{56A}$, 87 (1976).

16. V. A. Kovarskii, N. F. Perelman and S. S. Todirasku, Kvantovaya Elektron., $\underline{3}$, 1805 (1976) [Sov. J. Quantum. Electron., $\underline{6}$, 980 (1976)].

Multiphoton Laser Spectroscopy in Heavy Elements

R. W. SOLARZ, J. A. PAISNER, and E. F. WORDEN
Lawrence Livermore Laboratory
P. O. Box 808
Livermore, California

Multiphoton spectroscopy using either a collection of different optical fields or a single intense laser field has been vigorously applied to a number of systems within the last few years. Interest in multiphoton effects has been accelerated by its many applications to isotope separation, photochemistry, and the spectroscopy of highly excited atoms and molecules. Resonant multiphoton spectroscopy in particular has proven to be a very flexible and sensitive technique for the study of the excited states of complex atoms and molecules. While the use of a single intense laser field may be a convenient technique for inducing decomposition in a molecule or ionization in an atom, spectroscopic evaluation of these systems is best accomplished by stepwise resonant excitation experiments in weak fields. Our interest in multistep spectroscopy has been its application to the study of the structure of the lanthanides and actinides. In this paper we describe some recently discovered regularities in the spectra of heavy elements which are also applicable to the analysis of the spectra of lighter atoms.

Nearly all the previously reported spectra of the lanthanides and actinides were either from photoabsorption or emission experiments wherein the spectral lines of each atomic species are analyzed using the technique of finite differences [1]. These studies have generally provided a complete mapping of the low lying level structure (to about 3 ev) in the neutral and singly ionized species. In the few cases where the spectra are simple Yb, Tm, and Eu for example) states below the ionization limit and autoionizing levels in the continuum have been observed [2, 3]. However, time resolved stepwise laser photoionization permits study of even the most complicated atomic spectra in arbitrary energy regimes, permits preferential study of arbitrary groupings or classifications of states, and using the time

resolution and narrow linewidths of suitable lasers, also allows
the study of hyperfine structure, Stark effects, and radiative
lifetimes. Oscillator strengths may also be measured to high
accuracy using related methods.

To date, a dozen lanthanides and actinides have been studied
using a laser spectrometer which has been described in detail previously [4,5]. The important feature of the apparatus is the in-
corporation of several temporally related dye lasers in an atomic
beam spectrometer. The atomic beam is excited and in many cases
ionized by the output from these lasers. Often times the atoms
are brought to levels just below the continuum and are ionized by
collisional means, the application of an externally applied electric field, or IR photons from a fixed frequency laser. In each
instance the photoion yield is examined by a channeltron particle
multiplier. With sufficiently powerful tunable lasers, ionization can be achieved with very high efficiency. The apparatus is
shown in Fig. 1, and in Fig. 2 we have summarized the types of
experiments performed. In this figure λ_n denotes the wavelength
of the nth photon in the sequential excitation ladder.

We have previously reported methods for mapping excited state
level structure, measuring relative cross sections, and measuring
lifetimes in heavy atoms [4]. Excited state–excited state transitions are usually mapped by monitoring photoion yield as a func-
tion of λ_2 (varied) when λ_1 and λ_3 are held fixed as in Fig. 2b.
The existence of a resonance near 4 eV can be explored from a
variety of parent or initial levels near 2 eV of differing angular momentum J. Coupled with electric dipole selection rules
these experiments usually provide a value for the angular momentum
of the excited state. Even near 4 eV the level structure is dense
enough that frequently two or more states can be found within
0.5 cm^{-1} or less of each other. While single laser multiple photon resonance spectroscopy was used to obtain level structure in
simpler systems such as Ca [6], in the very heavy atoms the method
is clearly inapplicable.

The detection of Rydberg states in the heavy atoms is of
special interest. Except in the very simplest of the heavy atoms,
Rydberg states are not detected in conventional experiments. In
the most complicated lanthanides and actinides time resolved
methods are required. The great density of valence states near
the ionization limit, characterized by small orbital radii and
short radiative lifetimes, is sufficient to prevent the observation of a recognizable progression in stepwise photoionization
experiments as described above. We have previously reported the
preferential observation of Rydberg states in uranium wherein we
excited all levels near the ionization limit but delayed the final
ionization step out of these levels to the continuum by micro-
seconds [5]. During this time interval the Rydberg states retain

their energy and the valence states radiatively decay and are
lost from the detection or ionization step.

Rydberg states of cerium are shown in Fig. 3. These were
detected by two step resonant excitation followed by time delayed
field ionization of the excited states. This method was also
used to obtain the bound Rydberg states of Gd and Tb. Collisional
ionization of the bound states of holmium was used to obtain its
Rydberg spectrum. For the remainder of the lanthanides (except
for Pr and Pm) we observed autoionizing Rydberg states by direct
two or three step photoionization. Having identified Rydberg
series for these elements, we obtain the ionization limit by selecting that value which yields the smoothest and most constant
quantum defects for each observed Rydberg term. In this way we
obtain the precise values of the ionization limits listed in
Table I. This procedure for obtaining ionization limits ignores
perturbations in the spectra and we have adjusted our quoted mar-
gin of error accordingly. Currently we are using the methods of
quantum defect theory to analyze perturbations in the Rydberg
states of uranium. High resolution studies show that there are,
as expected, a great many interacting channels. However, preliminary results indicate that the perturbations in the spectra
are not sufficiently great to cause errors in the ionization
limit larger than we have quoted.

Figure 4 is a plot of the lanthanide ionization limits as a
function of atomic number. The values for Tm and Yb in this figure are taken from the work of Camus [7]. With the exception of
Gd and Ce, all ionization processes are the removal of an s electron from the $4f^N 6s^2$ (N = 2 to 14) configuration. For Ce (N = 2)
and Gd (N = 8) the lowest lying levels for the neutral and ion are
$f^{N-1}ds^2$ to $f^{N-1}d^2$ and $f^{N-1}ds^2$ to $f^{N-1}ds$ respectively. However,
the position of the lowest lying levels of the $f^N s^2$ and $f^N s$ configurations are accurately known from spectroscopic studies [1].
Using this information we normalize the ionization limits of Ce
and Gd to correspond to an $f^N s^2$ to $f^N s$ ionization mechanism as indicated in the figure. We have subsequently found that the piecewise linear variation of ionization limits with a slope change
occurring at the half filled shell as seen here, can be found for
any row of the periodic table when limits are normalized to correspond to ionization mechanisms of the type $\ell^N s^2$ to $\ell^N s$ or $\ell^N s$
to ℓ^N. Furthermore, we find that this is true for any stage of
ionization.

This observation can be explained simply in terms of the
Pauli principle and Hund's rule. For an ionization process of
the type $\ell^N s^2$ to $\ell^N s$, in the absence of electron correlations the
ionization limit increases monotonically and linearly with N as a
result of incomplete shielding of the added electrons as one moves
across a given row of the periodic table. However, the exclusion

principle constrains the various electronic motions to be correlated and as a result Coulomb repulsion between electrons is least
and binding energies the greatest when the spatial wavefunction
is completely antisymmetric with all electronic spins parallel.
From Hund's rule the term of highest multiplicity has the lowest
energy within a given configuration. Thus the ground state of
the ion experiences a substantial correlation energy between the
aligned spins of the f shell and the single s electron; this correlation does not exist in the neutral where the s shell is filled
with two antiparallel electrons. The greater the number of f
electrons, the greater the correlation term or binding energy of
electrons in the ion. Thus the deviation of ionization limits
(or the difference of the binding energy in the lowest level of
the ionic $\ell^N s$ and neutral $\ell^N s^2$ configurations) from the linear behavior increases as one moves to the half filled shell. At this
point additional electrons begin pairing spins and the deviation
diminishes.

Elementary application of the Slater-Condon semiempirical
methods to calculate the ionization energy, I_N, of the configurations $\ell^N s^2$ to $\ell^N s$ suggests that the pattern in ionization limits
should be

$$I_N = A + NB \qquad\qquad N \leqslant 2\ell + 1$$

$$I_N = A + N[B + G_\ell(\ell,s)] \qquad N \geqslant 2\ell + 1$$

that is the ionization limits are piecewise linear with a change
in slope occurring at the half filled shell. The constants A and
B are combinations of Slater parameters with the important point
being that the change of slope at the half filled shell is simply
the Slater exchange integral of the ion, $G_\ell(\ell,s)$. More elaborate
ab initio Hartree-Fock calculations confirm this interpretation.
As a consequence of this observation, the knowledge of any two
ionization limits plus an exchange integral within a row, or alternatively the knowledge of three limits, allows interpolation
of any other ionization limit within that row to a precision of
nearly 1 part in 10^3. We have applied this method to obtain precise values for the ionization limits of the neutral actinides
and to the first, second, third and fourth limits of the lanthanides and report the results elsewhere [8].

To summarize, stepwise resonant multiphoton methods are ir-
replaceable tools in the study of high lying states in complex
atomic systems. Systematic application of these methods has per-
mitted useful regularities to be observed which are also found to
hold for the lighter elements. It should be clear that greatly increased understanding of the excited state structure of heavy
atoms is now possible.

Laser spectroscopy apparatus
Photomultiplier
Boxcar Integrator
1.5m JY monochromator
Dual trace scope
ThI4
Microwave lamp
Fast monitor diode
Channeltron
Quadrupole mass filter
Detector
Vacuum chamber
Ion chamber
M oven
Nitrogen laser 1
Nitrogen laser 2
Nitrogen laser 3
Dye laser 1 (fixed)
Dye laser 2 (fixed)
Dye laser 3 (variable)
λ1
λ2
λ3
Scan control unit
Master trigger control unit
2 pen recorder
M = lanthanide or other metal

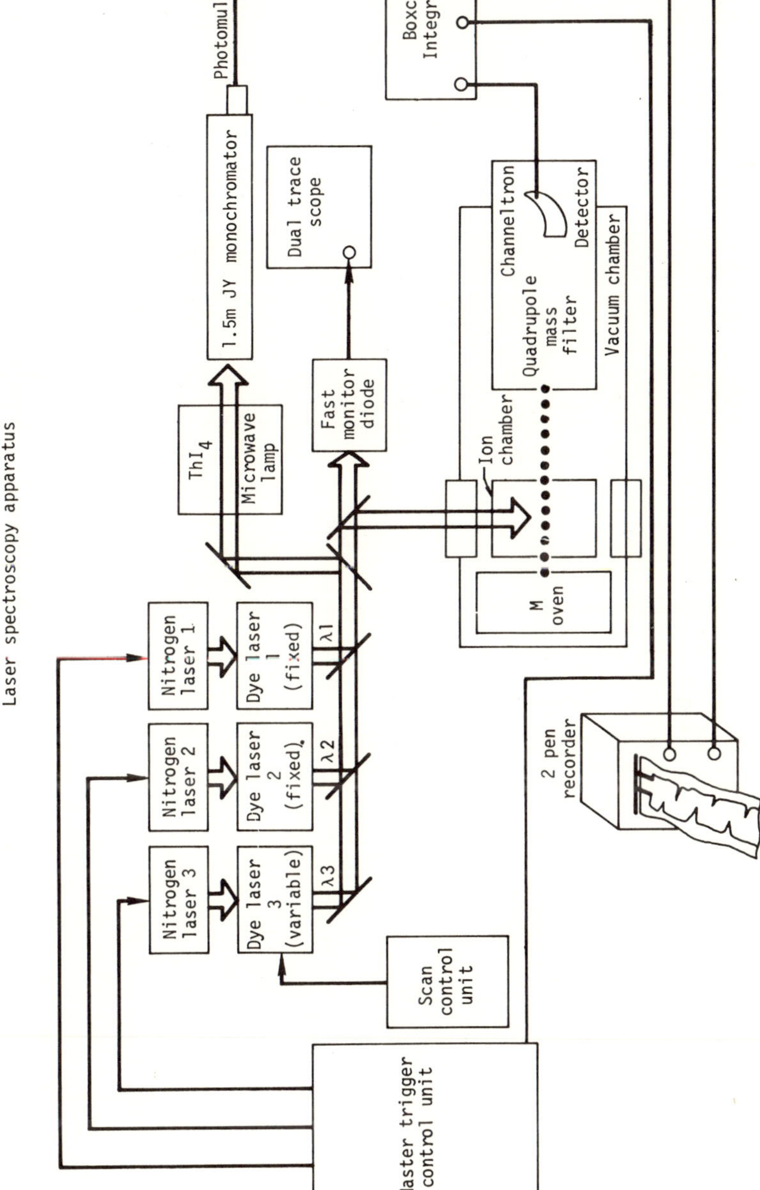

Figure 1

EXCITATION SCHEMES USED TO OBTAIN RYDBERG AND AUTOIONIZATION SPECTRA

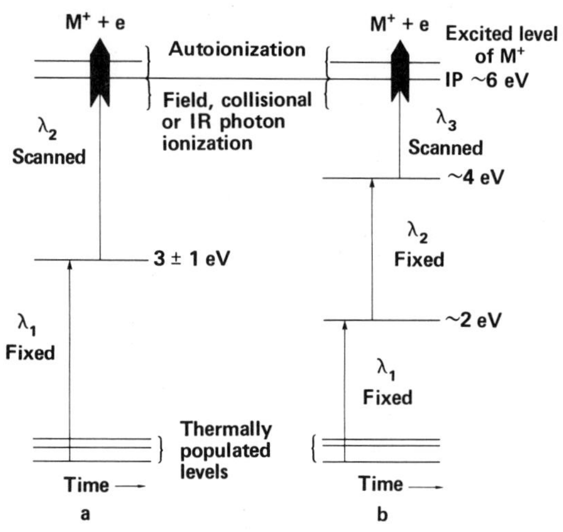

FIGURE 2

CERIUM I RYDBERG SERIES CONVERGING TO THE ION GROUND STATE

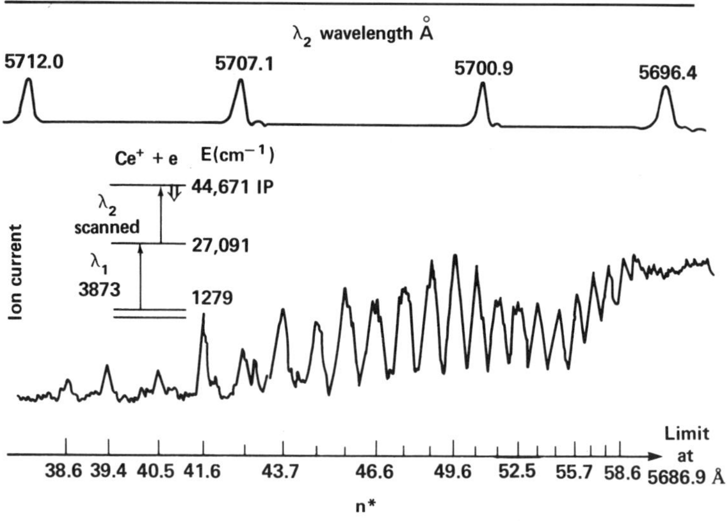

FIGURE 3

TABLE I. Some Lanthanide Rydberg Series Limits

Element	λ_1[a] (Å)	λ_2[a,b] (Å)	Excited Level Used (cm^{-1})	Convergence Energy (cm^{-1})	Convergence Level in Ion (cm^{-1})	First Ionization Limit[c] (cm^{-1})	(eV)
Ce	3793.83		26331.11	44674(3)	0.00	44674(3)	5.5389(4)
	3873.03		27091.56	44671(3)	0.00	44671(3)	5.5385(4)
Nd	4924.53		20300.84	45075(5)	513.32	44562(5)	5.5250(6)
Sm	4596.74		22041.02	45846(8)	326.64	45519(5)	5.6437(6)
Eu	4594.03		21761.26	47403(2)	1669.21	45734(2)	5.6703(3)
	4661.88		21444.58	47405(2)	1669.21	45736(2)	5.6706(3)
	6291.34	6787.48	30619.49	47403(2)	1669.21	45734(2)	5.6703(3)
Gd	5617.91	6351.72	33534.71	49603(5)	0.00	49603(5)	6.1501(6)
	5701.35	6573.83	32957.77	47604(5)	0.00	49604(5)	6.1502(6)
Tb	4139.06		24438.76	47295(5)	0.00	47295(5)	5.8639(6)
	4146.96		24392.75	47294(5)	0.00	47294(5)	5.8638(6)
Dy	4211.72		23736.60	48730(5)	828.31	47902(5)	5.9391(6)
	6259.09	6769.79	30739.79	48727(5)	828.31	47899(5)	5.9388(6)
Ho	4103.84		24360.55	49203(8)	637.4	48566(8)	6.0216(10)
	6305.36	6947.1	30246(2)	48567(5)	0.00	48567(5)	6.0216(6)
Er	4087.63		24457.15	49704(8)	440.43	49264(8)	6.1080(10)
	6221.02	6451.56	31565.94	49699(8)	440.43	49259(8)	6.1074(10)

[a] Excitation wavelengths (λ_1 and λ_2) and excited level values given to 0.01 are from NBS monograph 145 and Refs. therein.
[b] No value of λ_2 is given for two step observations when λ_2 is scanned. Values given are λ_2 for three step results.
[c] 8065.479 cm^{-1}/eV used to convert values from cm^{-1} to eV.

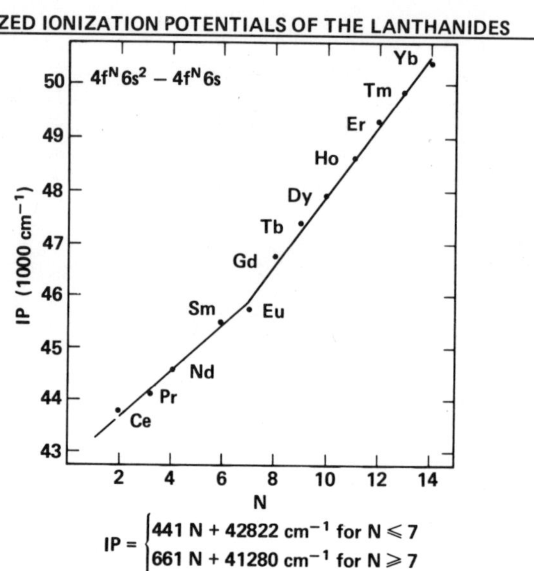

$$IP = \begin{cases} 441\,N + 42822 \text{ cm}^{-1} \text{ for } N \leqslant 7 \\ 661\,N + 41280 \text{ cm}^{-1} \text{ for } N \geqslant 7 \end{cases}$$

FIGURE 4

ACKNOWLEDGMENTS

We wish to acknowledge valuable discussions with B. Shore, K. Rajnak, J. Conway and L. Carlson.

REFERENCES

1.   J. Blaise, P. Camus, and J. F. Wyart, "Rare Earth Elements" in Gmelin Handbuch der Anorganishen Chemie, System 39-84 (Springer-Verlag, Berlin, 1976).

2.   F. S. Tomkins, J. Opt. Soc. Am. 65, 1188A (1975).

3.   G. Smith and F. S. Tomkins, Proc. Roy. Soc. Lond, A342, 149 (1975).

4.   L. R. Carlson, J. A. Paisner, E. F. Worden, S. A. Johnson, C. A. May, and R. W. Solarz, J. Opt. Soc. Am. 66, 846 (1976).

5.   R. W. Solarz, C. A. May, L. R. Carlson, E. F. Worden, S. A. Johnson, J. A. Paisner, and L. J. Radziemski, Phys. Rev. A14, 1129 (1976).

6.   P. Esherick, J. A. Armstrong, R. W. Dreyfus, and J. J. Wynne, Phys. Rev. Lett. 36, 1296 (1976).

7.   P. Camus, Thesis, University of Paris, Orsay, 1971.

8.   E. F. Worden, R. W. Solarz, J. A. Paisner, B. W. Shore, K. Rajnak, and J. G. Conway, International Conference on Atomic and Molecular States Coupled to a Continuum, June 13-17, 1977, Aussois, France.

Photon Statistics in Double Photon Experiments

J. KRASIŃSKI
Institute of Experimental Physics,
University of Warsaw,
Hoza 69, 00-681 Warsaw, Poland

The relation betwen the statistical properties of light and
the multiphoton absorption probability has been investigated by
many authors in theory as well as experiment. Lambropoulos and
others [1] found theoretically that the two-photon absorption
probability should depend on statistical properties of the light
and predicted that the absorption efficiency for thermal light
should be twice that for coherent light. Almost at the same time
the paper of Teich and Wolga was published with the same results
[2].

For a broad absorption level and narrow laser line, the prob-
ability of two-photon absorption $W^{(2)}$ can be written as $W^{(2)} = \alpha<I>^2 g^{(2)}(0)$ where I is the beam intensity and $g^{(2)}(0)$ is the
normalized second order correlation function of the light. Mollow
predicted [3] that the probability ratio $W^{(2)}_{chaotic}/W^{(2)}_{coherent}$ should
depend on the relation between laser and absorption band widths.
In the case of a broad laser line and narrow absorption line the
absorption efficiency for thermal light should be four times that
for coherent light.

The probability for two-photon absorption is rather low in
comparison with ordinary single photon absorption. The square
dependence of the probability on the intensity of the light forces
us to use high power pulse laser for two-photon experiments. However, the reproducibility of the peak power and spectrum of a
pulsed laser is not sufficient to investigate the rather weak dependence of the probability on the statistical properties of a
laser beam. In our laboratory we proposed and used an experimental setup for measuring two-photon absorption by means of medium power c.w. lasers [4]. The stability of such lasers is, of
course, much better.

If we focus the laser beam by means of a well corrected lens
we can obtain very high power density at the focal point. If we
use a 100x microscopic objective as a lens and 50 mW He-Ne laser,
the power density at the focus will be about 50 mW/cm^2 which is
almost the same as what we can obtain by means of a giant pulse

278   J. Krasiński

laser without focusing.

In our experiments we used fluorescence intensity as a measure of absorption. If we calculate the fluorescence intensity distribution along the focused light beam [5] we can see that almost 80% of the total intensity will be generated in small volume around the focal point of the lens. We can treat the focal point as a source of the fluorescent light.

Figure 1 shows two basic arrangements used in our measurements. The first of them is very simple. The L_F lens focused the laser beam in the sample. The L_C lens collected the fluorescence light on the PM photomultiplier cathode. The F filter cuts off the scattered laser light.

The apertures of the focusing lens L_F and the collimating lens L_C cannot be very large so we cannot obtain high power density at the L_F focus. Therefore we cannot collect all the fluorescence light from the sample. However, the required attenuation of the laser line in the F filter is not very high. The sensitivity of the system is sufficient if we use the argon ion laser as the light source.

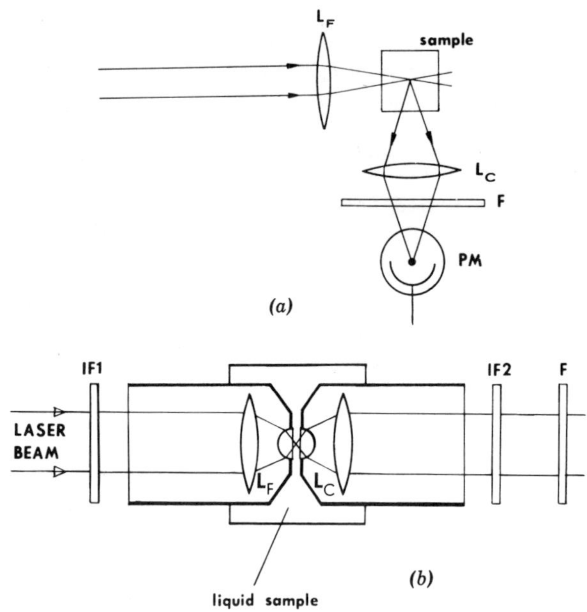

FIGURE 1.   Two basic arrangements for two-photon absorption investigation by means of a c.w. laser.

The second system [6] which we used recently enabled us to use large aperture immersion objectives L_F and L_C. Since the apertures of such objectives are about 1.3, we can collect by means of the L_C lens, about 25% of the whole fluorescent light. The fluorescence generated mainly at the L_F focus is recollimated by means of the L_C lens and propagates along the direction of the laser beam. Such high quality of the fluorescence light beam offers many advantages; for example, we can use a monochromator as a filter and the losses will still be very low.

Adding to the system a IF1 interference filter which can fully transmit the laser light and has high reflectivity in the region of the sample fluorescence, we can collect up to 50% of the whole fluorescence light. An additional IF2 interference filter transmits the fluorescence light but reflects the laser beam. The reflected beam is focused again by means of the L_C lens. In such a double pass configuration, using both IF1 and IF2 filters, the signal should be almost eight times stronger than that obtained without filters. In practical arrangements the gain will be something lower due to some losses in lenses and interference filters.

The only problem we have using such a system is connected with the F filter. The transmission of the filter for the laser line should be very low, about 10^{-16}, while the transmission for the fluorescence light should be as high as possible.

If we use a He-Ne or argon laser as a light source and organic scintillators as absorbers, we can use BG-12 or UG-11 Schott filters.

In such an arrangement we can measure double photon absorption in PPO scintillator solution starting from laser power as low as 10 μW. Our experiments have been done by means of such systems. Figure 2 shows a typical experimental setup. An argon ion laser operating in TEM_{∞} mode was used as a light source. The laser output spectrum was observed by means of a scanning interferometer. The laser beam power was regulated with the use of a polarization rotator and a polarizer. They were oriented in such a way as to obtain the constant polarization of the laser beam behind them. The Schott GG-14 filter (F_1) cuts off the wavelengths below 5.4 nm. A lens L_1 focuses the light in the center of the sample. The sample consisted of saturated toluene solution of an organic scintillator. All chemical compounds used here were of spectroscopic grade. The thickness of the sample cell was about 15 mm. The laser beam power was measured by a power meter located behind the sample. The amount of laser power absorbed in the cell may be neglected. The F_2 filter (Schott UG-11 or BG-12) cuts off the laser line as well as longer wavelengths. The fluorescence peak is in the region of 350-400 nm.

280   J. Krasiński

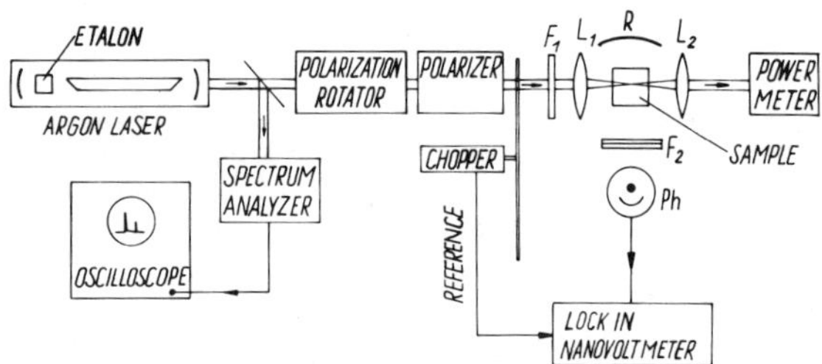

FIGURE 2. Experimental setup.

The fluorescence was detected by means of a photomultiplier and
the signal was measured by means of a lock in nanovoltmeter. In
our first experiment, we measured the two-photon absorption rate
in the case of single and multimode operation of the laser [7].
Figure 3 shows the intensity of the two photon excited fluorescence I_F versus laser power I_L. The slopes of these straight
lines (on log-log scale) sepend on the dye and take on the values
between 1.87 and 1.94 for tested POPOP, NPO and PPO. A small
deviation from the expected value of two may be attributed to
single photon absorption from higher vibrational levels in the
ground level. This process should be less probable for dyes with
higher energy of excited states. This in fact has been observed.
The best result 1.94, was measured in the case of PPO dye. As it
is seen, the fluorescence intensity I_F excited by the free running
laser was larger than that excited by the single frequency laser
by a factor of 1.86 in the case of PPO absorber. The small difference between this experimental factor and the predicted theoretically (equal 2) can be attributed to the contribution of
single photon process to the measured signal and to some discrepancies between theoretically assumed ideal coherent and thermal beams and their experimental realizations. If the laser generates two longitudinal modes which we can obtain inserting
special Fabry Perot etalon inside the laser cavity, we can easily
find the second order correlation function of such radiation as
a function of amplitudes of these modes. Denoting by $K^2 = I_1/I_2$,
where $I_{1,2}$ are intensities of the two modes and keeping output
power constant, i.e. $I_1 + I_2 = \text{const}$, we can find [8] the probability of two-photon absorption $I_F \sim 1 + 2K^2/(K^2 + 1)^2$.
    Figure 4 shows the experimental results. The continuous
curve was calculated from the theoretical formula. As it is seen,
there is good agreement between experiment and theory. The theory
has been recently extended for the case of N independent modes
[9]. However the experimental verification remains to be done.

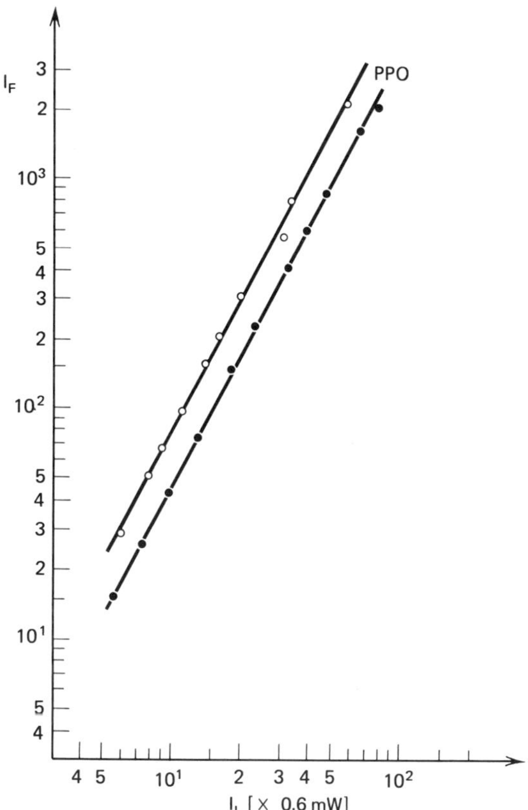

FIGURE 3.   Intensity of two-photon excited fluorescence versus
laser power.   "O"- multimode and "●" single mode
operation.

Up to now all the experiments concerning the influence of
the photon statistics on two-photon absorption have been perform-
ed for a broad atomic level, i.e. $\gamma \gg \Delta\omega$; γ being the decay rate
of excited atomic level, $\Delta\omega$ the bandwidth of the field.   In this
limit, the two-photon absorption rate is just proportional to
$G^{(2)}(0,0,0,0) = g^{(2)}(o)<I>^2$.   In general [3], when the width of
atomic level is comparable with the bandwidth of the field, two-
photon absorption efficiency may be written:

$$W^{(2)} = C \int_{-\infty}^{\infty} dt\ G^{(2)}(-t,-t,t,t)e^{2i\omega_f t-\gamma|t|}$$

where C=constant characteristic for a given atom (sum over the intermediate states of the atom), ω_f-frequency difference between atomic levels of interest.

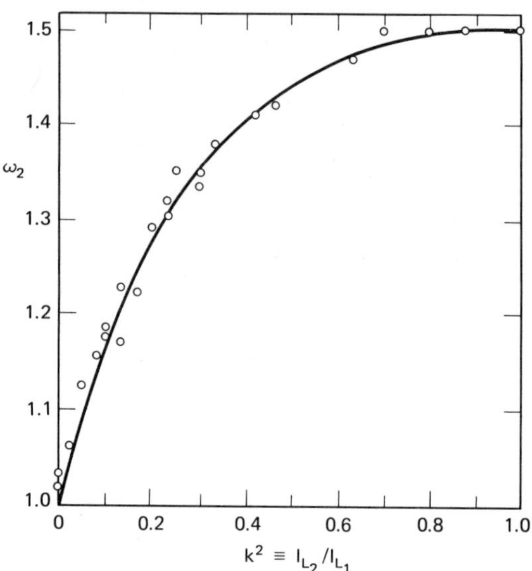

FIGURE 4.   Intensity of two-photon excited fluorescence versus intensity ratio of the two modes. The continuous curve was calculated from the theoretical formula.

It is interesting to compare the ratio between the rates from chaotic source and laser as a function of two photon detuning. The change of $W^{(2)}_{chaotic}/W^{(2)}_{laser}$ with relative detuning for different $\Delta\omega/\gamma$ is given in Fig. 5 [10].

There are also interesting effects in the region predicted by the theory [11] when two laser modes with phase diffusion are superimposed.

All these results are still not confirmed by the experiment.

The change of the statistical properties of a light beam passing through a two-photon absorber has been predicted by Weber [12]. For a beam propagating through a two-photon absorber the two-photon absorption law is

$$dI/dz = -\beta <I(z)>^2 \, g^{(2)}(\tau = 0, z)$$

Here β is the nonlinear absorption coefficient. For a fluctuating light beam we can observe smoothing of the fluctuations as a result of two-photon absorption. This is due, of course, to the

fact that the high intensity peaks are attenuated more than the
rest. As a consequence of such process, the second order correlation function of the light decreases.

In all of the experiments that I have discussed, the attenuation of the laser beam intensity due to double photon absorption
was negligibly small. If we want to investigate the change of
the statistical properties of the laser beam, we have to obtain
very strong two-photon absorption. In Weber's paper, we find
that for an attenuation of the laser beam intensity to 80% of its
initial value by means of a two-photon absorber, the second order
correlation function decreases by about 15%. In an experiment
one should obtain even higher absorption because such measurements are not precise enough.

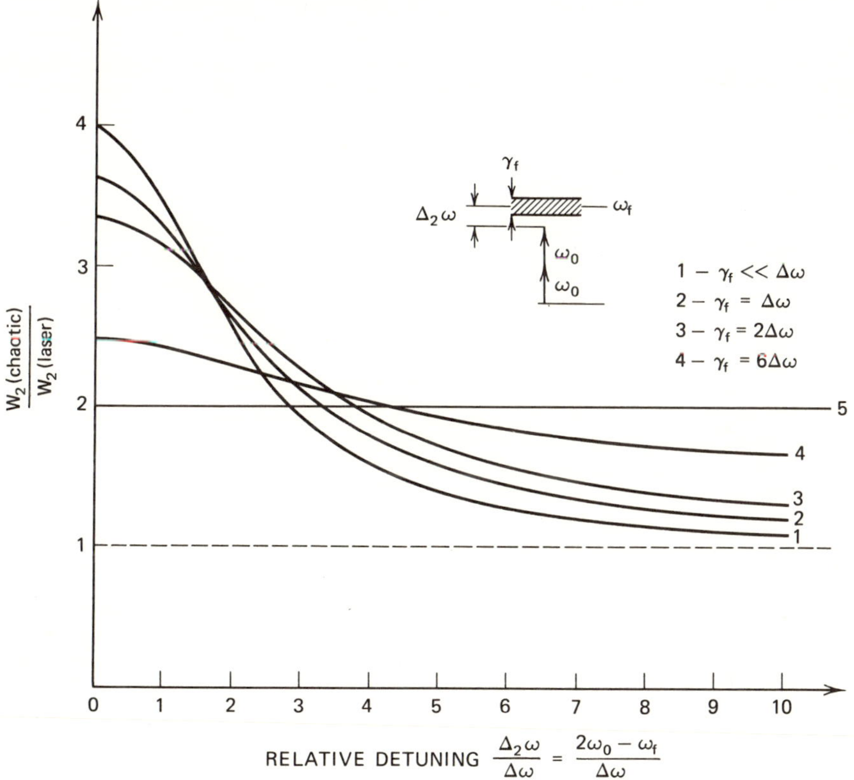

FIGURE 5. Change of $W^{(2)}_{\text{chaotic}}/W^{(2)}_{\text{coherent}}$ versus relative detuning for different $\Delta\omega/\gamma$.

It is not so easy to obtain such a high two-photon absorption rate. The power of the beam ΔP absorbed in two-photon process could be described as $\Delta P \approx P^2 \delta czs$, where P is the power of the beam (photon s^{-1}), c – absorber molecules concentration (molecule cm^{-3}), z – absorption length (cm), s – cross section of the beam (cm^2), δ – two-photon cross section (cm^4 s $photon^{-1}$ $molecule^{-1}$). The value of δ is approximately 10^{-50} for typical organic molecule.

The maximum concentration can reach 10^{22} molecule cm^{-3}. For $\Delta P/P = 0.2$ and absorption length $z = 1$ cm, we can find the power density $P/S = 10^{27}$ photons $s^{-1}cm^{-2}$. This number corresponds to 400 MW/cm^2 in the green part of the spectrum. In order to obtain higher absorption, the power density should be higher.

Such high power density can be obtained only by means of pulsed lasers. In the case of ruby or neodymium laser we have problems connected with nontunability, low repetition rate, irreproducibility of the power as well as duration of the pulse and irreproducibility of the mode structure. That is why we used a high repetition dye laser as a light source. The results of measurements of such statistical changes are reported in [13]. We used a second order correlation function $g^{(2)}(o) = <I^2>/<I>^2$ as a measure of statistical properties of the light beam. Figure 6 shows the experimental setup. A dye laser pumped by a 2 MW N_2 laser was used as a light source. P1 and P2 polarizers were used for laser beam attenuation. The beam was twice focused inside the C1 cell filled with the two-photon absorber. The wavelength of the laser beam was about 490 nm. This corresponds to the peak of the two-photon absorption spectra of α chloronaphtalene which we used as the two-photon absorber. In this double pass configuration the transmission of the cell measured by the use of full power of the laser was about 10% while small signal transmission reached almost 100%.

The central part of the transmitted beam recollimated by means of L2 lens and attenuated by means of P3 and P4 polarizers was focused again inside the other two-photon absorption cell C2, consisting of saturated solution of PBD scintillator. The intensity of this beam <I> was measured by means of PD1 photodiode and a sampling oscilloscope. The time resolution of this system is about 1 ns.

A fluorescence signal from the cell was a measure of $<I^2>$ of the light beam.

In order to improve the signal to noise ratio we also used double pass configuration here.

The fluorescence from the cell was detected by means of 1P28 photomultiplier in 5 dynods scheme and a sampling oscilloscope.

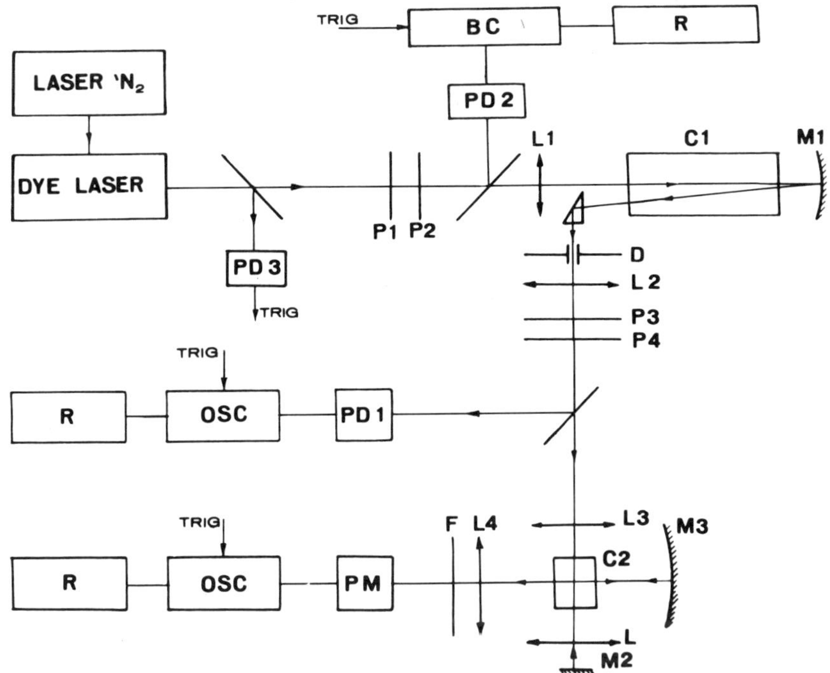

FIGURE 6.   Experimental setup for measuring changes of statistical properties of a light beam due to double photon
absorption.

The time resolution of this channel was about 2 ns and it
was sufficient for the observation of the 8 ns dye laser pulse.

Figure 7 shows the results of the measurements. As is seen,
the intensity correlation function decreases when the power of
the laser beam in C1 cell increases. This is in agreement with
the theoretical predictions.

However, we have recorded a weaker than theoretically predicted dependence of $g^{(2)}(o)$ on the nonlinear losses in the absorber. Even when more than 90% of the laser power were absorbed
in C1 cell the decrease of $g^{(2)}(o)$ was only ∿20%. From the
theory the change should approach 50%.

This difference could be explained by taking into account
other possible nonlinear processes which can take place at such
high power densities used in our measurements.

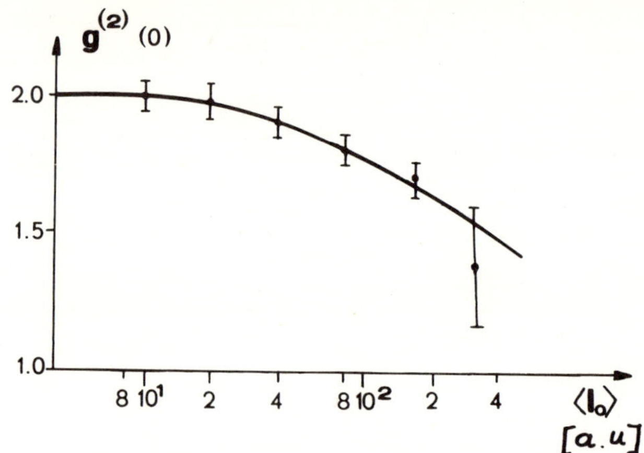

FIGURE 7.   The change of the normalized second order correlation function $g^{(2)}(o)$ versus laser power.

The maximum power density at the focus of the beam was $\sim$59 W/cm^2. This number is sufficient for the observation of strong two-photon absorption as well as other nonlinear processes like stimulated scattering or long lived absorption due to suspended submicron particles. The last process is slow as compared to the duration of the fluctuation of our broad line laser and might be responsible only for the high absorption but not for the changes of the statistical properties of the light beam.

REFERENCES

1.   P. Lambropoulos, C. Kikuchi, R. K. Osborn, Phys. Rev. <u>144</u>, 1081 (1966).

2.   M. C. Teich, G. J. Wolga, Phys. Rev. Lett. <u>16</u>, 625 (1966).

3.   B. R. Mollow, Phys. Rev. <u>175</u>, 1555 (1968).

4.   M. Głódź, J. Krasiński, Lett. al Nuovo Cimento, <u>6</u>, 566 (1974).

5.   E. H. A. Granneman, M. J. van der Wiel, Rev. Sci. Instrum. <u>46</u>, 332 (1975).

6.   P. Główczewski, Cz. Radzewicz, J. Krasiński, Proc. IX EGAS Conference, Cracow, July 12-15, 1977 (to be published).

7.   J. Krasiński, S. Chudzyński, W. Majewski, M. Głódź, Opt. Commun. <u>12</u>, 304 (1974).

8. J. Krasiński, B. Karczewski, W. Majewski, M. Głódź, Opt. Commun. $\underline{15}$, 409 (1975).

9. B. Karczewski, J. Krasiński, W. Majewski, Proc. EKON-76 Conference Poznań, 1976.

10. J. Chrostowski, private communication.

11. J. Chrostowski, T. Warenycia, Proc. Conference ILA-3 Dresden 1977 Sov. J. Q. Electronics (to be published).

12. H. P. Weber, IEEE JQE, $\underline{QE7}$, 189 (1971).

13. J. Krasiński, S. Dinev, Opt. Commun. $\underline{18}$, 424 (1976).

Resonant Two-Photon Processes
in the Fields of Modulated Waves

K. N. DRABOVICH, YU. G. GRIN, YU. N. KARAMZIN,
T. P. SHLEGEL, AND A. P. SUKHORUKOV
Moscow State University
Moscow, USSR

I. INTRODUCTION

Coherent resonant phenomena are of great interest in non-linear optics and laser spectroscopy. These include photon echo, optical nutation, self-induced transparency (SIT) etc. The coherent processes first began to be investigated at the one-photon resonance [1-6]. In recent years, the two-photon resonant phenomena have been the subjects of active research [7-17]. We note that the two-photon SIT was predicted in 1969 [7] and was then observed in the laser experiments [8,9].

In this paper, we develop the transient diffraction theory of optical frequency-mixing and coherent self-focusing (SF) in conjunction with two-photon SIT using a two-level model.

Transient coherent self-focusing in two-photon resonance was first studied in numerical experiments when SIT took place. Before, this kind of SF had been studied at one-photon resonance [16,17]. The computer simulation of short pulse propagation in a resonant absorber is based on the numerical solutions of Maxwell-Bloch equations. We investigate the following aspects of coherent SF: pulse breakup and peak amplification, pulse delays, and energy-density variation as functions of propagation distance. The critical energy of transient coherent SF is calculated in a simple manner.

It should be noted that the coherent SF has also been predicted due to phase-matched non-resonant interaction of three (or two) beams in a medium with second-order electronic non-linearity.

The other nonlinear effect investigated in our paper is the optical frequency conversion in a resonant isotropic medium. The resonant frequency-mixing is most important for high power short laser pulses [10-15], as in this case the saturation effect is negligible unlike the case of long pulses [15]. This process is closely connected with such coherent phenomena as optical nutation and SIT. In particular, we will consider the formation of third harmonic short pulses under SIT.

II.   FUNDAMENTAL EQUATIONS

We consider the coherent processes in an isotropic resonant absorber. In general, the input laser pulse excites the third harmonic, and the two waves then interact with the absorbing medium.

We assume the input pulse duration to be much shorter than the relaxation times T_1 and T_2. Thus, the transient propagation of the laser beam in a two-photon resonant absorbing medium can be described by Maxwell-Bloch equations. For slowly modulated waves, these equations simplify to the coupled truncated equations:

$$\frac{\partial A_1}{\partial Z} - \frac{i}{2K_1}\, \Delta_\perp A_1 = i\gamma_1\left[d_1(n-n_o)A_1/2 + 2qA_1^*\sigma_{21} + r\sigma_{12}A_3 e^{i\Delta Kz}\right], \quad (1)$$

$$\frac{\partial A_3}{\partial Z} - \frac{i}{2K_3}\, \Delta_\perp A_3 = i\gamma_3\left[d_3(n-n_o)A_3/2 + r\sigma_{21}A_1 e^{-i\Delta Kz}\right], \quad (2)$$

$$\partial\sigma_{21}/\partial\eta + i\left(d_1|A_1|^2 + d_3|A_3|^2\right)\sigma_{21} = i\left(qA_1^2 + A_3 A_1^* e^{i\Delta Kz}\right)n, \quad (3)$$

$$\partial n/\partial\eta = -4\,\mathrm{Im}\left[\left(qA_1^{*2} + rA_1 A_3^* e^{-i\Delta Kz}\right)\sigma_{21}\right], \quad (4)$$

where $A_{1,3}$ are complex amplitudes; σ_{21} is nondiagonal element of density matrix; n is the population inversion, z is coordinate along propagation; $\Delta_\perp = \partial^2/\partial x^2 + \partial^2/\partial y^2$; x,y are transverse coordinates; $\eta = t - z/u$, t is time, u is group velocity; $\Delta K = K_3 - 3K_1$; $d|A|^2$ is Stark line-shift; $\gamma_j = 2\pi N\omega_j^2\hbar/(K_j C^2)$; N is the molecular number density. The composite matrix elements equal

$$q = \frac{1}{\hbar^2}\sum_p \frac{d_{1p}d_{p2}}{\omega_{p1}-\omega_1}\,, \quad r = \frac{1}{\hbar^2}\sum_p d_{1p}d_{p2}\left(\frac{1}{\omega_{p1}+\omega_1} + \frac{1}{\omega_{p1}-3\omega_1}\right), \quad p \neq 1,2.$$

At the entrance, it has been assumed that the third harmonic is zero and the phase of the laser pulse is constant.

III. TWO-PHOTON SIT AND THIRD HARMONIC GENERATION

Let us consider the resonant interaction of the input pulse with phase-matched third harmonic, neglecting diffraction effects $(\Delta_\perp A_j = 0)$ and assuming the Stark line shift is small $(d_1 = d_2 = 0)$. Then, one can reduce Eqs. (1-4) to

$$\frac{\partial A_1}{\partial z} = -\gamma(2qA_1 - rA_3)\sin\psi, \tag{5}$$

$$\frac{\partial A_3}{\partial z} = -3\gamma\, rA_1\,\sin\psi, \tag{6}$$

where $A_j(\eta,z)$ is real amplitude, $\gamma = \gamma_1 n_o/2$,

$$\psi(\eta,z) = 2\int_{-\infty}^{\eta}\left[qA_1^2(\xi,z) + rA_1(\xi,z)A_3(\xi,z)\right]d\xi. \tag{7}$$

From Eqs. (5-6) we obtain the generalized area theorem [19]

$$\frac{d(W_1 + W_2)}{dz} = -2\gamma[1 - \cos0(z)], \tag{8}$$

where $W_j = \int_{-\infty}^{\infty} A_j^2\, d\eta$ is the pulse energy-density, $\theta(z)=\psi(\infty,z)$. The total energy of input and third harmonic pulses remains in the resonant absorbing medium at $\theta = 2\pi m$.

If third harmonic is not excited $(A_3 = 0)$, the input pulse evolution is described by a known formula [19]

$$A_1^2(\eta,z) = \frac{A_1^2(\eta,0)}{1 + 2\kappa z[\sin\psi(\eta,0) + \kappa z(1 - \cos\psi(\eta,0)]} \tag{9}$$

where
$$\kappa = \gamma_1\, n_o|q| \; . \tag{10}$$

By formula (9), the two-photon SIT threshold is

$$\theta(0) = 2q\int_{-\infty}^{\infty} A_1^2(\eta,0)d\eta \geq 2\pi \tag{11}$$

The input pulse with $\theta > 2\pi$ is broken into 2π-pulses at its tail. Their peak intensities grow proportionally to $(\kappa z)^2$ and their durations are shortened accordingly.

The influence of the third harmonic on two-photon SIT is determined by the parameter $\beta = r/q$. It should be first noted that the SIT threshold increases (compare with (11)):

$$\theta(0) \geqslant 2\pi \left\{ \left(2 - \sqrt{1-\beta^2} \right) \left| 1 - \left(\frac{q_2(q+q_1)}{q_1(q+q_2)} \right) \frac{q}{\sqrt{q^2-3r^2}} \right| \right\}^{-1} \tag{12}$$

where $q_{1,2} = q \pm \sqrt{q^2-3r^2}$.

If Raman connection is weak, $|\beta| \ll 1$, the SIT threshold is slightly more than (5), namely $\theta(0) \geqslant 2\pi(1+9\beta^2/4)$. However, the threshold goes to infinity at $|\beta| = 1/\sqrt{3}$. It means the 2π-pulse regime is impossible. In this case, the coupled stable 0π-pulses are formed on the long distance, $\kappa z \gg 1$. Here, the two-photon resonant excitation of the two-level system is suppressed by Raman interaction of the input pulse with the third harmonic. Their amplitudes are then connected as

$$q A_1 + r A_3 = 0 \tag{13}$$

The envelopes of 0π-pulses repeat the input pulse shape. The amplitudes of 2π-pulses are connected as

$$A_3 = A_1 (1 - \sqrt{1 - 3\beta^2})/\beta. \tag{14}$$

The peak intensities increase with distance $\sim [\gamma |q| (1 + \sqrt{1-3\beta^2})z]^2$ and the durations decrease accordingly. The input pulse with $\theta > 2\pi$, propagating through a resonant medium, forms 2π-pulses (14) at its back front and 0π-pulses (13) at the forward front.

The above consideration is illustrated with Fig. 1, where the evolution of an input pulse with $\theta = 2,2\pi$ is shown for three Raman connections: a) $|\beta| = 2$ – strong, b) $|\beta| = 0.5$ – middle, and c) $|\beta| = 0.06$ – weak.

In the case of weak Raman connection, $|A_3| \ll |A_1|$ at any distance. Then, one can derive the formula for the third harmonic taking into account the Stark line-shift:

$$|A_3|^2 = \frac{9}{4} \beta^2 \left\{ (A_1 - A_{10})^2 + \frac{d_1^2}{d_1^2 + 4q^2} \left[(A_1 - A_{10}) \operatorname{ctg} \frac{\psi_{st}}{2} + 2\kappa z A_1 \right]^2 \right\} \tag{15}$$

where $\psi_{st} = \psi(\eta,0)\sqrt{1 + d_1^2/(4q^2)}$; $A_{10} = A_1(\eta,0)$. The main pulse evolution is described by formula (9) with $\psi \to \psi_{st}$ and $\kappa \to \kappa_{st} = \kappa/\sqrt{1 + d_1^2/4q^2}$ instead of (10). It can be seen that the Stark shift essentially affects the third harmonic envelope.

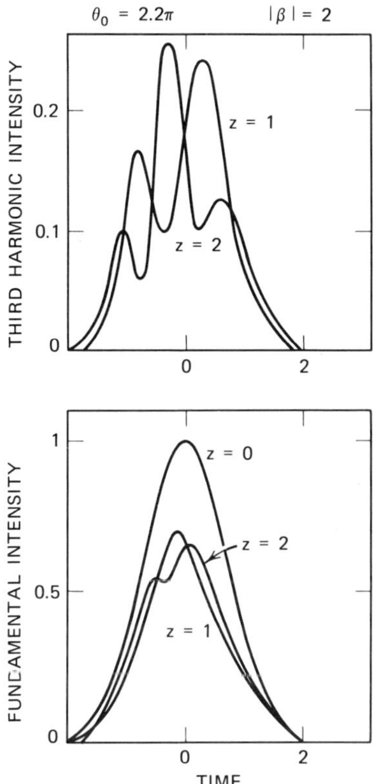

(a) $\beta = 2$ - strong Raman connection

FIGURE 1. Evolution of the input Gaussian and the third
harmonic pulses in the two-photon absorbing medium.
The propagation distance z is normalized to the nonlinear coherent length $L_{NL} = \kappa^{-1}$, $\theta(0) = 2.2\pi$.

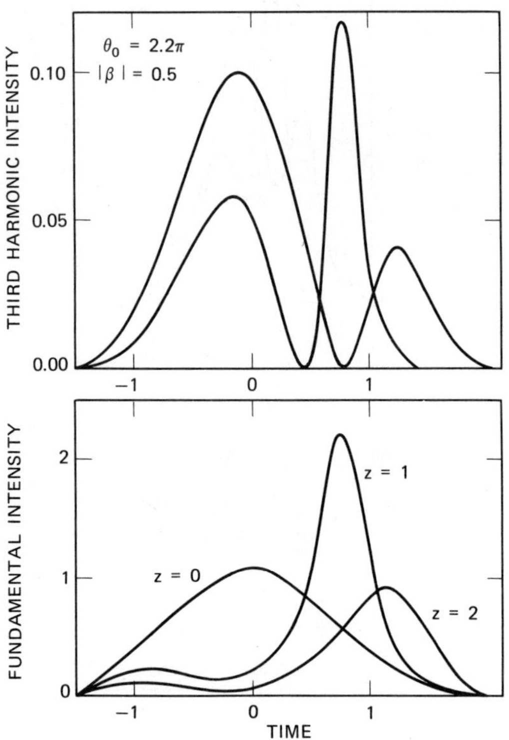

FIGURE 1.   (b) $\beta = 0.5$ - middle one

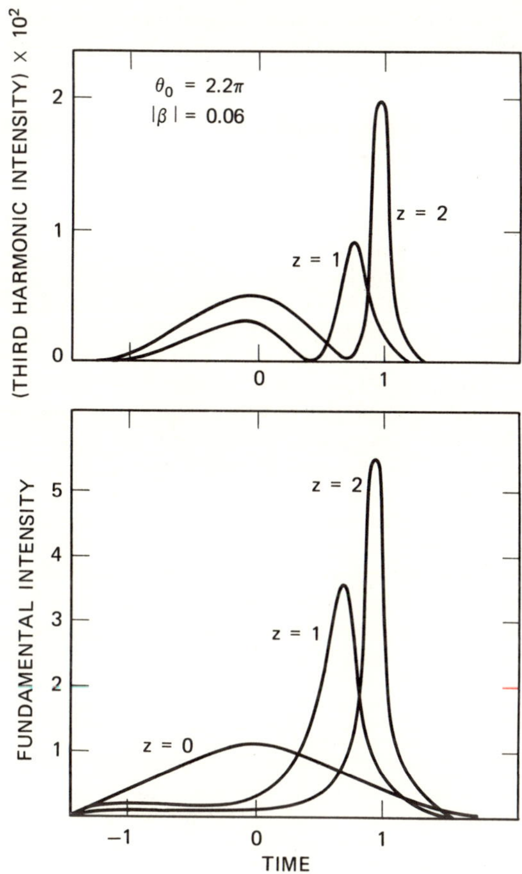

FIGURE 1.   (c) $\beta = 0.06$ - weak one.

The resonant interaction of three (or four) waves is more complicated. We only note, that ultra-short pulses can be formed, for example, on the frequencies ω_1, ω_2 and $\omega_3 = 2\omega_1 \pm \omega_2$.

IV. COHERENT TRANSIENT SELF-FOCUSING IN TWO-PHOTON RESONANCE

Now let us consider the coherent propagation of the input pulse taking into account the diffraction of the beam, $\Delta_\perp A_j \neq 0$. We assume that the third harmonic is not excited as a consequence of phase-mismatch, $\Delta K \gg \kappa$, i.e. $A_3 \equiv 0$, and put $d_1 = 0$. Under these conditions the coupled equations (1,3,4) have been solved with the help of the computer. The results of the numerical experiments are shown in Fig. 2 for a Gaussian beam $A = A \exp(-t^2/\tau^2 - \rho^2/a^2)$ with $\kappa z = 2$, $\theta < 2\pi$ and $\theta > 2\pi$. One can see that the diffraction of the beam has great influence on two-photon SIT. Moreover, the computer simulations predict the coherent transient self-focusing (SF) on two-photon resonance. The analogous phenomenon on single-photon resonance has recently been predicted [16] and has been investigated in laser experiments [17]. The coherent SF on single- or two-photon resonance is due to spatial distortion of the primary plane wave front of the travelling beam. The coherent transient SF arises when the confocal parameter exceeds the nonlinear coherent length by more than one order of magnitude:

$$D = \frac{1}{\kappa B} \lesssim 0.1; \quad Ka^2 \gtrsim 10/(\gamma_1 n_o |q|) \,. \tag{16}$$

This is the first condition of coherent SF. In addition, the normalized energy-density (the pulse area) has to be more or nearly 2π (2π-pulse). This gives the second condition of SF:

$$\theta(0) > 2\pi; \qquad W_1(0) > \pi/q \tag{17}$$

Excluding the beam radius from (16) and (17), one can estimate the critical total energy of coherent transient SF on two-photon resonance:

$$E_{cr} \simeq \frac{c\lambda^2}{1.6(2\pi)^3 N\hbar n_o q^2} \tag{18}$$

where $E_{cr} = ca^2 W_{cr}/16$. Typical values for E_{cr} are $10^{-4} - 10^{-3}$ joule.

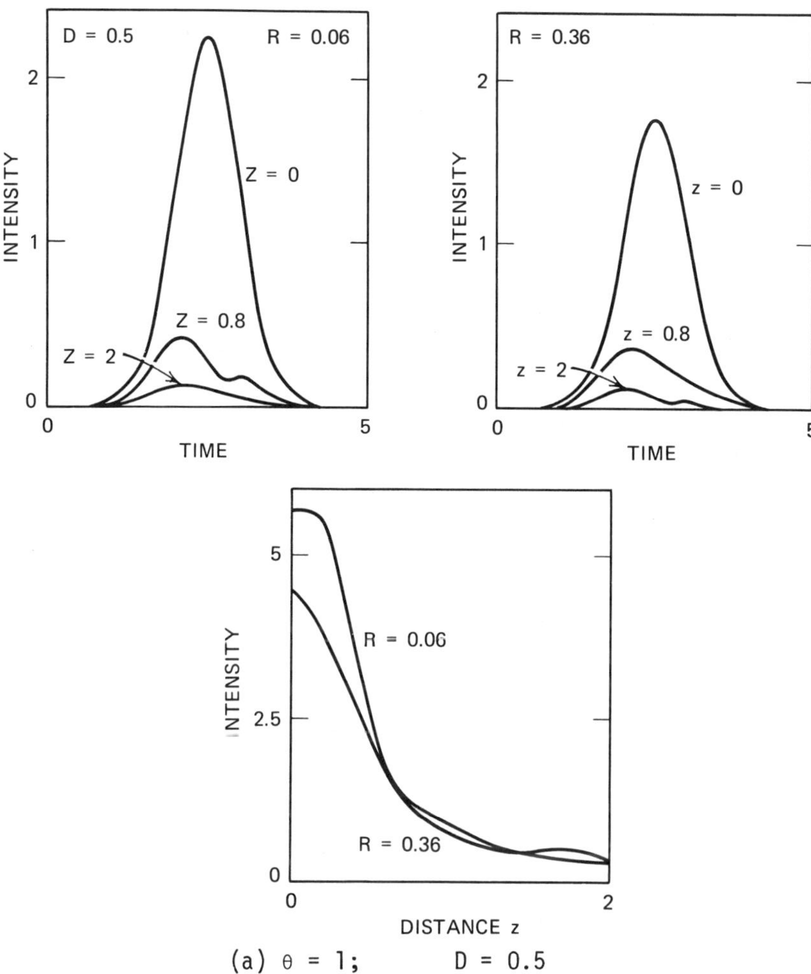

(a) θ = 1;   D = 0.5

FIGURE 2.   The results of computer simulations of the transient
            propagation of a focused Gaussian beam in a two-
            photon resonant absorber. The figures give the
            evolution of an input pulse at ρ = 0.06a (near beam
            axis) and ρ = 0.3a (half beam radius), pulse area,
            time delay at peak as functions of normalized dis-
            tance.

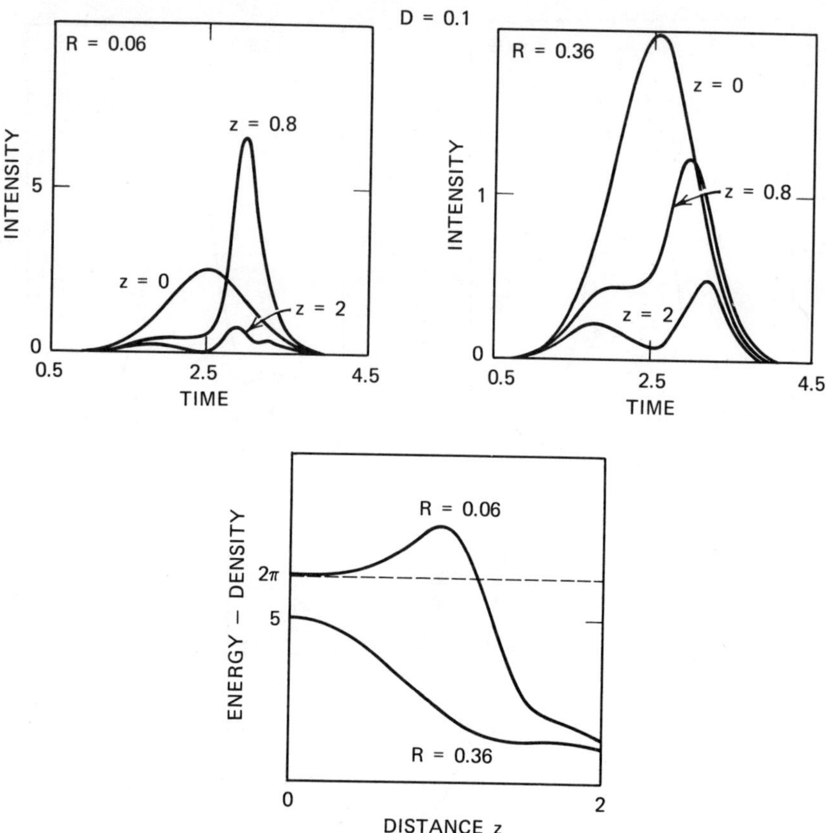

FIGURE 2.   (b) $\theta = 2\pi$; $D = 0.1$

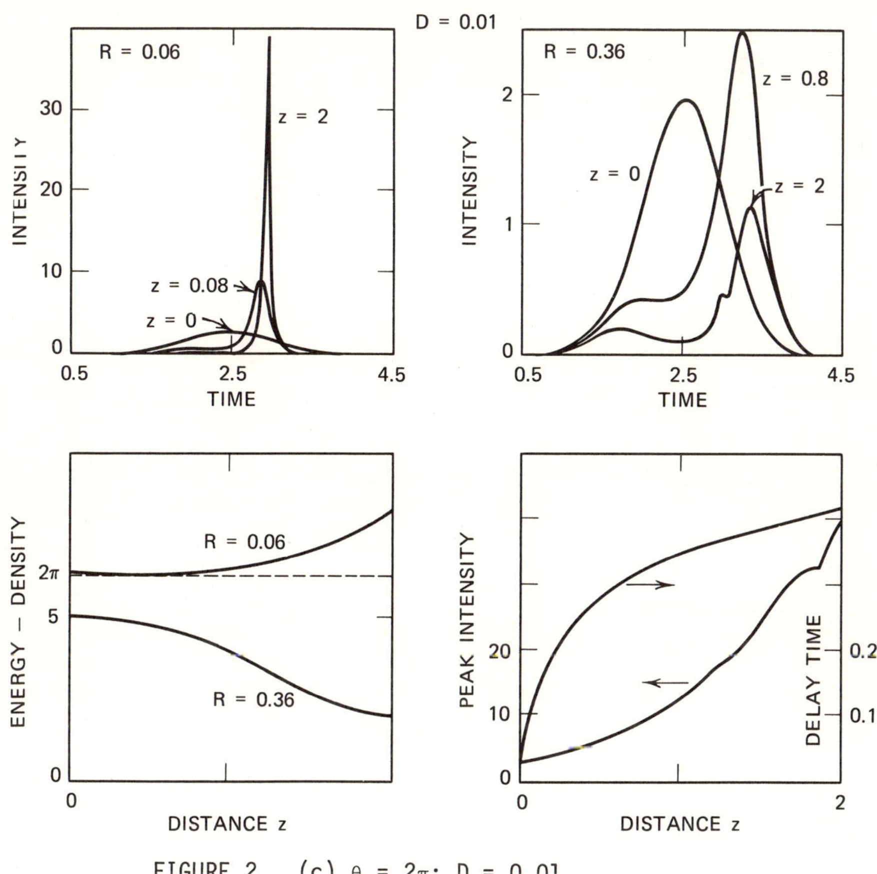

FIGURE 2.   (c) $\theta = 2\pi$; D = 0.01

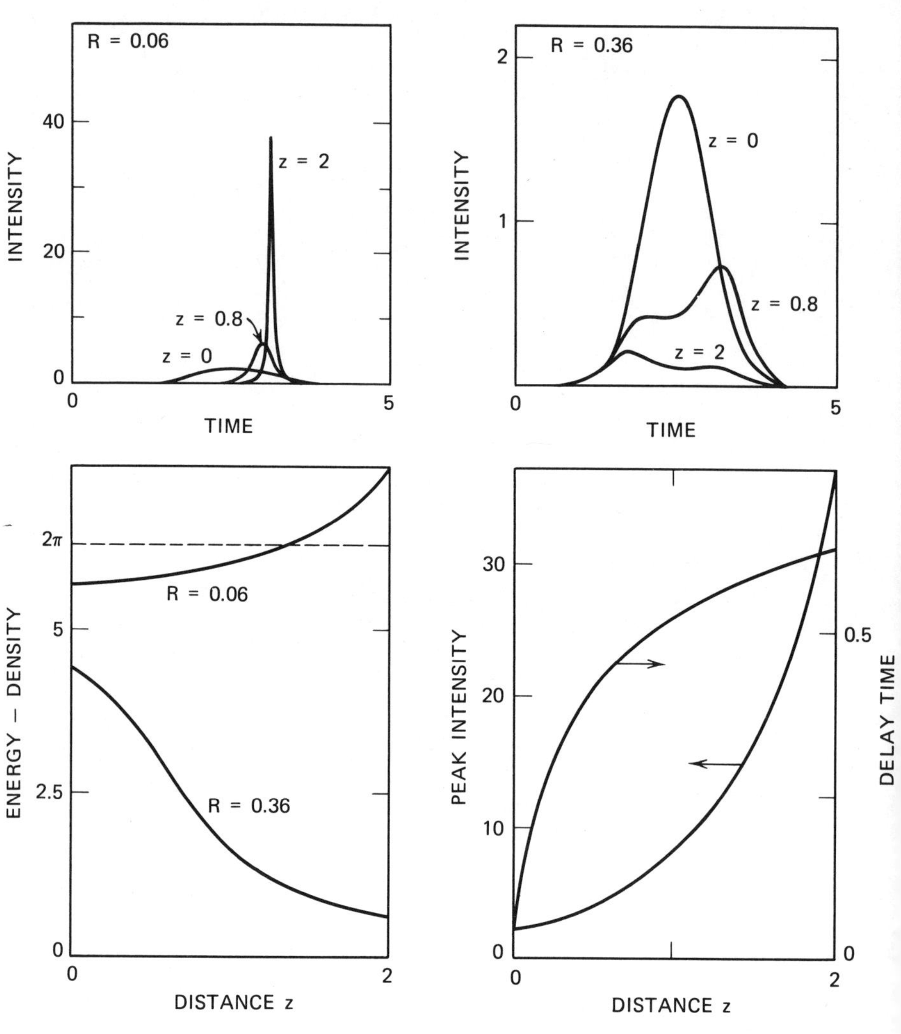

FIGURE 2.   (d) $\theta = 1.8\pi$; $D = 0.001$

The dependence of the coherent pulse propagation on the factor D is illustrated in Fig. 2. In the case of Fig. 2a,d, where $\theta < 2\pi$, the plane-wave theory predicts decreasing of the pulse area along the proportion distance. However, for $D = 0.001$ (Fig. 2d), we observe increasing of the pulse area on the axis due to coherent transient self-focusing on two-photon resonance. The same pulse is defocused at $D = 0.5$ due to diffraction (Fig. 2a). The 2π-pulse weakly self-focused at $D = 0.1$ (Fig. 2b, see (16)) and is essentially self-focused at $D = 0.01$ (Fig. 2c). Among the focused beams considered, the higher self-focusing is developed at $D = 0.001$ (see Fig. 2c).

It should be noted that the coherent self-focusing of the interacting beams have also been predicted in a medium with nonresonant electronic quadratic nonlinearity [18]. For example, such mutual SF can be observed in the phase-matched second harmonic generation. The interactions of diffracting beams of the first and the second harmonics are described in the quasioptical approximation by the parabolic equations

$$\frac{\partial A_1}{\partial z} - \frac{i}{2K_1} \Delta_\perp A_1 = i\gamma_q A_1^* A_2 \tag{19}$$

$$\frac{\partial A_2}{\partial z} - \frac{i}{2K_2} \Delta_\perp A_2 = i\gamma_q A_1^2 \tag{20}$$

where $\gamma_q = 2\pi\omega_1\chi/cn$, χ is the quadratic susceptibility. Here, the coherent nonlinear interaction of beams changes also the phase in the beam cross-section. These changes are such that coherent mutual self-focusing both of the fundamental radiation and of the harmonic takes place for a self-focusing factor $D_q = 1/\gamma_q A_{10} B = 10^{-2} - 10^{-4}$. This kind of SF is shown in Fig. 3. We note that according to plane-wave theory [20], the intensities of both waves cannot exceed the one of the incident wave $|A_{10}|^2$ (see dash-dot line in Fig. 3).

302   A. P. Sukhorukov et al.

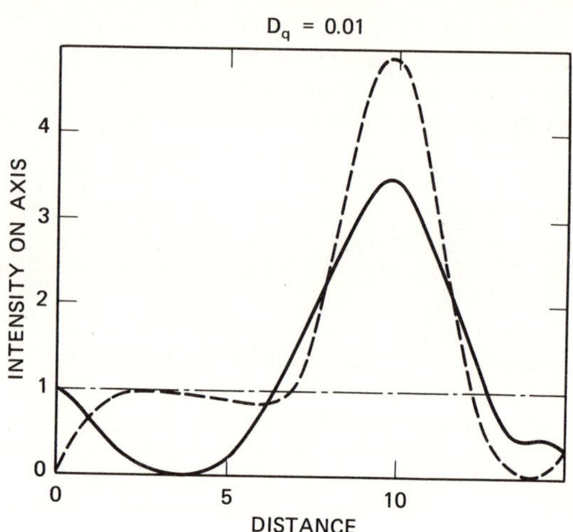

FIGURE 3.   The coherent quasistationary non-resonant self-
focusing of the first (solid) and second (dashed)
harmonics in a medium with quadratic nonlinearity.
The incidence beam is Gaussian. Factor $D_q = 10^{-2}$.

REFERENCES

1.   S. L. McCall, E. L. Hahn, Phys. Rev. Lett. $\underline{18}$, 908 (1967);
     Phys. Rev. $\underline{183}$, 457 (1969).

2.   C. K. N. Patel, Phys. Rev. A$\underline{1}$, 979 (1970).

3.   I. M. Asher, M. O. Scully, Opt. Comm. $\underline{3}$, 395 (1971).

4.   R. E. Slusher, H. H. Gibbs, Phys. Rev. A$\underline{5}$, 1634 (1972);
     A$\underline{6}$, 2326 (1972).

5.   E. Courtens, Laser Handbook, Ed. F. Arrechi (North-Holland
     Publishing Co. 1972) Vol. 2, p.1259.

6.   I. A. Poluektov, J. M. Popov, V. S. Roijtberg, Usp. Fiz.
     Nauk $\underline{114}$, 97 (1974).

7.   E. M. Belenov, I. A. Poluektov, Zh. Eksp. Teor. Fiz. $\underline{56}$,
     1407 (1969).

8.   T. L. Gvardjaladze, A. Z. Grasjuk, I. K. Zubarev, P. K.
     Krjukov, O. B. Shatberashvili, ZhETF Pis. Red. $\underline{13}$, 159 (1971).

9.   N. Tan-no, K. Yokoto, H. Ihaba, Phys. Rev. Lett. 29, 1211
     (1972).

10. A. M. Afanasjev, E. A. Manikin, Zh. Eksp. Teor. Fiz. 48,
     931 (1965).

11. A. H. Kung, Appl. Phys. Lett. 25, 653 (1974).

12. K. N. Drabovich, D. Metchkov, S. Mitev, L. Pavlov, K.
     Stamenov, Opt. Comm. 20, N 3 (1977).

13. V. I. Anikin, V. D. Gora, K. N. Drabovich, A. N. Dubovik,
     Kvantovaja Elecktronoka 2, 330 (1976).

14. V. I. Anikin, K. N. Drabovich, A. N. Dubovik, Zh. Eksp.
     Teor. Fis. 72, N 5 (1977).

15. R. B. Miles, S. E. Harris, IEEE Journal of Quantum Electr.
     QE-9, 470 (1973).

16. N. Wright, M. C. Newstein, Opt. Comm. 9, 8 (1973).

17. H. M. Gibbs, B. Bölger, F. P. Mattar, M. C. Newstein,
     G. Forster, P. E. Toschek, Phys. Rev. Lett. 37, 1743 (1976).

18. Yu. N. Karamzin, A. P. Sukhorukov, ZhEIF Pis. Red. 20, 734
     (1974), JETP Lett. 20, 339 (1974); Zh. Eksp. Teor. Fiz. 68,
     834 (1975), Sov. Phys. -JETP 41, 414 (1975).

19. I. A. Poluektov, Yu. M. Popov, V. S. Rojitberg, ZhETP Pis.
     Red. 18, 638 (1973); 20, 533 (1974).

20. N. Bloembergen, Nonlinear Optics (W. A. Benjamin, New York
     1965).

Part 5

MOLECULAR MULTIPHOTON PROCESSES

On the Influence of Molecular Structure
Upon the Collisionless
Laser Photodissociation of SF_6

C. D. CANTRELL, H. W. GALBRAITH AND J. ACKERHALT
Theoretical Division
University of California, Los Alamos Scientific Laboratory
Los Alamos, New Mexico

I. INTRODUCTION

Attempts at understanding the phenomena of collisionless multiple-photon excitation (CMPE) and collisionless multiple-photon dissociation (CMPD) of SF_6 and other molecules [1-5] should, we believe, be based on physical principles rather than phenomenological models.** Such a first-principles theory must be founded on qualitatively physically correct theories for both the structure (i.e., energy levels, transition moments and selection rules) of the molecular system, and the dynamics of the interaction of the molecules with laser light.

Among the many theoretical or phenomenological models which have been suggested as explanations for CMPE and CMPD [6-22], most have considered only the rather simple structure of a non-rotating, one-dimensional anharmonic oscillator. Such one-dimensional models are strictly applicable only to heteronuclear diatomic molecules, in which CMPD has not been observed experimentally [20,21]. The excited vibrational states of highly symmetric polyatomic molecules such as SF_6 are not simply *shifted* by the anharmonic part of the vibrational potential energy; they are also *split* into as many levels as are allowed by the symmetry

---

* Research supported by USERDA.

** We eschew the use of the term "superexcitation" to describe the process by which CMPE and CMPD occur, as suggested in [12]. Superexcitation is currently understood to denote the excitation of a molecule to levels above the ionization threshold by ionizing radiation [30].

of the molecule [14,23,24]. Anharmonic splitting (AS) has been known theoretically for a long time [25,26]; its existence, and importance for spectroscopy, have been confirmed by detailed assignments of vibration-rotation spectra in CH_4 and CD_4 [27-29]. Only one published dynamical calculation of CMPD [24] includes the effects of AS. In that case, it was shown that the dependence of the number of SF_6 molecules excited to $v_3 = 4$ or 5 upon the frequency of the laser is qualitatively different with AS than with a simple anharmonic shift such as occurs in diatomic molecules.

In this talk we shall concentrate primarily on the influence of molecular structure on CMPE, and shall discuss in some detail how AS arises and in what ways it affects CMPE in the well-studied SF_6 molecule. Our discussion of the dynamics of excitation of SF_6 would not be complete, however, without an indication of the influence and importance of vibrational modes other than the triply-degenerate v_3 mode. Only two of the six vibrational modes of SF_6 (v_3 and v_4) interact with infrared radiation, and of these only v_3 is strongly coupled to CO_2 laser light. However, the anharmonic portion of the vibrational potential energy is expected to lead to the interaction between the excited states of the v_3 mode alone, which we shall denote $[vv_3]$, and a multitude of other vibrational states containing quanta of more than one mode. This interaction will broaden and shift the excited states $[vv_3]$, in a manner which can be described by an equation of motion for the density matrix of the v_3 mode alone [16,17], as we shall describe briefly below. It is commonly believed [1,8,11,13,16,17,22] that the rapid increase with excitation of the total density of vibrational states in a molecule with many vibrational modes provides a quasi-continuum in which a resonant, dipole-allowed absorption of a laser photon is always possible. The quasi-continuum, if it exists as supposed, would allow the resonant absorption of many CO_2 laser photons [4], following an initial absorption of $\lesssim 5$ laser photons by low-lying vibrational states which are not significantly broadened by interaction with other vibrational states.

II. ANHARMONIC SPLITTING OF $[vv_3]$ IN SF_6

The Cartesian displacements of the S and F nuclei in the v_3 and v_4 vibrational modes of SF_6 are shown in Fig. 1 [31]. Each of these modes is triply degenerate; the displacements shown occur in one of the three degenerate v_3 or v_4 normal modes. If we expand the vibrational potential energy V in a power series in the v_3 normal coordinates $q_i^{(3)}$ (i=1,2,3), we find that the central symmetry of the octahedral SF_6 molecule and the fact that the coordinates $q_i^{(3)}$ change sign under inversion imply that no terms of odd degree in the $q_i^{(3)}$ can occur in V. The anharmonic terms must, also, occur in groupings which are invariant under the operations of the octahedral point group [32]. There are three such terms which are quartic in the $q_i^{(3)}$:

$$\left[r^{(3)}\right]^4 = \left[\left(q_1^{(3)}\right)^2 + \left(q_2^{(3)}\right)^2 + \left(q_3^{(3)}\right)^2\right]^2 \tag{1}$$

which is spherically symmetric (i.e., invariant under all rotations);

$$0_1^{(3)} = \left(q_1^{(3)}\right)^4 + \left(q_2^{(3)}\right)^4 + \left(q_3^{(3)}\right)^4 \tag{2}$$

$$0_2^{(3)} = \left(q_1^{(3)}\right)^2\left(q_2^{(3)}\right)^2 + \left(q_1^{(3)}\right)^2\left(q_3^{(3)}\right)^2 + \left(q_2^{(3)}\right)^2\left(q_3^{(3)}\right)^2 \tag{3}$$

which are invariant only under the operations of the octahedral group. The Hamiltonian $H^{(3)}$ for the ν_3 vibrational mode, expanded to fourth order in the ν_3 normal coordinates, is allowed by symmetry to contain only the above potential-energy terms in addition to the harmonic potential energy. It is a convenient and reasonably accurate approximation to construct a Hamiltonian which is equal to $H^{(3)}$ on the manifold states [$\nu\nu_3$], but which does not couple states with different values of v. This means that we first expand (1)-(3) in terms of the annihilation and creation operators

$$a_i = \frac{1}{\sqrt{2}}\left(q_i^{(3)} + ip_i^{(3)}\right) \tag{4}$$

$$a_i^+ = \frac{1}{\sqrt{2}}\left(q_i^{(3)} - ip_i^{(3)}\right) \tag{5}$$

for the three degenerate ν_3 modes. Then we shall discard terms such as $(a_1^+)^3 a_1$, which change the value of the sum

$$v = n_1 + n_2 + n_3 \tag{6}$$

of the number of quanta (n_i) in the three ν_3 modes, in the Cartesian basis

310   C. D. Cantrell et al.

$$|n_1 n_2 n_3\rangle = \frac{(a_1^+)^{n_1}(a_2^+)^{n_2}(a_3^+)^{n_3}}{\sqrt{n_1! n_2! n_3!}}\,|0\rangle \qquad (7)$$

(where $|0\rangle$ is the vibrational ground state of the SF_6 molecule).
The correction to the energy eigenvalue $E(n_1,n_2,n_3)$ due to the
term $(a_1^+)^3 a_1$ is approximately $(1/16)n_1^2(n_1+1)(n_1+2)$ (coefficient
of $0_1^{(3)})^2/(E(n_1,n_2,n_3)-E(n_1+2,n_2,n_3))$. We shall see below that
the coefficient of $0_1^{(3)}$ is approximately -3 cm^{-1}; hence for $n_1=4$,
the correction is approximately 0.1 cm^{-1}, appreciably less than
the anticipated uncertainty in our calculations of the energies
$E(n_1,n_2,n_3)$ for $v=4$ or greater. We shall refer to the neglect of
operators which change v as *truncation* of the Hamiltonian. (This
procedure is rather well known to spectroscopists, see [25] and
[26].)

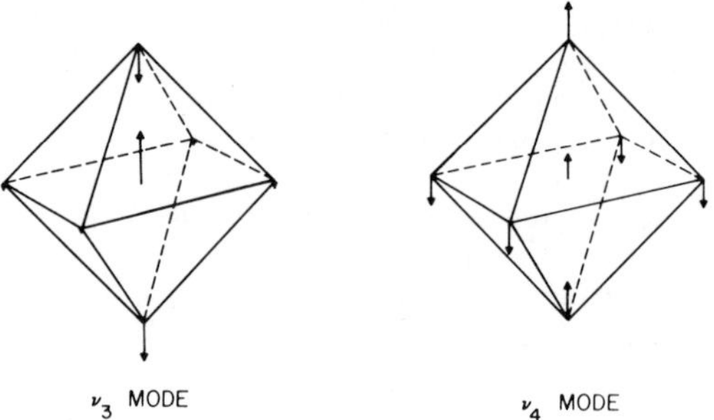

FIGURE 1. Cartesian displacement coordinates for the infrared-
active fundamental vibrations of SF_6.

Before we express $(r^{(3)})^4$, $0_1^{(3)}$ and $0_2^{(3)}$ in terms of creation and annihilation operators, it is convenient to subtract the
average of $0_1^{(3)}$ and $0_2^{(3)}$ over all orientations , leading to a re-
definition of $0_1^{(3)}$ and $0_2^{(3)}$

$$0_1^{(3)} = \sum_{i=1}^{3}\left[q_i^{(3)}\right]^4 - \frac{3}{5}\left[r^{(3)}\right]^4 \qquad (8)$$

$$O_2^{(3)} = \sum_{i<j} \left[q_i^{(3)} \right]^2 \left(q_j \right)^2 - \frac{1}{5} \left[r^{(3)} \right]^4 . \tag{9}$$

We also define the operator for the number of quanta in the ith ν_3 mode,

$$n_i = (a_i^+) \, a_i \, . \tag{10}$$

Straightforward algebra using (4), (5) and (8–10) shows that the truncated anharmonic contributions to the potential energy are

$$\left\{ r^{(3)4} \right\}_t = \frac{1}{4} \left[6 \sum_1^3 (n_i)^2 + 14 \sum_i^3 n_i + 2 \sum_{i<j} \left\{ a_i^2 a_j^{+2} + a_i^{+2} a_j^2 \right\} \right.$$
$$\left. + 8 \sum_{i<j} n_i n_j + 15 \right] \tag{11}$$

$$\left\{ O_1^{(3)} \right\}_t = \frac{3}{10} \left[2 \sum_1^3 (n_i)^2 - 2 \sum_1^3 n_i - 4 \sum_{i<j} n_i n_j \right.$$
$$\left. - \sum_{i<j} \left\{ a_i^2 \left(a_j^+ \right)^2 + \left(a_i^+ \right)^2 a_j^2 \right\} \right] \tag{12}$$

$$\left\{ O_2^{(3)} \right\}_t = - \frac{1}{2} \left\{ O_1^{(3)} \right\}_t \, . \tag{13}$$

Since the harmonic-oscillator portion of the ν_3 vibrational Hamiltonian $H^{(3)}$ is

$$H_{har}^{(3)} = \omega_3 \sum_1^3 (n_i + \frac{1}{2}) \tag{14}$$

we see from (11–14) that to represent the expansion of the truncated Hamiltonian $\{H^{(3)}\}_t$ to fourth order in the ν_3 normal coordinates, we need only the number operators n_i and the operator $\sum_{i<j} \{a_i^2(a_j^+)^2 + (a_i^+)^2 a_j^2\}$. The latter operator arises most naturally from the square of the vibrational angular momentum operator [33], whose Cartesian components are given by

$$\vec{\ell}^{(3)} = - i \, \vec{a}^+ \times \vec{a} \, . \tag{15}$$

312   C. D. Cantrell et al.

In (15), $\vec{a}$ is a vector with components (a_1, a_2, a_3). Simple algebra leads to the relation

$$\left(\vec{\ell}^{(3)}\right)^2 = 2\sum_1^3 n_i + 2\sum_{i<j} n_i n_j - \sum_{i<j}\left\{a_i^2\left(a_j^+\right)^2 + \left(a_i^+\right)^2 a_j^2\right\} . \tag{16}$$

From (11–14) and (16) it is obvious that one may express $\{H^{(3)}\}_t$ in terms of the number operators n_i and the square of the vibrational angular momentum $(\vec{\ell}^{(3)})^2$ [34], multiplying each operator by a parameter which in principle can be calculated from the vibrational force field [35], but in practice must be evaluated experimentally:

$$\begin{aligned}
\left\{H^{(3)}\right\}_t &= (\omega_3 + 3X_{33} - 8T_{33})N \\
&\quad + (X_{33} - 6T_{33})N^2 + 10T_{33}M \\
&\quad + (G_{33} + 2T_{33})\left[\vec{\ell}^{(3)}\right]^2 + \text{constant}
\end{aligned} \tag{17}$$

where we have defined [34]

$$N = \sum_1^3 n_i \tag{18}$$

$$M = \sum_1^3 (n_i)^2 . \tag{19}$$

The somewhat arbitrary expressions for the parameters multiplying the operators in (17) were introduced in order to conform with the notation established by Hecht [26], who expressed $\{H^{(3)}\}_t$ in the form

$$\left\{H^{(3)}\right\}_t = H^{(3)}_{har} + X_{33}\left[H^{(3)}_{har}/\omega_3\right]^2 + G_{33}\left(\vec{\ell}^{(3)}\right)^2 + T_{33}0_{33} \tag{20}$$

where

$$0_{33} = \frac{20}{3}\left\{0_1^{(3)}\right\}_t . \tag{21}$$

The physical interpretation of our representation (17) of the Hamiltonian $\{H^{(3)}\}_t$ is as follows [34]: The first term (proportional to N) gives the harmonic-oscillator energy levels, corrected for anharmonic changes to the oscillator frequency. The second term (proportional to N^2) gives a level shift which depends only on the eigenvalue v of N [see (6)]; this is the analog of the anharmonic shift familiar from diatomic molecules. The third and fourth terms in (17) split the shifted level into as many levels as the octahedral symmetry of SF$_6$ allows.

From (17) it is clear that if the coefficient $(G_{33}+2T_{33})$ of $\left(\vec{\ell}^{(3)}\right)^2$ is negligible, the eigenvalues of $H^{(3)}$ in the manifold $[vv_3]$ may be written down by inspection [34]. The eigenvalue of N is simply v; the eigenvalues of M are all the possible values of $n_1^2+n_2^2+n_3^2$, given that $n_1+n_2+n_3=v$. The levels for which n_1,n_2 and n_3 are all different will be six-fold degenerate, and those for which two of n_1,n_2,n_3 are equal will be three-fold degenerate. Adding the $\left(\vec{\ell}^{(3)}\right)^2$ term prevents one from obtaining a solution quite so trivially, although analytic solutions have been given by Hecht through $v=4$ [26]. In general, an analytic solution (obtained by solving a polynomial secular equation) will be possible only when the maximum degree of the secular equation for a given v, which is equal to the maximum number of levels in $[vv_3]$ with the same octahedral symmetry type, is no greater than 4. Thus, the energy eigenvalues for $v=3$ and 4 involve only a quadratic secular equation; the energy levels for $v=5$ and 6 involve a quartic secular equation; and those for $v=7$ or greater must be obtained numerically [4,24]. The eigenvalues of (17) or (19) for $v=1$ to $v=5$ are shown in Fig. 2 [24] for the parameter values $X_{33} = -2.8\text{cm}^{-1}$, $G_{33}= 1.34\text{cm}^{-1}$, $T_{33} = -0.44\text{cm}^{-1}$, which were obtained from a model of the SF$_6$ force field [35]. The symmetry types A_1, A_2, E, F_1, F_2 in Fig. 2 are the labels of the irreducible representations of the octahedral point group 0, and were assigned by determining the transformation properties of the numerically calculated eigenvectors under the elements of 0. We also show in Fig. 2 that positions of the levels obtained by neglecting the $\left(\vec{\ell}^{(3)}\right)^2$ term in (17), and (by arrows) the levels predicted by the simple shift

$$E(v) = vv_3 + X_{33}v(v-1) \ . \tag{22}$$

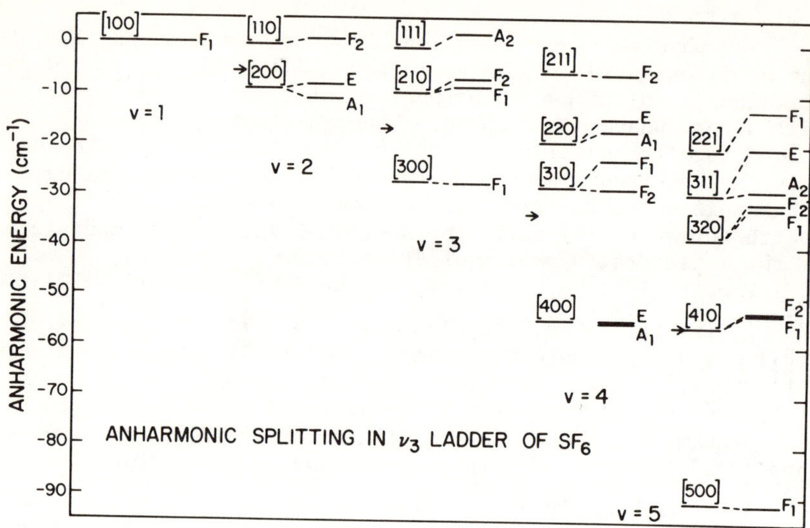

FIGURE 2.   Anharmonic splitting in ν_3 ladder of SF_6.

Summary. It is obvious from Fig. 2 that the AS is as large in order of magnitude as the anharmonic shift and scales similarly with v. In [14], [23], and [24] it was argued that this may happen not only for SF_6, but for a large number of other spherical-top molecules. As a result, we believe it is relatively easy to understand qualitatively how a small number (~4) of CO_2 laser photons can be absorbed resonantly by SF_6 and other molecules. In addition to the vibrational states discussed in this section, it is also necessary to consider the contribution of rotational energy levels, which undoubtedly play a role in further compensating whatever detunings from resonance remain after AS is taken into account [8,18,33], and thus facilitate the rapid absorption of laser photons at low laser intensities.

Although the AS of some molecules has been measured experimentally, that of SF_6 has not been. The observation [36] of a single peak in the "forbidden" transition from v=0 to v=3 in gaseous SF_6 (where the two final states of F_1 symmetry should give rise to two peaks) could be interpreted as evidence that the AS of SF_6 is actually much smaller than is shown in Fig. 2. However, a recent calculation [37] of the relative strengths of the two peaks to be expected in the v=0 to v=3 transition has shown that one peak (corresponding to the state designated [210] in Fig. 2) should be about 10^3 times weaker than the other. Thus, only one peak should ordinarily be observed, in agreement with

the results of [36]. The spectroscopic parameters of [14] or
[35] do not correctly predict the experimentally observed frequency [36] of the v=0 to v=3 ([300]) transition assigned in [37],
however. If the AS parameters T_{33} and G_{33} are left fixed at the
values found in [35], but X_{33} is adjusted to give the observed
position [36] of v=3 ([300]), then the resulting value of X_{33} is
$-0.50 \mathrm{cm}^{-1}$. This value of X_{33} conflicts with that quoted in [38],
but should be regarded as speculative since the parameters T_{33}
and G_{33} calculated in [35] cannot, at the moment, be checked
against other data. The value $X_{33}= -0.50 \mathrm{cm}^{-1}$ (rather than $X_{33}=$
$-2.8 \mathrm{cm}^{-1}$ [35]) predicts energy levels of [vν_3] which are shifted
upwards by $v(v-1)\Delta X_{33}$ with respect to those shown in Fig. 2, and
overlap the harmonic-oscillator energy levels at least up to
v=10. If the smaller value for X_{33} turns out to be more nearly
correct than the one reported in [35], it would appear that AS
alone can account for nearly resonant laser-driven transitions
from [(v-1)ν_3] to [vν_3] up to v=9 at least, without the necessity
for invoking a vibrational quasi-continuum [36a].

   The final aspects of molecular structure we shall mention
which have significance for CMPE are the transition dipole moments
and selection rules for transitions from [(v-1)ν_3] to [vν_3]. To
first approximation (setting $G_{33}+2T_{33}\cong 0$ in (17)) the transition
dipole moments may be calculated between the elements of the
Cartesian basis (7). When this is done [34,39], it is found that
some transition moments (which are not required by symmetry to be
zero) vanish because of a selection rule on the allowed changes
of the eigenvalues of M (Eq. (19)). The exact matrix of dipole
transition moments may, of course, be calculated numerically by
a similarity transformation after the matrix of $H^{(3)}$ has been
diagonalized. The latter method was used in the dynamical calculations of CMPE described in the next section [24].

III. DYNAMICS OF COLLISIONLESS, LASER-DRIVEN VIBRATIONAL EXCITATION

   The exact numerical and analytical study of laser-driven excitation of closed (i.e., undamped) multilevel systems has recently been the subject of intensive study, with some elegant
analytical results having been obtained for transitions in non-
degenerate systems [18,40]. Unfortunately, the problem of interest to us, the excitation of the triply-degenerate ν_3 mode in
SF$_6$, does not appear to lend itself well to analytical solutions.
We shall give a brief description of the solution of the
Schrödinger equation for the levels shown in Fig. 2, and the results for the fraction of the total population of SF$_6$ molecules

316   C. D. Cantrell et al.

found in the v=4 and v=5 states as a function of the laser frequency. We have chosen v=4 and v=5 because there is spectroscopic
evidence [36] that the [vv_3] states remain discrete up to v=3
(i.e., the broadening due to anharmonic interactions with other
levels is much less than the apparent rotational width at v=3
[36]), and because v=4 and v=5 are believed [13] to be the levels
at which the quasicontinuum begins. Other authors [18] have put
the beginning of the quasicontinuum at a lower value of v.

A system which is coherently driven and lacks significant
damping may be described by a wave function. The Schrödinger
equation for the unitary time-development operator $U(t,t_0)$ is

$$i\hbar \frac{\partial}{\partial t} U(t,t_0) = \left[H^{(3F)} + H^{(3)} + H^{(F)} \right] U(t,t_0) \qquad (23)$$

where $H^{(3F)}$ is the interaction Hamiltonian between a (quantized)
laser field and the v_3 mode, using the transition dipole moments
calculated in [34,39]; and $H^{(F)}$ is the Hamiltonian of a monochromatic laser field with frequency ω. We shall find it convenient
in this section to denote the eigenstates of $H^{(3)}$ as $|v\alpha\rangle$, where
v is given by (6) and α denotes the remaining indices required to
specify the states shown in Fig. 2:

$$H^{(3)} |v\alpha\rangle = E(v\alpha) |v\alpha\rangle \; . \qquad (24)$$

The states of the system composed of (laser field + molecule)
will be denoted $|n;v\alpha\rangle$, so that

$$\left[H^{(3)} + H^{(F)} \right] |n;v\alpha\rangle = \left(E(v\alpha) + n\hbar\omega \right) |n;v\alpha\rangle \qquad (25)$$

Here, and elsewhere in this section, we draw on the work of
Mukamel and Jortner [11]. In the rotating-wave approximation,
only the states $|n-v;v\alpha\rangle$ are coupled together. For large n, the
off-diagonal matrix elements of $H^{(3F)}$ are well approximated by
the molecular transition matrix elements for a classical laser
field. The diagonal matrix elements of

$$H = H^{(3F)} + H^{(3)} + H^{(F)} \qquad (26)$$

are

$$\langle n-v;v\alpha|H|n-v;v\alpha\rangle = n\hbar\omega + \left(E(v\alpha) - v\hbar\omega \right) . \qquad (27)$$

The solution of the Schrödinger equation for such a system
with a laser pulse which starts at time t_0 and has constant intensity thereafter may be expeditiously accomplished by a *single*

diagonalization of the matrix H [11]. The matrix elements of
$U(t,t_0)$ in the basis $|n-v;v\alpha\rangle$ are

$$\langle n-v;v\alpha|U(t,t_0)|n-v';v'\alpha'\rangle$$

$$= \sum_\lambda c(v\alpha;\lambda)\ c(v'\alpha';\lambda)^* \exp\left[-\frac{i}{\hbar} E_\lambda (t-t_0)\right] \tag{28}$$

where $|\lambda\rangle$ is an eigenvector of H,

$$H|\lambda\rangle = E_\lambda|\lambda\rangle \tag{29}$$

and

$$|\lambda\rangle = \sum_{v\alpha} c(v\alpha;\lambda)|n-v;v\alpha\rangle\ . \tag{30}$$

One of the advantages of this approach is that the optical Stark
shifts are automatically included in the levels E_λ [11]. Equation (28) may be used to find the probability for the (molecule
+ field) to be in the state $|n-v;v\alpha\rangle$ at time t, given that it was
in the state $|n-v';v'\alpha'\rangle$ at time t_0:

$$p(v\alpha,t|v'\alpha',t_0) = |\langle n-v;v\alpha|U(t,t_0)|n-v';v'\alpha'\rangle|^2\ . \tag{31}$$

The probability averaged over many cycles of the lowest frequency,
$\min(E_\lambda/\hbar)$, is

$$\overline{p(v\alpha|v'\alpha')} = \sum_\lambda |c(v\alpha;\lambda)|^2\ |c(v'\alpha';\lambda)|^2\ . \tag{32}$$

<u>Results</u>. Figure 3 shows the dependence of the fraction of the
population in v-4 and v=5,

$$p(\omega) = \sum_\alpha p(4,\alpha|0) + \sum_{\alpha'} p(5,\alpha'|0) \tag{33}$$

upon the laser frequency ω, calculated by diagonalizing an
84 × 84 matrix H using the parameters $X_{33}=-2.8\mathrm{cm}^{-1}$, $G_{33}=1.34\mathrm{cm}^{-1}$,
$T_{33}=-0.44\mathrm{cm}^{-1}$ [35] and a laser intensity of 3 GW/cm^2. The laser
field was assumed to be polarized along the (111) axis of a non-
rotating SF$_6$ molecule. Other choices of polarization direction
produce slightly different curves of $p(\omega)$, but the differences
are unimportant for the rather qualitative use we shall make of
$p(\omega)$. Also shown by a dotted line in Fig. 3 is the curve of $p(\omega)$

318   C. D. Cantrell et al.

obtained for a three-dimensional oscillator with the energy levels
(22), representing a simple anharmonic shift of a type found in
less symmetrical molecules. Evidently, there is a major difference between the p(ω) found in these two different models: the
presence of AS causes p(ω) to peak around 944 cm^{-1}, while for an
anharmonic oscillator with a simple shift p(ω) peaks at lower frequencies and achieves a lower average excitation. These qualitative differences do not depend on the specific AS parameter values,
but persist for a wide-range of parameters, as was suggested in
[14].

  For comparison with experiment, we turn to the two-frequency
CMPD experiments of Ambartzumian et al. [41], in which SF_6 molecules were excited by a weak, nearly resonant CO_2 laser and dissociated by a strong, non-resonant CO_2 laser. The experimental
curve of the dissociation rate per molecule, per laser pulse,
sampled at the discrete set of available CO_2 laser frequencies,
shows a peak near ω = 944cm^{-1}, in qualitative agreement with Fig.
3. Owing to the uncertainties inherent in the interpretation of
this experiment, and the deficiencies of this calculation, a
quantitative comparison between theory and experiment in this
case is probably unwarranted. The comparison does appear clearly
to support the importance of AS as one of the major determinants
of the dependence of CMPE upon laser frequency.

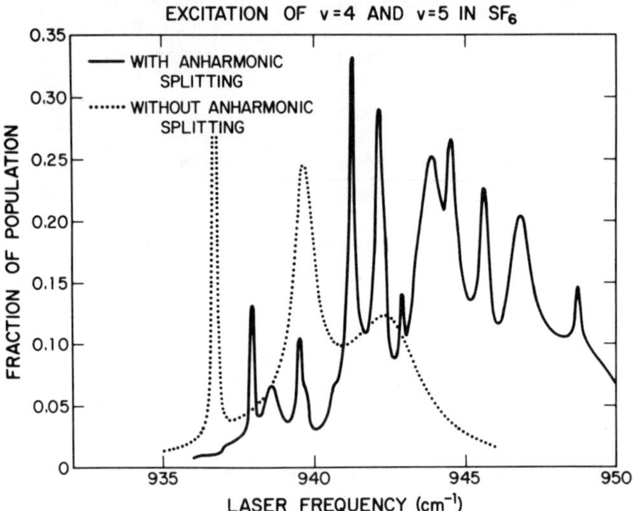

FIGURE 3. Laser frequency (cm^{-1}).

The   difference between the excitation achieved with and without AS can be vividly seen by means of a color computer-generated movie showing the dependence upon time t of the population in each of the states $|n-v;v\alpha\rangle$ for v=0 to 5,

$$p(v,t) = \sum_\alpha p(v\alpha,t|0,t_0) \ . \tag{34}$$

A sample frame from the movie is shown in Fig. 4.

Two major ingredients which affect the magnitude of $p(\omega)$ are lacking in the calculation we have described: molecular rotation, and the interaction of the states $[v\nu_3]$ with other vibrational states. Numerical calculations have shown that for both a non-degenerate anharmonic oscillator [19] and the triply-degenerate ν_3 mode of SF$_6$ [42] the major effect of molecular rotation is to increase the order of magnitude of $p(\omega)$. The increase can amount to several orders of magnitude at low laser intensity. When AS is included, adding molecular rotation does not qualitatively change the overall shape of $p(\omega)$ as a function of ω [42].

The influence of the interaction of $[v\nu_3]$ with other vibrational levels will be considered briefly in the next section.

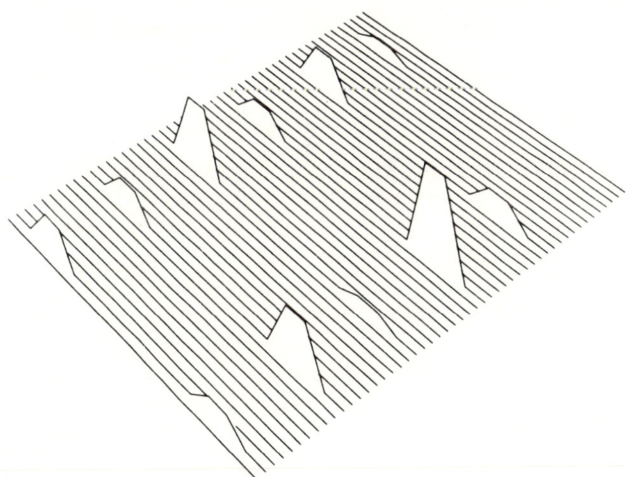

FIGURE 4.   Right (left) column represents excitation of molecule without (with) AS. Peak height corresponds to population.   Each successive peak moving toward the right corresponds to a successively higher energy level: v = 0-5 are shown.

IV. INTERACTION BETWEEN $[v\nu_3]$ AND OTHER VIBRATIONAL STATES: SHIFTING AND BROADENING

The vibrational normal modes of a molecule provide a basis in terms of which it is possible to expand an arbitrary displacement of the nuclei in the molecule, apart from rotations and translations. The fact that only the ν_3 and ν_4 modes of SF_6 are infrared-active depends only upon the octahedral symmetry of the molecular electronic-ground-state potential energy surface. The frequencies of the ν_3 and ν_4 modes are sufficiently different (948 and 615cm^{-1}, respectively) that only ν_3 interacts strongly with CO_2 laser light. Therefore it is convenient to reduce the mammoth dimensions of the problem by focusing attention upon the ν_3 mode, as we have done in the previous sections, while allowing for the fact that anharmonic terms in the Hamiltonian couple the ν_3 mode to all the other modes at large vibrational amplitudes. This coupling results in a leakage of energy to the other modes, and is the origin of the vibrational quasicontinuum [1,8,11,13, 16,17,22]. For the description of a phenomenon in which coherent excitation, complex energy-level structure, and interaction with a large set of unobserved internal states must all be taken into account, it is natural to use the reduced density matrix [43,44]. Accordingly, we introduce the reduced density matrix for the ν_3 modes, $\rho^{(3)}$:

$$\rho^{(3)}_{ss'} = \sum_r \rho_{sr;sr'} \tag{35}$$

where s,s' denote states $|s\rangle$ of the *system* (the ν_3 mode) and r,r' denote states $|r\rangle$ of the *reservoir* (the remaining modes), and ρ is the density matrix for the entire molecule. Likewise, the reduced density matrix for the reservoir is

$$\rho^{(R)}_{rr'} = \sum_s \rho_{sr;sr'} \; . \tag{36}$$

Since the diagonal matrix element $\rho_{sr;sr'}$ is the fraction of the SF_6 molecules in the state $|s\rangle|r\rangle$, it is clear that the "population" $\rho^{(3)}_{ss}$ in the state $|s\rangle$ (e.g., $|s\rangle = |3\alpha\rangle$) is not just the population in the state $|s\rangle|0\rangle$ (e.g., $|3\alpha\rangle$ with *no* other vibrational excitation), but is instead the sum of the populations in all the vibrational basis states $|s\rangle|r\rangle$ which include $|s\rangle$. A ladder of states with increasing ν_3 excitation begins at every state $|0\rangle|r\rangle$ (e.g., $\nu_3 = 0$, $\nu_6 = 1$) which lacks ν_3 excitation. The reduced density matrix $\rho^{(3)}$ is capable of describing completely the interaction of this enormous set of ν_3 ladders with an external field. Clearly $\rho^{(3)}$ depends upon $\rho^{(r)}$, for as the laser field pumps the ν_3 mode, which transfers energy to the reservoir, the fact that reservoir states $|r\rangle$ which initially were empty are acquiring population, can qualitatively alter $\rho^{(3)}$.

The derivation of the equation of motion for $\rho^{(3)}$ and $\rho^{(r)}$ in the Markovian approximation has been described by Lax [45]. It is a direct consequence of the derivation that the equation of motion for $\rho^{(3)}$ may be cast in a form that splits off the effect of the laser from that of the reservoir:

$$\frac{\partial \rho^{(3)}_{ss'}}{\partial t} = \left(\frac{\partial \rho^{(3)}_{ss'}}{\partial t}\right)_{laser} + \left(\frac{\partial \rho^{(3)}_{ss'}}{\partial t}\right)_{res} \quad , \tag{37}$$

where

$$\left(\frac{\partial \rho^{(3)}_{ss'}}{\partial t}\right)_{laser} = -\frac{i}{\hbar} \sum_{s''} \left\{ H^{(3F)}_{ss''}(t)\rho^{(3)}_{s''s'} - \rho^{(3)}_{ss''}H^{(3F)}_{s''s'}(t), \right\} \tag{38}$$

$$\left(\frac{\partial \rho^{(3)}_{ss'}}{\partial t}\right)_{res} = -\frac{1}{2}\left(\gamma_{ss} + \gamma_{s's'}\right)\left(1 - \delta_{ss'}\right)\rho^{(3)}_{ss'}$$

$$- \delta_{ss'} \sum_{s''} \left[\rho^{(3)}_{ss} w(s \to s'') - \rho^{(3)}_{s''s''}w(s'' \to s)\right]$$

$$- i\Delta\omega_{ss'}\, \rho^{(3)}_{ss'} \quad . \tag{39}$$

The solution of (38), where $H^{(3F)}(t)$ is in an interaction picture based on $H^{(3)}+ H^{(F)}$, is described in Section III. In (39), the three terms on the right-hand side represent, respectively, the decay of molecular coherence (represented by the off-diagonal elements of $\rho^{(3)}$); the relaxation of the populations $\rho^{(3)}_{ss}$ to an equilibrium distribution; and a second-order shift of the transition frequency from s' to s, $\omega_{ss'}$. All of these terms may be evaluated in terms of the coupling Hamiltonian $H^{(3R)}$ between the ν_3 mode and the reservoir, which has been assumed to be of the separable form [44,45]:

$$H^{(3R)} = \sum_i \left(Q_i F_i + Q_i^+ F_i^+\right) \quad . \tag{40}$$

In [40], Q_i is an operator function of the ν_3 normal coordinates; F_i is an operator function of the normal coordinates of the remaining vibrational modes. Equations of the form of (39) have

322   C. D. Cantrell et al.

been used previously by Goodman et al. [17] and by Hodgkinson and Briggs [16] to discuss CMPE.

The constants in (39) may be explicitly evaluated in terms of the operators appearing in $H^{(3R)}$ as follows [46]:

$$\gamma_{ss} = \sum_{s''} \sum_{i} \left\{ |Q_i(0)_{ss''}|^2 \gamma_i^+(\omega_{ss''}) + |Q_i(0)_{s''s}|^2 \gamma_i^-(\omega_{s''s}) \right\} \quad (41)$$

$$w(s \rightarrow s'') = \sum_{i} \left\{ |Q_i(0)_{ss''}|^2 \gamma_i^+(\omega_{ss''}) + |Q_i(0)_{s''s}|^2 \gamma_i^-(\omega_{s''s}) \right\} \quad (42)$$

$$\Delta\omega_{ss'} = \hbar^{-1} \left[\Delta H^{(3)}_{ss} - H^{(3)}_{s's'} \right] \quad (43)$$

where

$$\Delta H^{(3)}_{ss} = \hbar \sum_{s''} \sum_{i} \left\{ |Q_i(0)_{ss''}|^2 \Delta_i^+(\omega_{ss''}) - |Q_i(0)_{s''s}|^2 \Delta_i^-(\omega_{s''s}) \right\} \quad (44)$$

and

$$\frac{1}{2} \gamma_i^{+(-)}(\omega) + i\Delta_i^{+(-)}(\omega)$$

$$= \sum_{r,r'} \rho_{rr}^{(R)} |<r| (F_i^{(+)} |r'>)|^2 \int_0^\infty e^{i(\omega + (-)\omega_{rr'})\tau} \, d\tau. \quad (45)$$

In (45), the symbols in parentheses are to be used together as a group; and it has been assumed that the reservoir density matrix is diagonal in a basis of energy eigenstates.

<u>Physical discussion</u>. In principle, it is necessary to solve (37) together with a similar equation for $\rho^{(R)}$. However, we shall assume that $\rho^{(R)}$ is a thermal density matrix; Lax [45] has shown that this is the first of a hierarchy of approximations to $\rho^{(R)}$. Physically, this amounts to assuming that the energy transferred to the reservoir from the ν_3 mode is completely thermalized, the reservoir temperature being determined by energy conservation. In making this assumption, we have, of course, neglected the coherent properties of the excitation induced in the reservoir by interaction with the ν_3 mode. We expect such coherent reservoir effects to constitute some of the finer details of the process of CMPE. At the moment, we are interested in the gross features, which we expect to be well represented by a thermal reservoir density matrix $\rho^{(R)}$.

At the present time, one of the major questions in the theory of CMPE in SF_6 concerns the relative importance of excitation of

the ν_3 mode *versus* excitation of the reservoir: does the excitation process in SF$_6$ consist of exciting many, say, $\gtrsim 20$, ν_3 quanta
[*12*], or does the excitation consist of exciting a few, $\lesssim 5$, ν_3
quanta, up to a level of excitation at which leakage of energy to
the reservoir essentially balances the uptake of energy from the
laser field by the ν_3 mode [*8,10,11,13,16,17,18*]? The available
experimental evidence on the process of dissociation [*47,48*] is
consistent with a complete randomization of energy in the SF$_6$
molecule prior to dissociation. Unfortunately, these experiments
appear to us to say little about the process of excitation prior
to dissociation.

   While we shall not attempt a definitive answer to these
equations about the process of excitation, it is worth noting
that the reduced-density-matrix equation of motion (37–39), together with the relations (40–43) which define the input of
molecular structure to (37–39), includes, and yet can distinguish
between, two quite different types of reservoir-induced broadening and shifting of the energy levels [$\nu\nu_3$]: (a) The tendency of
anharmonic interactions with nearby vibrational states to create
many levels in the near vicinity of [$\nu\nu_3$], each with some non-
zero amplitude of [$\nu\nu_3$] in its wave function. (b) The tendency of
an increase in reservoir temperature to broaden any optical absorption feature, including the transition from [$(v-1)\nu_3$] to
[$\nu\nu_3$], i.e., the growth of "hot bands" as the reservoir temperature increases. In the reduced-density-matrix formalism, only
(a) will occur when the reservoir temperature is zero. As the
reservoir temperature increases, (b) will become important also.
In the remainder of this section, we shall evaluate (45) (the
reservoir spectral density [*49*]) for a general term in the interaction Hamiltonian (40) between the ν_3 mode and the reservoir.
Armed with explicit expressions for (45), we shall then comment
on the qualitative changes in the nature of the process of excitation which can develop as the reservoir temperature increases,
as evidenced by the dependence of (41)–(43) on the reservoir
temperature.

Evaluation of reservoir spectral densities, damping rates, and
frequency shifts. For the sake of obtaining analytic expressions
for the reservoir matrix elements appearing in (45) and the system matrix elements appearing in (41–42), we shall use the following normally-ordered forms for F_i and Q_i:

$$Q_{i,j_i,k_i} = \left(a_i^+\right)^{j_i} \left(a_i\right)^{k_i} \qquad (i = 1,2,3) \qquad\qquad (46)$$

$$F_{i,j_i,k_i} = \sum_{\{m\}} \sum_{\{p\}} \kappa(i,j_i,k_i; \{m\},\{p\}) \prod_n \left(b_n^+\right)^{m_n} \left(b_n\right)^{p_n} . \tag{47}$$

In (47), b_n^+ and b_n are creation and annihilation operators for the nth reservoir normal mode, and κ is a coupling constant which can, in principle, be calculated or at least estimated from a model of the SF_6 force field. We shall assume that

$$j_i > k_i , \qquad p_n < m_n , \tag{48}$$

so that Q_i and F_i in (41-42) and (45) are fully off-diagonal [45]. With the choice expressed in (48), the reservoir spectral density [49] displayed in (45) has the physical significance of a spectral density for the *emission* of reservoir phonons if the + sign is chosen in (45), or for the *absorption* of reservoir phonons if the − sign is chosen.

We shall use a Cartesian basis (7) for both the ν_3 mode (denoted $|\{n\}>$) and the reservoir (denoted $|\{v\}>$) to evaluate the matrix elements of Q_{i,j_i,k_i} , F_{i,j_i,k_i} . The matrix elements of (46) are

$$<\{n\}|\left[Q_{i,j_i,k_i}|\{n'\}>\right] \delta_{n_i',n+k-j_i} \prod_{j\neq i} \delta_{n_j',n_j}\left[\frac{(n_i+k_i-j_i)!n_i!}{\{(n_i-j_i)!\}^2}\right]^{1/2}$$
$$\tag{49}$$

and those of (47) are

$$<\{v\}|\left[F_{i,j_i,k_i}|\{v'\}>\right] = \sum_{\{m\}} \sum_{\{p\}} \kappa(i,j_i,k_i; \{m\},\{p\})$$

$$\cdot \prod_n \left\{\left[\frac{(v_n+p_n-m_n)!v_n!}{\{(v_n-m_n)!\}^2}\right]^{1/2} \delta_{v_n',v_n+p_n-m_n}\right.$$

$$\left. \cdot \prod_{n'\neq n} \delta_{v_{n'}',v_{n'}}\right\} . \tag{50}$$

The reservoir transition frequency $\omega_{rr'}$ in (45) is

$$\omega_{rr'} = - \sum_n \omega_n(p_n - m_n) \tag{51}$$

where ω_n is the harmonic frequency of the nth reservoir mode. The thermal average of the square of (50), which enters into (45), may be evaluated by a generating-function technique. The thermal average of the square of any one of the quantities in square brackets in (50) is

$$\left\langle \frac{(v_n + p_n - m_n)! \, v_n!}{\{(v_n - m_n)!\}^2} \right\rangle_R = m_n! \, p_n! \, <v_n>^{m_n} \sum_{q_n=0}^{p_n} \binom{p_n}{q_n} \binom{m_n + p_n - q_n}{m_n} <v_n>^{p_n - q_n} \tag{52}$$

where the mean number of quanta in the nth reservoir mode is

$$<v_n> = \left[e^{\hbar \omega_n / kT_R} - 1 \right]^{-1} . \tag{53}$$

At zero reservoir temperature $<v_n> = 0$, and the thermal average in (52) vanishes unless $m_n = 0$. When $m_n = 0$ and $<v_n> = 0$, the right-hand side of (52) becomes $p_n!$. If the reservoir temperature is nonzero but low in the sense that $\hbar \omega_n / kT_R \ll 1$ then (52) becomes

$$\left\langle \frac{(v_n + p_n - m_n)! \, v_n!}{\{(v_n - m_n)!\}^2} \right\rangle_R \cong m_n! \, p_n! \, <v_n>^{m_n} \left\{ 1 + (m_n + 1) p_n <v_n> \right\}. \tag{54}$$

In the opposite limit of very high reservoir temperature, only the term $q_n = 0$ will contribute to (52), so that

$$\left\langle \frac{(v_n + p_n - m_n)! \, v_n!}{\{(v_n - m_n)!\}^2} \right\rangle_R \cong (m_n + p_n)! \, <v_n>^{m_n + p_n} . \tag{55}$$

We now substitute our results for the thermal average of the matrix elements of F_i, and the matrix elements of Q_i, into (45) and (41) to obtain the decay constant $\gamma_{\{n\}}^+$ in the ν_3 state $|\{n\}>$ (Eq. (7)) for the emission of reservoir phonons:

$$\gamma_{\{n\}}^+ = 2\pi \sum_{i,j_i,k_i} \sum_{\{m\}} \sum_{\{p\}} \left| \kappa(i,j_i,k_i;\{m\},\{p\}) \right|^2 \delta\left(\omega_3 (j_i - k_i) \right.$$
$$\left. - \sum_n \omega_n (p_n - m_n) \right)$$

$$\cdot \frac{(n_i + k_i - j_i)! \, n_i!}{\{(n_i - j_i)!\}^2} \prod_n m_n! \, p_n! \, <v_n>^{m_n} \sum_{q_n=0}^{p_n} \binom{p_n}{q_n} \binom{m_n + p_n - q_n}{m_n} <v_n>^{p_n - q_n} . \tag{56}$$

<u>Physical interpretation</u>. We have derived a decay rate $\gamma_{\{n\}}^{+}$ which increases very rapidly with excitation of either the ν_3 mode or the reservoir. The quantity shown in (56) is one of the two contributions to the rate at which coherence decays, as shown in (39), and is also the sum of one set of terms in the rate constant (42) which describes the flow of population out of the level $\{n\} \equiv s$. Therefore, we may regard (56) as being proportional to the width in frequency of the level $|\{n\}>$; we have derived an excitation-dependent level width. At zero reservoir temperature, the terms in which $m_n = 0$ for all reservoir modes (n) in (56) give the contribution of the anharmonic mixing of other states (with the level $|\{n\}>$) to the width of $|\{n\}>$. The extremely rapid increase of both the number of terms which contribute significantly to (56) and the magnitude of each term as $<v_n>$ increases is due to the stimulation of the remission of reservoir phonons by the presence of reservoir excitation.

A few aspects of (56) which should not be taken literally deserve mention. First, the transition frequencies appearing in the energy-conserving delta function in (56) have not been corrected for anharmonicity, which is relatively simple to do in the case of the ν_3 mode [34] and the reservoir [45] to a reasonable level of approximation. In deriving (39) and subsequent equations, we have tacitly assumed that this has been done, so that certain terms which contribute only for harmonic oscillators have already been deleted from (39) [45]. Second, there is an equally important contribution to the ν_3 mode transition frequencies owing to the reservoir excitation itself, i.e., owing to the growth of "hot bands" as the reservoir temperature increases. This manifests itself as a dependence of the redefined Hamiltonian for the ν_3 mode [45]

$$H^{(3)} = \mathrm{tr}_R(\rho^{(R)} H_{mol}) \qquad (57)$$

upon the state of excitation of the reservoir. In (57), H_{mol} is the total molecular vibrational Hamiltonian. Finally, in (56) the energy-conserving delta function itself should not be taken too literally, since the upper limit of the time integration in (45) is really the coarse-graining time interval Δt [49]. As long as $\Delta t << (\gamma_{\{n\}})^{-1}$ and $\Delta t >> $ (width in ω of $\gamma_i^{\pm}(\omega))^{-1}$ [49], the resulting equation of motion (39) does not depend strongly on the choice of Δt. All of these corrections to (56) may be made trivially at the expense of an increase in complexity of the resulting equations.

Hodgkinson and Briggs have suggested [16] that the CMPE of SF_6 proceeds in two phases: a long "slow heating" phase in which the reservoir is not strongly excited, and energy flows in a quasi-steady-state manner from the laser field through the ν_3

mode into the reservoir; and a short, final "explosive" phase.
Our expression (56) for the excitation-dependent width appears
to support this explanation. The rate at which energy flows from
the ν_3 mode to the reservoir is increased rapidly as soon as any
significant amount of energy is deposited in the reservoir. Since
(in a very crude sense) the excitation of the reservoir depends
exponentially upon this rate, the cutoff of the "slow heating"
phase should be rather sharp. In the limit of very high reservoir temperature, only a very few ν_3 "levels" (modulo all the
other modes which have been traced out in (35)) will be involved
in the absorption of energy from the laser field, since the width
$\gamma_{\{n\}}^+$ (56) may already be large when any $n_i = 1$. Thus, the formalism we have described (in an admittedly cursory fashion) appears
to allow one to describe the transition from fully coherent excitation of discrete levels in the initial phase of CMPE, to an
essentially fully statistical molecular excitation after large
amounts of energy have been absorbed.

In future studies we hope to apply the methods developed by
Goodman and Thiele [50] for the numerical solution of density-
matrix equations of motion to the solution of (39), using parameters derived from a realistic model of the force field of SF$_6$.
Such an approach should be able to address some of the major
questions which have arisen in the past year concerning the roll
of the ν_3 mode *versus* the role of the reservoir in the CMPE of
SF$_6$.

ACKNOWLEDGEMENTS

We are indebted to R. V. Ambartzumian, N. Bloembergen, J. H.
Eberly, K. Fox, W. E. Lamb, Jr., V. S. Letokhov, G. N. Makarov,
C. Paul Robinson and E. Yablonovitch for helpful and stimulating
discussions. Special thanks are due to S. M. Freund and J. L.
Lyman for their comments on the Section 4 of this paper, and to
R. S. McDowell for permission to use Fig. 1.

REFERENCES

1.   N. R. Isenor and M. C. Richardson, Appl. Phys. Lett. <u>18</u>,
     224 (1971); N. R. Isenor and M. C. Richardson, Optics Commun.
     <u>3</u>, 360 (1971); N. R. Isenor and M. C. Richardson, Proc.
     Tenth Intern. Conf. on Ionization Phenomena in Gases (Oxford,
     Donald Parsons, 1971); N. R. Isenor, V. Merchant, R. S.
     Hallsworth, and M. C. Richardson, Can. J. Phys. <u>51</u>, 1281
     (1973); R. S. Hallsworth and N. R. Isenor, Chem. Phys. Lett.
     <u>22</u>, 283 (1973).

2.   R. V. Ambartzumian, V. S. Letokhov, E. A. Ryabov and N. V. Chekalin, ZhETF Pis. Red. $\underline{20}$, 597 (1974) [JETP Lett. $\underline{20}$, 273 (1974)]; R. V. Ambartzumian, Yu. A. Gorokhov, V. S. Letokhov and G. N. Makarov, ZhETF Pis. Red. $\underline{21}$, 375 (1975) [JETP Lett. $\underline{21}$, 171 (1975)].

3.   J. L. Lyman, R. J. Jensen, J. Rink, C. P. Robinson and S. D. Rockwood, Appl. Phys. Lett. $\underline{27}$, 87 (1975).

4.   R. V. Ambartzumian, Yu. A. Gorokhov, V. S. Letokhov, G. N. Makarov, E. A. Ryabov and N. V. Chekalin, pp. 114–121 in *Laser Spectroscopy*, edited by J. C. Pebay-Peyroula, T. W. Hansch and S. E. Harris (Springer-Verlag, Berlin, 1975).

5.   N. G. Basov, V. T. Galochkin, A. N. Oraevsky and N. F. Starodubtsev, ZhETF Pis. Red. $\underline{23}$, 569 (1976) [JETP Lett. $\underline{23}$, 521 (1976)].

6.   N. Bloembergen, Optics Commun. $\underline{15}$, 416 (1975).

7.   C. J. Elliott and B. J. Feldman, paper OA-2, 28th Annual Gaseous Electronics Conference, Rolla, Missouri (1975).

8.   D. M. Larsen and N. Bloembergen, Optics Commun. $\underline{17}$, 254 (1976).

9.   F. H. M. Faisal, Optics Commun. $\underline{17}$, 247 (1976).

10.   R. V. Ambartzumian, Yu. A. Gorokhov, V. S. Letokhov, G. N. Makarov and A. A. Puretzky, ZhETF Pis. Red. $\underline{23}$, 26 (1976) [JETP Lett. $\underline{23}$, 221 (1976)].

11.   S. Mukamel and J. Jortner, Chem. Phys. Lett. $\underline{40}$, 150 (1976).

12.   T. P. Cotter, W. Fuss, K. L. Kompa and H. Stafast, Optics Commun. $\underline{18}$, 220 (1976).

13.   N. Bloembergen. C. D. Cantrell and D. M. Larsen, pp.162–176 in *Tunable Lasers and Applications*, edited by A. Mooradian, T. Jaeger and P. Stokseth (Springer-Verlag, Berlin, 1976).

14.   C. D. Cantrell and H. W. Galbraith, Optics Commun. $\underline{18}$, 513 (1976).

15.   V. I. Gorchakov and V. N. Sazonov, ZhETF $\underline{70}$, 467 (1976) [JETP $\underline{43}$, 241 (1976)].

16.   D. P. Hodgkinson and J. S. Briggs, Chem. Phys. Lett. $\underline{43}$, 451 (1976).

17.   M. F. Goodman, J. Stone and D. A. Dows, J. Chem. Phys. $\underline{65}$, 5052 (1976); J. STone, M. F. Goodman and D. A. Dows, J. Chem. Phys. $\underline{65}$, 5062 (1976).

18.   V. S. Letokhov and A. A. Makarov, Optics Commun. <u>17</u>, 250 (1976); *Coherent Excitation of Multilevel Molecular Systems in Intense Quasi-Resonant Laser IR Field* (Moscow, 1976).

19.   D. M. Larsen, Optics Commun. <u>19</u>, 404 (1976).

20.   J. P. Aldridge, J. H. Birely, C. D. Cantrell and D. C. Cartwright, pp. 57–144 in *Laser Photochemistry, Tunable Lasers and Other Topics* (Physics of Quantum Electronics, Vol. 4), edited by S. F. Jacobs, M. Sargent III, M. O. Scully and C. T. Walker (Addison–Wesley Publishing Co., Reading, Mass., 1976).

21.   V. S. Letokhov and C. B. Moore, Kvant. Elekt. <u>3</u>, 247 and 485 (1976) [Sov. J. Quant. Elect. <u>6</u>, 129 and 259 (1976)].

22.   V. M. Akulin, S. S. Alimpiev, N. V. Karlov, B. G. Sartakov, and L. A. Shelepin, ZhETF <u>71</u>, 454 (1976).

23.   V. M. Akulin, S. S. Alimpiev, N. V. Karlov and B. G. Sartakov, ZhETF <u>72</u>, 88 (1977).

24.   C. D. Cantrell and H. W. Galbraith, Optics Commun. (in press).

25.   W. H. Shaffer, H. H. Nielsen and L. H. Thomas, Phys. Rev. <u>56</u>, 895 and 1097 (1939).

26.   K. T. Hecht, J. Mol. Spectrosc. <u>5</u>, 355 (1960).

27.   K. Fox, J. Mol. Spectrosc. <u>9</u>, 381 (1962).

28.   E. J. Hurley, Ph.D. Thesis, Florida State University (1974) (unpublished).

29.   K. Fox, Proc. Orbis Scientiae (1977) (to be published).

30.   R. L. Platzman, Radiation Research <u>17</u>, 419 (1962).

31.   R. S. McDowell, private communication.

32.   J. D. Louck and H. W. Galbraith, Rev. Mod. Phys. <u>48</u>, 69 (1975).

33.   A. Messiah, *Quantum Mechanics* (North–Holland Publishing Co., Amsterdam, 1960), Vol. I, Ch. 12.

34.   H. W. Galbraith, Proc. Orbis Scientiae 1977 (to be published).

35.   C. C. Jensen, W. B. Person, B. J. Krohn and J. Overend, Optics Commun. <u>20</u>, 275 (1977).

36.   H. Kildal (to be published).

36a. Since ICOMP it has been brought to our attention that a new calculation of the parameters G_{33}, T_{33} and X_{33} may be required based on other possible interpretations of [*36*], private communication K. Fox.

37.   C. Marcott, W. G. Golden and J. Overend, J. Chem. Phys. (to be published).

38.   R. S. McDowell, private communication (1976); R. S. McDowell, J. P. Aldridge and R. F. Holland, J. Phys. Chem. <u>80</u>, 1203 (1976).

39.   J. Overend, H. W. Galbraith, W. B. Person and C. D. Cantrell, J. Chem. Phys. (to be published).

40.   J. H. Eberly, B. W. Shore, Z. Bialynicka-Birula, and I. Bialynicki-Birula, Phys. Rev. A. (in press, 1977); Z. Bialynicka-Birula, I. Bialynicki-Birula, J. H. Eberly and B. W. Shore, Phys. Rev. A. (in press, 1977).

41.   R. V. Ambartzumian, N. P. Furzikov, Yu. A. Gorokhov, V. S. Letokhov, G. N. Makarov and A. A. Puretzky, ZhETF Pis. Red. <u>23</u>, 217 (1976); Optics Commun.

42.   H. W. Galbraith and C. D. Cantrell (unpublished).

43.   J. von Neumann, *Mathematische Grundlagen der Quantenmechanik* (Julius Springer, Berlin, 1931).

44.   U. Fano, Rev. Mod. Phys. <u>29</u>, 74 (1957).

45.   M. Lax, J. Phys. Chem. Solids <u>25</u>, 487 (1964).

46.   J. R. Ackerhalt, C. D. Cantrell and H. W. Galbraith (unpublished).

47.   J. G. Black, E. Yablonovitch, N. Bloembergen and S. Mukamel, Phys. Rev. Lett. <u>38</u>, 1131 (1977).

48.   E. R. Grant, M. J. Coggiola, Y. T. Lee, P. A. Schulz and Y. R. Shen, (this volume), and M. J. Coggiola, et al., Phys. Rev. Letts. <u>38</u>, 17 (1977).

49.   W. H. Louisell, *Quantum Statistical Properties of Radiation* (John Wiley & Sons, Inc., New York, 1973).

50.   M. F. Goodman and E. Thiele, Phys. Rev. A<u>5</u>, 1355 (1972).

Multiphoton Excitation and Dissociation of Molecules and Isotope Separation by Intense Infrared Laser Radiation

V. S. LETOKHOV
The Institute of Spectroscopy of the
Academy of Sciences of the USSR,
Moscow, USSR

I. INTRODUCTION

In this review report I'll try to give a short outline of
the present state of the selective multiphoton excitation and dissociation of polyatomic molecules by the intense infrared radiation and application of this phenomenon for isotope separation.
A great interest to the phenomenon of multiphoton molecule dissociation by the infrared radiation, has, obviously two reasons:
an untrivial interaction mechanism of the intense infrared field
with a multilevel quantum system and perspectives of the practical application of this phenomenon in isotope separation and
chemical technology.

It is worth accentuating, that for the first time we have
got the possibility to control the excitation of the high vibrational levels and selective dissociation of molecules in the
ground electronic state. Figure 1 illustrates in the simplified
form the basic types of molecule photoexcitation: a) a single-
photon excitation of electronic or vibrational state; b) a two-
step excitation of the electronic state via intermediate state;
c) a multiphoton excitation and dissociation via vibrational
levels of the ground electronic state. The first (a) single-
quantum approach has been well-known in atomic and molecular
physics and chemistry before the laser invention. The last two
approaches (b,c) have been carried out by means of laser radiation, as they require, at least, absorption saturation of the
intermediate transitions. In principle, it's impossible to get
this by means of usual incoherent light sources because of the
low radiation temperture. Just because of this, beginning from
1970, we have been researching at the Institute of Spectroscopy
a selective stepwise [1,2] and selective multiphoton [3] photo-
processes. To our mind, just they in principle open new possibilities of selective action on the substance by laser radiation.

331

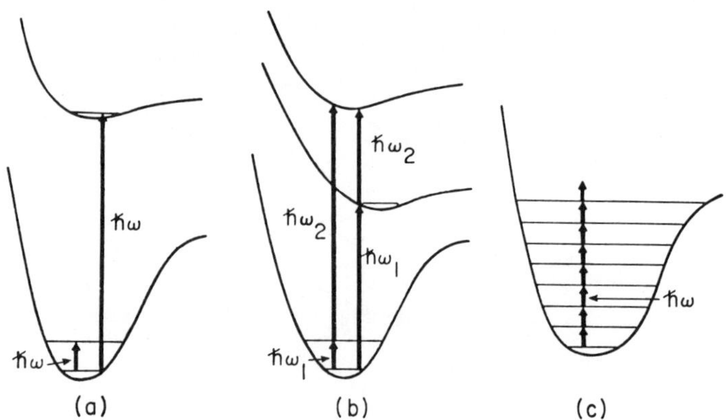

FIGURE 1.   Types of selective molecular photoexcitation pro-
            cesses: a) Single-step excitation of an electronic or
            vibrational states; b) two-step excitation of electron
            state through intermediate vibrational or electron
            state; c) multiphoton excitation by intense IR radia-
            tion.

By means of multiphoton excitation of high vibrational levels
all the basic photoprocess types have been successfully demonstrated: 1) a photochemical reaction with a proper acceptor
($BCl_3^{**} + O_2$); 2) a photodissociation ($SF_6^{**} \rightarrow SF_5^{*} + F$); 3) a photo-
isomerization (trans $- C_2H_2Cl_2^{**} \rightarrow$ cys $C_2H_2Cl_2$).   Therefore, we can
already speak about the appearance of a new field – multiphoton
infrared photophysics and photochemistry (see review [4]).   In
this field for the last two years so many works have been carried
out, that I have no chance to mention even a small part of them.
I'll be guided mainly by the results we have obtained at the
Institute of Spectroscopy.   The main contributors in this work
are R. V. Ambartzumian, V. N. Bagratashvili, N. V. Chekalin,
U. A. Gorokhov, V. S. Dolzikov, N. P. Fursikov, I. N. Knyasev,
U. R. Kolomijaski, V. V. Lobko, V. N. Lokhman, G. N. Makarov,
A. A. Makarov, A. A. Puretski, E. A. Ryabov, A. S. Shibanov.
First, I will talk about the main features of multiphoton
molecular excitation and dissociation phenomenon by the infrared
radiation.

II.   MAIN FEATURES OF THE PHENOMENON

The multiphoton dissociation is a nonlinear process with a
sharp intensity threshold.   Figure 2 illustrates the experimental
dependence of SF_6 molecule dissociation yield in relative units

per one pulse on the CO_2 laser radiation intensity, the frequency
of which is tuned either to the ν_3 fundamental vibrational band
or to the $\nu_2 + \nu_6$ weak compound vibrational band. A dotted horizontal line shows a measurement sensitivity. The dissociation
threshold region of about 23 ± 2 MW/cm^2 is shaded. We shall remark, that in some experiments with pure SF_6 other authors did
not observe a sharp threshold. It is explained by "dirty" experiment conditions, when the homogeneity of the laser beam intensity
is not controlled. The dissociation threshold exists for many
other polyatomic molecules as well. For example, the appearance
of C_2 radicals during C_2H_4 molecule dissociation, which are detected by means of a dye laser, has also a distinct threshold [7].

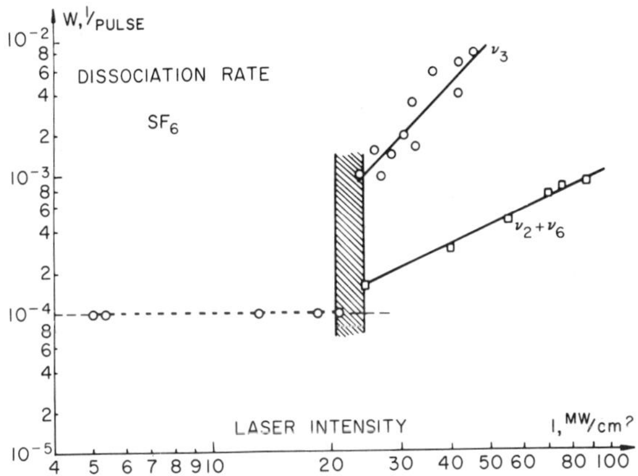

FIGURE 2.   The dependence of the dissociation yield (in relative
            units) in SF_6 on laser intensity. Collimated beam
            measurements, no scavanger added. The laser pulse
            length was always the same - 90 nsec FHWH, the SF_6
            pressure - 0,2 torr (from [5,6]).

     Unlike the multiphoton dissociation, a multiphoton SF_6 mole-
cule excitation is a nonlinear phenomenon without any intensity
threshold (Fig. 3). This pecularity has been discovered in our
first experiments [6,8]. In the subsequent experiments [9] by
means of an opto-accoustic method we have found that the threshold is absent when the intensity is reduced up to 10^5 W/cm^2. The
same experiments have shown that three-atom molecules of OCS and
D_2O absorb only a few photons and at the intensity of about
10^7 W/cm^2 absorption saturation takes place. In our last experiments [10] we found that in the large range of intensities from
10^2 to 10^7 W/cm^2 the number of absorbed quanta per one molecule
increases almost in proportion to radiation intensity (Fig. 3).

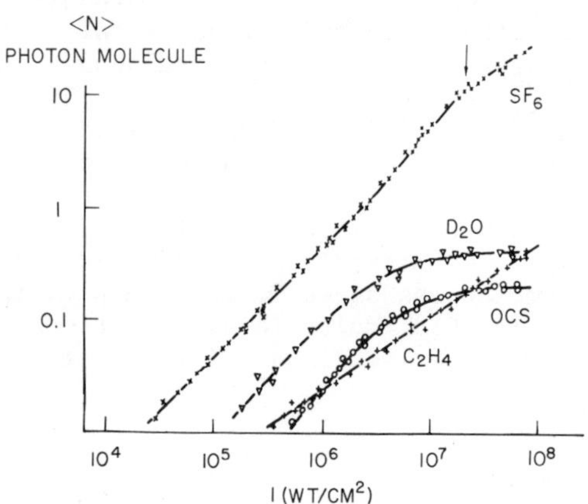

FIGURE 3.   The dependence of the number of absorbed photons per molecule <n> in irradiated volume on laser intensity measured by opticoacoustical method for several different molecules (from [9]).

SF$_6$ molecule dissociation yield during frequency scanning of laser radiation has a distinct resonance character [6]. The maximum of the resonance curve is shifted into the "red" side relative to the spectrum maximum of the linear infrared absorption in the ν_3 band.   The width and the "red" shift of the resonance of the dissociation yield correlated well with the dependence of isotopic selectivity of ^{32}SF$_6$ and ^{34}SF$_6$ molecule dissociation on the radiation frequency, which was observed in the work [6].

Multiphoton absorption rate measured by the number of the absorbed photons per one molecule also has a distinct resonance character.   The resonance width and its shift into the red side increases with the radiation intensity increase (Fig. 4) [9].

A polyatomic molecule in the strong infrared field absorbs energy noticeably exceeding the dissociation energy of the weakest bond [11]. This conclusion resulted from the two indirect considerations.   Firstly, polyatomic molecule dissociation products consisted of many simple fragments, i.e. photolysis products in the strong infrared field greatly differed from the ultraviolet photolysis products [12].   Secondly, if we take into account the molecule distribution by rotational sublevels, we may suppose that only some part of the molecules can really absorb a large number of photons.   Then the measured average number of absorbed photons <n> $\simeq$ 20 at SF$_6$ dissociation threshold (see Fig. 3) in reality corresponds to the absorption of much larger number of quanta <n> $\simeq$ 60–100 by a small part of molecules.

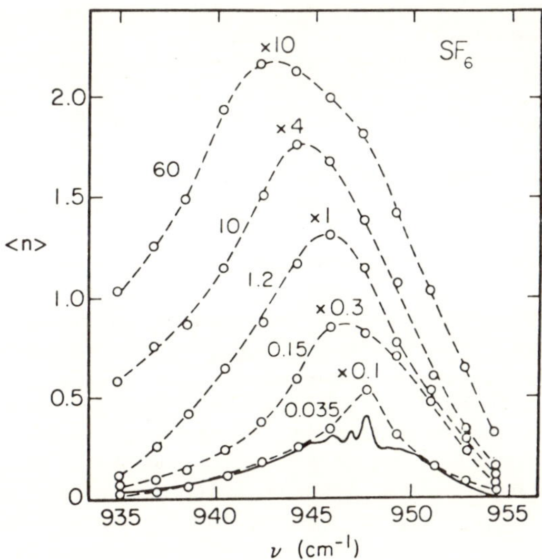

FIGURE 4. The dependence of the number of absorbed photons per molecule <n> for SF$_6$ on pump frequency of CO$_2$ laser, measured by opticoacoustical method. The parameter of the curves is the laser intensity (in MW/cm^2 units) (from [9]).

III. MODEL OF THE PROCESS AND ITS CHECK UP

Having learned the main characteristics of multiphoton absorption phenomenon and polyatomic molecule dissociation, we tried to create a simple model of the process that could at least qualitatively explain these characteristics.

The absorption of comparatively large number of quanta in the field of moderate intensity (less than 1 MW/cm^2), which was observed by us for many molecules (SF$_6$, OsO$_4$ and others) proved, that the anharmonicity, at least at low transitions, is not an essential obstacle for molecule excitation. Therefore we supposed that some kind of mechanism of "soft" anharmonicity compensation exists at low transitions, this compensation doesn't require a "rough force", that is a very strong field. As a means of such a mechanism we have researched a rotational anharmonicity compensation at three consequent P-Q-R transitions (triple-rotational-vibrational resonance) [5,6].

As a result of a stepwise excitation at the sequence of three vibrational transitions v=0 → 1 → 2 → v = 3 the molecule gets

to the region where the vibrational level density is rather high.
In this region of so called vibrational quasicontinuum the frequency of the strong infrared field almost always coincides with
one of the numerous, but weak vibrational-rotational transitions.
Thus, a sufficiently intense infrared field can excite the molecule up to the energy levels sufficient for its dissociation.
Canadian researchers have paid attention to the possible role of
high density of polyatomic molecule vibrational levels in the
multiphoton dissociation mechanism in one of the first works on
this problem [13]. Then Bloembergen has made estimates of stimulated transition probability in vibrational quasicontinuum [14].

The molecule excitation at the transitions in the vibrational quasicontinuum leads to the excitation of many vibrational
modes. The distribution of vibrational energy among many modes
makes difficult its dissociation at the energy, which is exactly
equal to the dissociation energy of the weakest bond. This can
explain the effect of vibrational molecule superexcitation, that
is absorption of the energy, exceeding the energy of its dissociation [8,11].

Thus, this very simple model gives the ideas about the existence, firstly, of resonance excitation at low vibrational transitions in the field of moderate intensity and, secondly, about the
existence of nonresonance excitation at transitions in vibrational quasicontinuum in sufficiently intense field. To check this
model we studied the molecule excitation and dissociation in the
two-frequency infrared field (Fig. 5) [15]. In these experiments
ω_1 is the radiation frequency of the first laser with moderate
intensity (10^4 –10^6 W/cm^2) scanned in the ν_3 absorption band of
SF$_6$ molecule. The ω_2 radiation frequency of the second laser with
much larger power (10^7 –10^8 W/cm^2) scanned at some range far away
from the absorption band, i.e. in the region of the supposed absorption at vibrational quasicontinuum transitions. Our experiments have shown that this simple model is correct but only in
the first approximation.

Figure 6 illustrates the dependence of SF$_6$ molecule dissociation yield during ω_1 radiation frequency tuning of the first
laser in the presence of a strong field at ω_2 frequency far from
the resonance. Frequencies and intensities of both laser beams
are chosen in such a way that the dissociation is possible only
at joint action of the two laser pulses. In fact, it turned out
that the dissociation takes place when ω_1 resonance radiation intensity is at the level of 10^4-10^6 W/cm^2, that is much lower than
the threshold dissociation intensity in the single frequency field.
This safely proved the existence of the "soft" compensation mechanism of anharmonicity at low vibrational transitions. However,
the resonance width at frequency scanning of the resonance field

has turned out to be much larger than it should be from the triple
P-Q-R resonance model. Having supposed that this broadening is
connected with the "hot" bands contribution, we have carried out
the experiments at lower temperature and gotten some narrowing of
the resonance curve. Nevertheless, the resonance width exceeds
the width of $v = 1 \rightarrow v = 2$ transition Q-branch required for the
triple P-Q-R resonance.

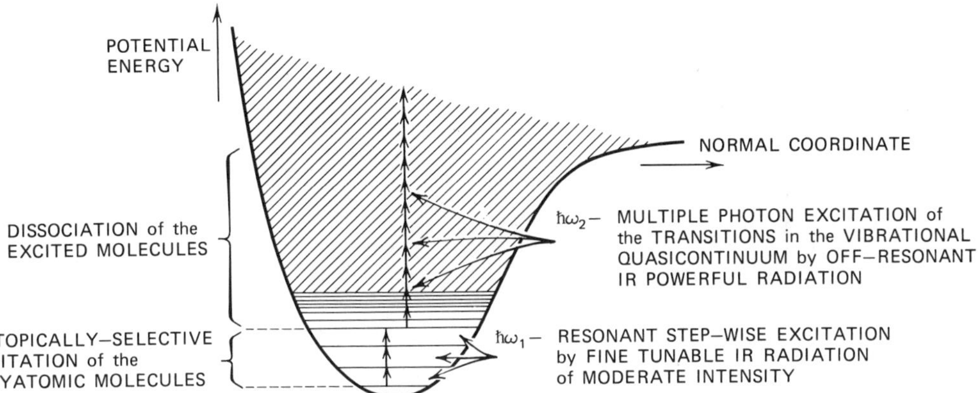

FIGURE 5.   Dissociation model of polyatomics by two-frequency IR
field based on resonant selective excitation in the
system of few lower transitions and nonresonant excitation of transitions in the "vibrational quasicontinuum".

Figure 7 illustrates the dependence of the absorbed photon
number from the second beam with ω_2 frequency on first beam energy density at E_1 resonance frequency. This dependence has recently been measured in the experiments by the optoacoustic technique and two CO_2 lasers. Energy density E_1 increase of the
resonance field causes a proportional (almost linear) absorption
increases in vibrational quasicontinuum. Hence, at the power
level of about 3 MW/cm^2 absorption saturation in the low level
systems doesn't take place yet. Therefore, coherent oscillation
with Rabi frequency in the low vibrational level system seems to
us hardly probable for SF_6. It is probable that the molecule leakage in vibrational quasicontinuum takes place with the rate close
to the Rabi oscillation frequency in the low vibrational level
system.

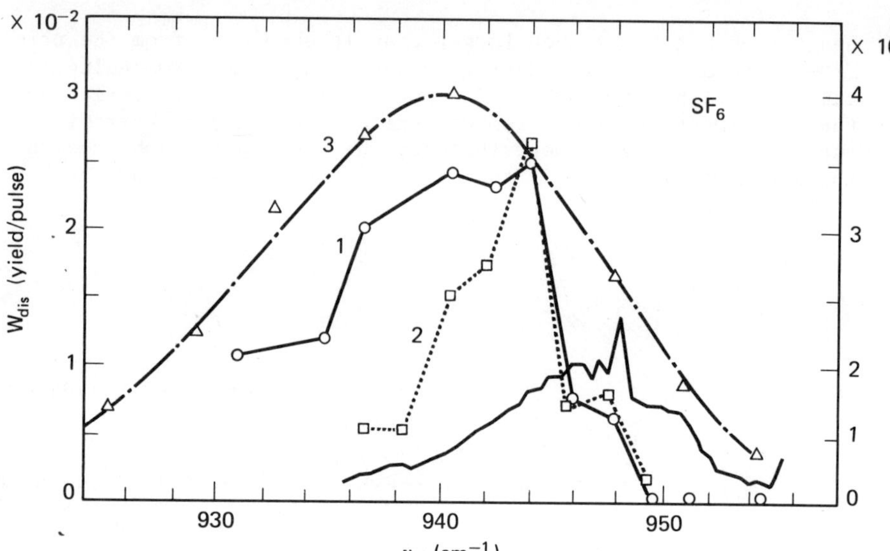

FIGURE 6.   Dissociation yield of SF_6 (left side scale) in two-frequency irradiation as a function of frequency ν_1 of resonant field at fixed frequency of nonresonant field (ν_2=1084cm^{-1}). Curves 1 and 2 correspond to two different temperatures of SF_6 - 300 and 190 °K. The linear absorption of SF_6 (lower curve) and the dissociation yield in the case of single-frequency irradiation (right side scale) at T=300°K are also shown (curve 3) (from [15]).

Figure 8 illustrates SF_6 molecule dissociation yield dependence on ω_2 frequency of the strong nonresonant infrared field at ω_1 fixed frequency of the "weak" resonance field. In fact, in accordance with the vibrational quasicontinuum model, dissociation takes place even with the strong field frequency detuned far from the abosrption band. This is a proof of the absorption existence at the transitions in the vibrational quasicontinuum. However, the dissociation yield increases when the strong field frequency approaches the absorption band, pointing at the probability of the existence of a wide resonance, shifted to the red side relative to the ν_3 SF_6 absorption band [15]. This wide resonance has recently been measured in the experiments on Si-F$_4$ two-frequency excitation by CO_2 TEA lasers [17]. The dependence of ω_2 frequency absorbed quanta number in the vibrational quasicontinuum on the E_2 energy density of the strong field is illustrated in Fig. 9. This dependence has recently been measured by the opto-acoustic technique in a two-frequency field [16]. The observed dependence of $\sqrt{E_2}$ type correlates well with the analogous dissociation yield dependence in a single frequency field with the intensity higher than that of the threshold.

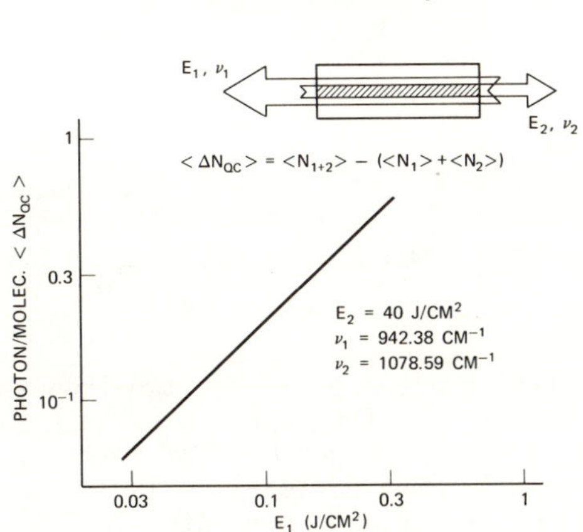

FIGURE 7.   The number of absorbed photons $\langle \Delta N_{qc} \rangle$ per molecule on transitions in the vibrational quasicontinuum of SF_6 as a function of energy density E_1 of resonant field at frequency ν_1. The value $\langle \Delta N_{qc} \rangle$ is measured as a difference on absorbed energy from the two joint laser pulses in opticoacoustical cell $\langle N_{1+2} \rangle$ and the absorbed energy from the independent laser pulses $\langle N_1 \rangle$ and $\langle N_2 \rangle$ (from [16]).

Thus, the experiments in the two-frequency infrared field confirm the simple model of resonant excitation at low transitions and quasiresonant exictation at transitions in the vibrational quasicontinuum. However, there are essential additions to this simplest model, at least in two aspects: 1) The resonance absorption width at low transitions is noticeably larger than it should be from the three-step P-Q-R resonance model; 2) the absorption in the vibrational quasicontinuum is uniform and has a wide but quite distinct resonance, shifted to the "red" side. It is already clear now how to correct our simple model in order to explain these pecularities.

340   V. S. Letokhov

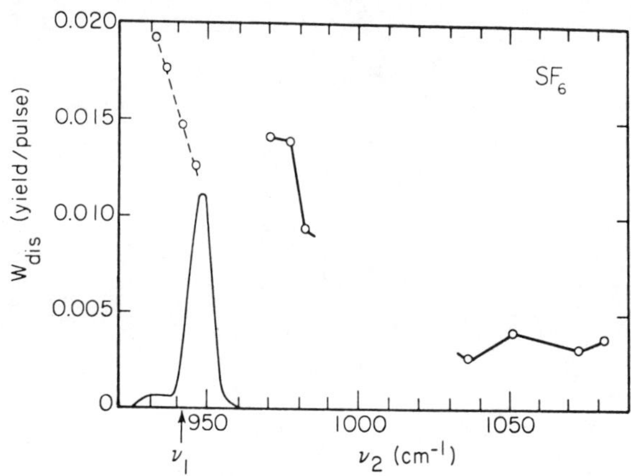

FIGURE 8.   Dissociation yield of SF_6 in two-frequency irradiation
as a function on frequency ν_2 of nonresonant strong
field at fixed frequency ν_1 of resonant field (ν_1 =
942.1 cm^{-1}).   The dotted line is the result of calculations of the data presented in Fig. 6 (from [15]).

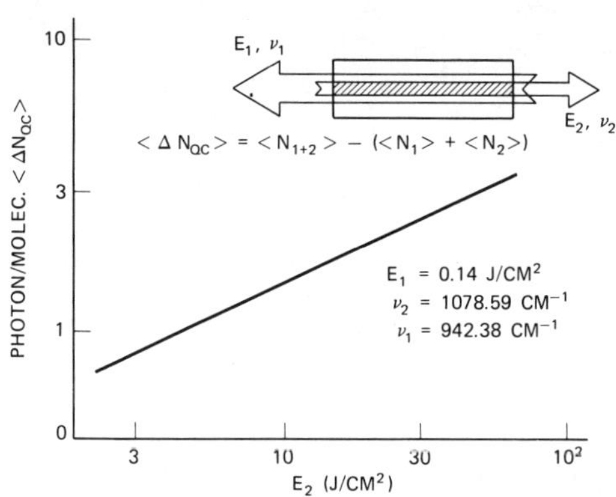

FIGURE 9.   The number of absorbed photons $\langle\Delta N_{qc}\rangle$ per molecule on
transitions in the vibrational quasicontinuum of SF_6
as a function of energy density E_2 of nonresonant
field at frequency ν_2. Definition of value $\langle\Delta N_{qc}\rangle$ is
given on Fig. 7 (from [16]).

Firstly, we must take into consideration the possibility of
two or three-photon transitions in the system of low levels with
some field frequency detuning relative to the resonance with intermediate transitions [18,19]. This makes the requirements of
the exact resonance for triple P-Q-R transition easier and gives
additional two or three-photon resonances [18-20]. The consideration of multiphoton transitions in the system of low vibrational
levels can obviously explain the power broadening of the resonance
during the increase of infrared field intensity. Multiquantum
transitions unlike the pure three-step P-Q-R transition can touch
a large number of rotational states of the ground vibrational
state. This is proved by the recent experiments on probing by a
weak infrared field at $v=0 \rightarrow v=1$ transition of rotational states
during the strong infrared radiation pulse irradiation on SF_6
molecule [21]. The observed data may be interpreted as the result of a depletion of a large number of rotational sublevels of
SF_6 molecule by a strong infrared pulse.

Secondly, we must take into consideration the possibility of
anharmonic splitting of excited degenerate vibrational levels,
which increases the number of possible intermediate resonances
for "soft" anharmonicity compensation [22] and can form a relatively wide absorption resonance in the vibrational quasicontinuum [23]. Wide resonances in the vibrational quasicontinuum
spectrum are obviously a common feature of polyatomic molecules.
Our experiments [24] on multiphoton absorption and dissociation
of CH_3NO_2 molecule in ν_{13} band, having no isotopical shift in a
linear infrared spectrum , testify to this fact. However, a distinct shift which probably appears at the expense of a wide resonance in the vibrational quasicontinuum absorption spectrum, appears in the multiphoton absorption spectrum.

To research the last step of the process dissociation of the
highly excited molecules we have used the detection technique of
simple dissociation fragments by dye laser radiation [7]. We have
detected, for example, that C_2 radical is formed during C_2H_4 molecule dissociation mostly in the ground electronic state and only
0.1% of C_2 radicals are in the electronically excited state and
therefore gives a fluorescence. The laser radical detection
technique can give the information about the kinetics of vibrational and rotational radical temperature and on Doppler broadening of spectral lines. One can also get the information about
the translational temperature of primary dissociation products.
The recent experiments [25] with molecular beams have given more
reliable information about the primary dissociation products.
The inertial character of a strongly excited molecule dissociation and the consequent dissociation of primary polyatomic fragments by infrared field [25] gives the possibility to understand
the experimental fact, the absorption of energy, noticeably

exceeding the dissociation energy of the weakest bond [*11*].

Thus, by relying upon the numerous experiments carried out and the theoretical considerations, it becomes convenient to consider the process of multiphoton excitation and dissociation of polyatomic molecules as a sequence of three processes: 1) a resonance (isotopically selective) multistep and multiphoton absorption on the sequence of several low transitions; 2) a wide resonant absorption in vibrational quasicontinuum; 3) a dissociation of the super-excited molecule and a consequent multiphoton absorption by polyatomic dissociation products. However, such a division of the process at three consequent stages is,to a considerable extent, very conventional. The illustrated (Fig. 3) proportional dependence of the number of <n> absorbed photons on the intensity proves that the accumulation of the energy by SF_6 molecule has no determined stages. The single determined point is the dissociation threshold, after which the dependence slope changes slightly. To my mind all of these three stages overlap one another and are hardly distinguishable.

Though we have no quantitative picture of the polyatomic molecule dissociation, the information we have obtained about the main properties of this phenomenon is quite sufficient to formulate the principles of optimal isotope separation by this method.

IV. OPTIMAL SCHEME OF ISOTOPICALLY-SELECTIVE DISSOCIATION

Selective dissociation of polyatomic molecules by the infrared radiation at single frequency has already been successfully employed in the enrichment of a number of isotopes.As a method of isotope separation, the dissociation of molecules by a single IR frequency suffers from several disadvantages [*4*]. The requirement of a strong IR field for the dissociation contradicts the requirement of high selectivity of the excitation process. Though the dissociation rate is more sensitive to the pulse energy rather than the intensity, the V-V transfer rate determines the maximal pulse length, and therefore the minimal intensity of the field. The selectivity of the excitation falls with the rise of field intensity due to power broadening, and this makes the method not applicable to the separation of heavy isotopes, i.e. when the isotope shift is small compared to the width of the absorption band. The next disadvantage is that it is difficult from the technological point of view to create high field intensity in a large volume. This makes the process of the separation difficult to scale.

The above considered dissociation by two IR pulses of different frequencies avoids both of these mentioned problems. In this method a resonant low intensity IR pulse at the frequency ω_1 performs selective excitation of the molecules. The second pulse at ω_2 , which is more intense, dissociates the excited molecules. Here we return, as a matter of fact, to the selective stepwise dissociation method in a two-frequency laser field, where the selective excitation of molecules was achieved by the infrared radiation, while the dissociation of selectively excited molecules by the radiation tuned to resonance with the electronic transition (Fig. 10). Five years ago we used this approach in the first experiments on laser isotope separation [1,2,26]. The essential difference of the present approach is in the fact that in the case of polyatomic molecules, the presence of the vibrational quasicontinuum allows the selective dissociation of excited molecules by the infrared radiation within the ground electronic state without excitation of the electronic transitions. This makes the method much more universal.

The two-frequency dissociation method with separation of the functions of the selective excitation and selectively excited molecule dissociation allows essentially the decrease of the power broadening of the resonance on the excitation selectivity. This, firstly, allows the use of the method for the separation of heavy elements, isotopes with isotopic shift of only $\Delta\nu_{isotop} \simeq 10^{-3} \nu_{vib}$, and, secondly, the essential decrease of the requirement on the energetics of the softly tunable frequency laser.

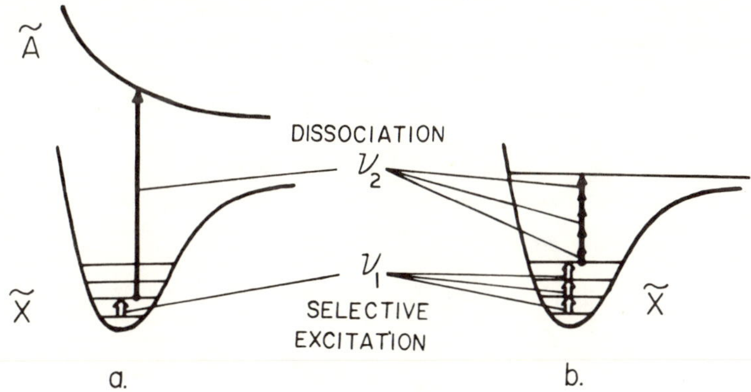

FIGURE 10.   The schemes of two-frequency dissociation by IR-UV excitation of excited electronical state (a) and by IR-IR multiphoton excitation in the ground electronical state (b) with separated functions of selective excitation and dissociation of excited molecules.

As it has been shown above (Fig. 8), the multiphoton absorption spectrum at the transitions in the vibrational "quasicontinuum" through which the dissociation is performed represents a broad maximum shifted to the red from the absorption band of the mode. The rise of the dissociation cross-section means the decrease of the dissociation threshold. Therefore the proper choice of the frequency ω_2 at which the dissociation is performed allows the use of beams of much lower intensity compared with the threshold value in the single frequency case.

For the first time this was realized in the experiments on separation of osmium isotopes by the radiation of two IR pulses [27,28]. In these experiments the intensities of both fields at ω_1 and at ω_2 were significantly lower compared with the threshold value, and the dissociation of OsO_4 took place in *unfocused* laser beams of moderate intensities. Figure 11 illustrates OsO_4 linear absorption spectrum in the CO_2 laser oscillation region, optimum ω_1 field frequency location providing the maximum dissociation selectivity and two possible frequency positions of ω_2 strong field in the "red" and "blue" side relative to the absorption band. During the ω_2 frequency tuning into the low frequency side of the absorption band, the quantum dissociation yield sharply increases, which testifies to the existence of the multiphoton absorption cross-section maximum in the vibrational quasicontinuum. Table I illustrates the results of the measurements of various osmium isotope enrichment in the natural abundance, in the experiments with single-frequency and two-frequency dissociations. In the case of single-frequency dissociation the enrichment is completely absent, but has a noticeable magnitude for the two-frequency method.

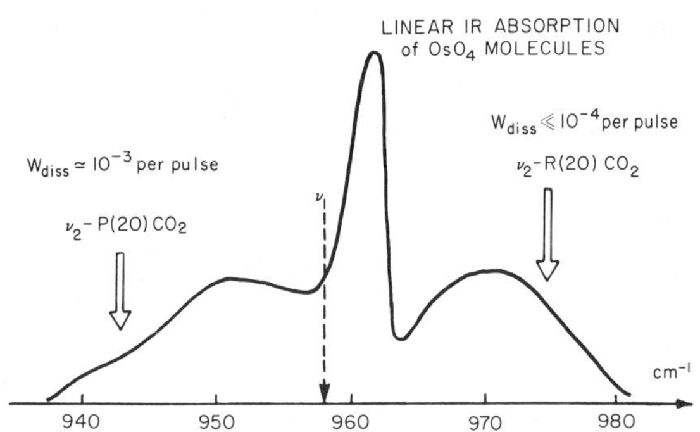

FIGURE 11. Two-frequency dissociation of OsO_4 molecule has a great difference of dissociation yield for a "red" and "blue" shifts of ν_2 frequency of intense IR field relative to the absorption band (from [28]).

TABLE I. Separation of Os isotopes by multiphoton dissociation of OsO_4 [28].

Regime of irradiation	Enrichment coefficient $\alpha=K \, (^{i}O_s/^{j}O_s) - 1$				
	$^{192}/190$	$^{192}/189$	$^{192}/188$	$^{192}/187$	$^{192}/186$
Single-Frequency $\nu - P(14)$	<1%	<1%	<1%	<1%	<1%
Two-Frequency $\nu_1 \quad \nu_2$					
P(6) + P(20)	11%	13%	24%	48%	58%
P(6) + P(12)	4%	4%	8%	4%	2%
P(20)+ P(20)	2%	2%	8%	1%	1%

These experiments confirm the conclusion that the optimal scheme for isotope selective dissociation of molecules is the dissociation by two IR pulses. The first pulse at ω_1 of very moderate intensity ($10^4 - 10^5$ W.cm^{-2}) should be tuned to the frequency of maximum isotope selective excitation. The second one at ω_2 with the intensity $10^6 - 10^7$ W.cm^{-2} (or energy density 0.1-1 J/cm^2) should be tuned to the frequency of maximum dissociation cross-section of excited molecules, i.e., the minimum of the threshold intensity.

The progress in the development of the selective multiphoton dissociation of polyatomic molecules and its application to isotope separation is so rapid that in the near future one can expect the creation of large-scale pilot installations for isotope separation in the practical scale. We have carried out the first successful experiments in this direction together with the I. V. Kurchatov Institute of Atomic Energy [29]. Thus, multiphoton processes in polyatomic molecules turned out to be not only an extraordinarily interesting field of fundamental scientific research, but they are undoubtedly of great practical interest. Perhaps this is the first case of practical application of multiphoton processes to which our conference is devoted.

REFERENCES

1. R. V. Ambartzumian, V. P. Kalinin, V. S. Letokhov, Pis'ma
    Zh.ETF, 13, 305 (1971); R. V. Ambartzumian, V. S. Letokhov,
    Appl. Optics, 11, 354 (1972); R. V. Ambartzumian, V. S.
    Letokhov, G. N. Makarov, A. A. Puretzky, "Laser Spectroscopy",
    Proceedings of Intern. Laser Spectroscopy Conf. Vail (June
    24-27, 1973, USA), ed. by R. Brewer, A. Mooradian (Plenum
    Press, N.Y. and London, 1974) p. 611.

2. V. S. Letokhov, Science 180, 451 (1973).

3. R. V. Ambartzumian, V. S. Letokhov, E. A. Ryabov, N. C.
    Chekalin, Pis'ma Ah.ETF, 20, 597 (1974); R. V. Ambartzumian,
    Yu. A. Gorokhov, V. S. Letokhov, G. N. Makarov, Pis'ma
    Ah.ETF, 21, 375 (1975).

4. R. V. Ambartzumian, V. S. Letokhov, in "Chemical and Bio-
    chemical Applications of Lasers", ed. by C. B. Moore, Vol.2,
    (Academic Press, 1977).

5. R. V. Ambartzumian, in "Tunable Lasers and Applications",
    Proceedings of Intern. Conf. in Norway, 7-11, June 1976
    (Springer-Verlag, 1976), p. 150.

6. R. V. Ambartzumian, Yu. A. Gorokhov, V. S. Letokhov, G. N.
    Makarov, A. A. Puretzkii, Pis'ma Zh.ETF, 23, 26 (1976);
    Zh.Eksp. i Teor. Fix. 71, 440 (1976).

7. N. V. Chekalin, V. S. Doljikov, V. S. Letokhov, V. N.
    Lokhman, A. N. Shibanov, Appl. Phys. 12, 191 (1977).

8. R. V. Ambartzumian, Yu. A. Gorokhov, V. S. Letokhov, G. N.
    Makarov, Zh.Eksp. Teor. Fiz., 69, 1957 (1975).

9. V. N. Bagratashvili, I. N. Knyazev, V. S. Letokhov, V. V.
    Lobko, Optics Comm. 18, 525 (1976).

10. I. N. Knyazev, V. S. Letokhov, V. V. Lobko (in press).

11. R. V. Ambartzumian, N. V. Chekalin, Yu. A. Gorokhov, V. S.
    Letokhov, E. A. Ryabov, in "Laser Spectroscopy", Proceed-
    ings of Second Intern. Laser Spectroscopy Conf. (Megeve,
    France, 23-27 June 1975), Lecture Notes in Physics, Vol. 43
    (Springer-Verlag, 1975), p. 121.

12. R. V. Ambartzumian, N. V. Chekalin, V. S. Letokhov, E. A.
    Ryabov, Chem. Phys. Lett. 36, 301 (1975).

13. N. R. Isenor, V. Merchant, R. F. Hallsworth, M. C.
    Richardson, J. Canad. Phys. 51, 1281 (1973).

14. N. Bloembergen, Optics Comm. $\underline{15}$, 416 (1975).

15. R. V. Ambartzumian, Yu. A. Gorokhov, V. S. Letokhov, G. N. Makarov, A. A. Puretzky, N. P. Furzikov, Pis'ma Zh.ETF, $\underline{23}$, 217 (1976); R. V. Ambartzumian, N. P. Furzikov, Yu. A. Gorokhov, V. S. Letokhov, G. N. Makarov, A. A. Puretzky, Optics Comm. $\underline{18}$, 517 (1976).

16. V. N. Bagratashvili, V. S. Letokhov, V. V. Lobko (in press).

17. V. M. Akulin, S. S. Alimpiev, N. V. Karlov, A. M. Prokhorov, B. G. Sartakov, E. M. Khokhlov, Pis'ma Ah.ETF (in press).

18. V. S. Letokhov, A. A. Makarov, Optics Comm. $\underline{17}$, 250 (1976).

19. N. Bloembergen, C. D. Cantrell, D. M. Larsen, in "Tunable Lasers and Applications", Proceedings of Intern. Conf. in Norway, Loen, 7-11 June 1976 (Springer-Verlag, 1976),p.162.

20. D. M. Larsen, N. Bloembergen, Optics Comm. $\underline{17}$, 254 (1976); D. M. Larsen, Optics Comm. $\underline{19}$, 401 (1976).

21. S. S. Alimpiev, V. N. Bagratashvili, N. V. Karlov, V. S. Letokhov, V. V. Lobko, A. A. Makarov, B. G. Sartakov, E. M. Khokhlov (in press).

22. C. D. Cantrell, H. W. Galbraith, Optics Comm. $\underline{18}$, 513 (1976).

23. V. M. Akulin, S. S. Alimpiev, N. V. Karlov, B. G. Sartakov, Zh.Eksp. i Teor. Fiz. $\underline{72}$, 88 (1976).

24. N. V. Chekalin, V. S. Doljikov, Yu. R. Kolomiisky, V. S. Letokhov, V. N. Lokhman, E. A. Ryabov, Phys. Lett. $\underline{59A}$, 243 (1976).

25. M. J. Coggiola, P. A. Shulz, Y. T. Lee, Y. R. Shen, Phys. Rev. Lett. $\underline{38}$, 17 (1977).

26. R. V. Ambartzumian, V. S. Letokhov, G. N. Makarov, A. A. Puretzky, Pis'ma Zh.ETF, $\underline{15}$, 709 (1972); $\underline{17}$, 91 (1973).

27. R. V. Ambartzumian, Yu. A. Gorokhov, G. N. Makarov, A. A. Puretzky, N. P. Furzikov, Kvantovaya Elektronika (Russian) $\underline{4}$, N7 (1977).

28. R. V. Ambartzumain, N. P. Furzikov, Yu. A. Gorokhov, V. S. Letokhov, G. N. Makarov, A. A. Puretzky, Optics Letters (in press).

29. V. N. Bagratashvili, V. Yu. Baranov, E. P. Velikhov, S. A. Kazakov, Yu. R. Kolomiisky, V. S. Letokhov, V. G. Niz'ev, V. D. Pis'menny, E. A. Ryabov, A. I. Starodubtzev, Appl. Phys. (in press); Invited Report on CLEA-77, 1-2 June 1977, Washington, D.C., USA.

Quasiresonant Multiphoton Processes in Molecules

I. I. TUGOV*
P. N. Lebedev Physical Institute
Moscow, USSR

Multiphoton dissociation of molecules by intense infrared and visible laser radiation was considered theoretically in 1964 by Askar'yan [1] and by Bunkin et al. [2] and was discovered experimentally and interpreted at Saclay in 1971-1972 during a study on interaction of infrared Nd-laser radiation and its second harmonic with diatomic molecules by Mainfray et al. [3,4].

In view of the strong interest shown in this phenomenon in many laboratories nowadays let us characterize the main features of multiphoton dissociation and review briefly the related theoretical and experimental results and formulate a general theoretical approach developed for multiphoton transition probability calculations for molecules. This approach is based on application of the time-dependent perturbation theory (worked out in detail over the last few years by Gontier and Trahin [5], Rapoport et al. [6], by Lambropoulos [7] and other authors for multiphoton transition in atoms) to diatomic and polyatomic molecules, the states of which are considered in terms of mixed states of zero order model approximation. Such mixed states may be due for example, to interactions between different molecular vibrational modes, configurational interactions or dynamic coupling between electron motion and nuclear motion, and are considered in a general way in terms of expansions of the molecular wave functions over wave functions of one or another model approximation.

Studies [1] and [2] led to experimental investigations on photochemical and photophysical processes arising from excitation and dissociation of molecules by infrared and visible laser radiation. The first works in this direction were accomplished mainly with condensed matter or with gases under sufficiently high pressure. Prokhorov et al. [8] observed two photon and three-photon

---

*Present Address: Service de Physique Atomique, Centre d'Etudes Nucléaires de Saclay, B. P. No. 2 - 91190, Gif-Sur-Yvette (France).

absorption processes due to interaction of intense ruby laser
radiation with a polycrystalline sample of complex organic molecules (transparent at $\lambda = 0.69$ μm) followed by excited state luminescence. They also observed a rise in the vibrational temperature with respect to the translational temperature and photolysis
of α-nitronaphtholene which absorbs strongly at $\lambda = 2.36$ μm during
interaction with the continuous dysprosium laser infrared radiation (stepwise absorption of 2 or 3 quanta of radiation). In
contrast with multiphoton excitation the probability of multistep (cascade) processes depends strongly on the V–V and V–T re-
laxations which determine the nature of the interaction between
the radiation and the dense medium. Multistep dissociation de-
veloping through vibrational sub-levels was observed under different conditions summed up in [9]. Visible luminescence over a
wide spectral band of heavy organic molecules (possessing a high
vibrational state density) was observed in [10] in a crystalline
diphenyl sample and in some liquids irradiated by a ruby laser
pulse under very fast V–V relaxation conditions. The multistep
dissociation of H_2O in the liquid state under infrared radiation
of an Nd-laser was observed in [11].

A series of papers was devoted to the investigation of heating [12,13] and cascade dissociation [14,15] of a number of molecular gases which resonantly absorb infrared CO_2 laser radiation (pulsed or continuous) at pressures around a few tens or
hundreds torr. Visible luminescence from the thermal dissociation products of NH_3, SF_6 and other gases caused by heating under
continuous CO_2 laser radiation was observed by Borde et al. [12]
and by Losev et al. [13]; dissociation of BCl_3 due to vibrational
heating of the molecule by resonant IR radiation (resulting from
the cascade population of high vibrational levels by V–V relaxation processes) was studied by Karlov et al. [14]. Cascade dissociation of N_2F_4 was observed in [15]. Note that the absorption
frequencies of N_2F_4 fuse to an absorption band of width about
100 cm^{-1} at pressure p > 15 torr [16].

Visible luminescence from electronically excited dissociation
fragments was observed by Isenor and Richardson [17] at p ~ 50 –
1000 torr and by Letokhov et al. [18] in a number of resonantly absorbing gases under the action of CO_2 laser pulses (flux
density F ~ $10^{27} - 10^{29}$ cm^{-2} s^{-1}) of frequency coinciding with vibrational-rotational absorption lines. It was found in [18] that
the luminescence appeared at a definite, threshold value of pulse
energy. The value of the threshold energy was strongly dependent
upon gas pressure. It was suggested that visible luminescence
probably occurs by a certain avalanche mechanism induced by effective vibrational heating of molecules similar to that considered in [14].

The luminescence observed in [17] had no laser intensity threshold. This type of luminescence has been shown to occur in gases irradiated with continuous CO_2 lasers of low power [12-14]. Considerable enhancement of luminescence of excited electronic states of the dissociation products was observed with the addition of a small proportion of admixture [17]. In [18] it was also noted that the presence of different admixture which were not controlled in the experiment, may essentially facilitate the operation of the avalanche mechanism. Consequently the luminescence observed in [17,18] results from the important part played by collisions and is not related to multiphoton collisionless dissociation.

The multiphoton collisionless dissociation of the molecule was established and studied experimentally for the first time during research on the interaction between the infrared radiation of Nd laser (and visible radiation of its second harmonic) and some diatomic molecules, especially hydrogen molecules [3,4]. [Experimental studies of laser interactions with low-pressure gases were also carried out by the group of Delone at the Lebedev Institute. The interaction of H_2 with Nd laser was studied in [19] and with ruby laser in [20]. Since H_2^+ ions were mainly observed $N(H_2^+)/N(H^+) \simeq 10^3$ [19] the authors concluded that ionization occurs ($H_2 + 15.4$ eV $\rightarrow H_2^+ + e$) but not dissociation ($H_2 + 4.5$ eV $\rightarrow H + H$). As a matter of fact, the multiphoton dissociation of H_2 suggested [19] being the result of vibrational transition in the ground electronic state is forbidden by the Frank-Condon principle. (Similar misleading propositions prevented interpretation of the laser radiation interaction with the molecules also in some other studies.) Under experimental conditions similar to those of [19] other results were obtained in [4]: the H^+ ions were observed mainly ($N(H^+) >> N(H_2^+)$) as in the case of the second harmonic Nd-laser radiation [3]. The origin of H^+ was accounted for by dissociation of H_2 being the multiphoton electronic vibrational transition which leads to atoms in the ground state H(n=1) and the excited state H(n=2). In [19,20] the dissociation products were analyzed by a time-of-flight mass spectrometer. Note that the use of a rather long time-of-flight path for good signal separation can lead to considerable ion loss. In [3,4] the masses were analyzed with a magnetic mass spectrometer.] Owing to the fact that the multiphoton dissociation study was carried out with H_2, which has well-known spectral properties, it was possible to reveal from this example the main characteristics of the phenomenon observed.

The features of multiphoton dissociation established in [3,4] are of a general nature and are common to any interaction between intense radiation and isolated molecules:

352   I. I. Tugov

1)   because the experiment was carried out at low pressure
      ($p \simeq 10^{-4}$ torr) it was shown that purely radiative collision-
      less dissociation under intense laser radiation (flux den-
      sity $F \sim 10^{30}$–10^{31} cm^{-2}s^{-1} and photon energy $\hbar\omega \ll D$) is pos-
      sible as a result of multiphoton processes (D is the dis-
      sociation energy).
2)   the multiphoton dissociation probability is characterized by
      the intensity threshold and reaches saturation when the in-
      tensity is high enough.
3)   multiphoton dissociation is a resonant process. In contrast
      to multistep processes this property is generally not re-
      lated to the coincidence of the laser frequency with that of
      allowed transition. Resonances appear at the nuclear struc-
      ture of intermediate states when the energy of transition
      from the initial state $|o\rangle$ to some vibrational-rotational
      level $|n\rangle$ coincides with the energy of some quanta of radia-
      tion, i.e., they are multiphoton resonances ($E_n - E_o = n\hbar\omega$).

The resonant nature of intense laser interaction with the
molecule is, on the condition that $\hbar\omega \ll D$, a general feature of
multiphoton dissociation. After the generally non-resonant absorption of some first quanta at the initial stage of the multiphoton process, the region of high density of states is reached:
quasi-continuum of vibrational-rotational levels. (It is obvious
that the passage to quasi-continuum absorption behavior takes
place all the sooner as the molecule is more complicated and as
the number of electronic or vibrational degrees of freedom is
larger.) In the case of the multiphoton dissociation of H_2 the
resonances at the vibrational-rotational levels of excited electronic states play an important part. Since the half-width of
the laser line in [3,4] was of the order of the excited state rotational constant B_e, and since the band of the possible rotational energy values of these resonance states was of the same
order as its vibrational frequencies ω_e, the effective quasicontinuum started to spread out at the beginning of the excited
terms region (at the 5th intermediate state for $\lambda = 0.53$ μm and
at the 11th intermediate state for $\lambda = 1.06$ μm).

4)   The quasi-continuum for multiphoton absorption corresponds
      to the energy region where the molecular structure and
      spectrum cannot be considered in terms of approximations
      connected with the treatment of the degrees of freedom of
      the molecule independently.

Resonances were observed in [3] and [4] on the states related to the $E,F^1\Sigma_g^+$ term, which results from the strong interaction between covalent and ionic configurations. A similar situation must arise in polyatomic molecules, the vibrational levels

of which cannot be considered in terms of anharmonic oscillators
related to different vibrational modes considered independently
(especially if the resonance conditions are considered or the
quasi-resonance excitation probabilities are calculated).

5)   As a result of the passage to the quasi-continuum (which was
fast under the experimental conditions used in [3,4]),the molecule irreversibly disintegrates into fragments in the ground
and electronically excited states. The dissociation may take
place either directly or through the predissociation of highly
excited states strongly dynamically coupled with dissociative
continuum [3].

These general features of multiphoton dissociation of molecules established in [3] and [4] using the example of H_2 are entirely confirmed by experiments accomplished recently with CO_2
lasers. The observations and correct interpretation of multiphoton dissociation at 10.6 µm became possible owing to considerable reduction of the gas pressures which ensured the collisionless nature of the molecule-radiation interactions. The first
work in which multiphoton dissociation was really observed at
10.6 µm was carried out by Ambartzumian et al. [21] in 1974.*
The instantaneous luminescence of electronically excited fragments
of BCl_3 at pressures down to 0.05 torr was observed (at flux density $F \sim 10^{28} - 4.10^{29}$ $cm^{-2}s^{-1}$, without accounting for possible
self-synchronization of axial laser modes, and at full pulse duration $\tau_p = 400ns$). Note that using a shorter pulse duration similar
to that used in [3], the flux densities necessary for observation
of the effect overlap with those used in [3]. At similar intensities ($F \sim 5.10^{28} - 10^{29}$ $cm^{-2}s^{-1}$) the selective dissociation of
SF_6 molecules under the action of CO_2 laser was observed in [24].
Initially the observed dissociation of polyatomic molecules was
related to the need to tune the laser frequency on the peak of

---

*The fast fluorescence of SiF radicals was observed in [22] during
irradiation of SiF_4 by a laser flux $\sim 10^{29}$ $cm^{-2}s^{-1}$; this phenomenon
appears with a delay of $\sim 100ns$ at p=5 torr. The presence of this
delay and the absence of intensity threshold led the authors to
the conclusion that SiF formation was due to an energy build-up
process rather than a multiphoton process. In [23] it was noted
that the fast s age of luminescence observed in [22] differed
essentially from the instantaneous luminescence observed in BCl_3
[21] by, for example, its dependence on gas pressure. The addition of a foreign gas led in [22] to damping of the fast fluorescence, a fact characteristic of phenomena where the role of collisions is essential. In [23] it was also noted that the in-
stantantaneous luminescence [21] probably had a threshold
character.

the absorption band of the molecule. It was therefore suggested
that only a small fraction of the molecules was excited, those
situated in the small number of rotational levels [24] and that
the phenomenon observed is not related to the multiphoton dissociation processes considered in [1] and [2].

Let us recall some of the theoretical results obtained. The
idea of coherent resonant excitation of high vibrational levels
and dissociation of molecules by intense IR radiation was proposed in [1]. For the estimation of the effect the harmonic bond
model was used. The excitation probability W of the Vth level of
quantum oscillator with frequency ω_e in field $E_o \sin \omega t$ is determined by classical work accomplished by the field $\varepsilon(t) =
2\hbar\omega_e(dE_o/\hbar\Delta)^2(\sin \Delta t/2)^2$, $\Delta = \omega_e - \omega$. From quasi-classical esti-
mates [1a] it follows that at resonance $(t \cdot \Delta/2 \ll 1)$, the time τ
for build-up of the oscillation amplitude $a(t)$ to the critical
value a_o about of molecular dimensions, $\varepsilon(t) = m\omega_e a_o^2/\hbar$, at field
intensity $I = 10^9$ Wcm^{-2} will be 10^{-11}s; m is the reduced mass related to the vibrational mode, a_o taken as the Bohr radius; the
molecular bond parameters used in [1a] correspond in quantum
terms to the dipole moment matrix element $d \simeq 0.3D$ and to absorption by the molecule, at a time small compared with the pulse
duration, $\tau \ll \tau_p$ of about 50 quanta of radiation.

Independently the possibility of multiphoton dissociation
of molecules was considered in [2] using the example of the isolated anharmonic vibrational bond (quasi-diatomic model). The
role of level broadening for compensation of the detuning of
resonances was mentioned (see also [1]), together with the importance of the irreversible radiative disintegration of the
resonance levels in eliminating resonant population oscillations.
A method of multiphoton probability calculations for diatomic
molecules was given in [25], based on the use of the model potential sets proposed for the homopolar and for the ionic bonds.

The basis of any theoretical consideration of the molecular
systems and their interaction with radiation lies in the determination of the transition matrix elements. In spite of the
obvious importance of matrix element calculations, a satisfactory
general approach to this problem does not exist nowadays even for
transitions determined by first-order perturbation theory.

For the multiphoton transition probability calculations considered in the Nth order perturbation theory, the problem becomes much more difficult. For (N-1) summations over the intermediate state spectrum it is necessary, in principle, to know
transition matrix elements between all molecular states coupled
during a multiphoton transition. The known methods of calculating matrix elements are not adequate for this purpose. For
example, considering multiphoton dissociation it is necessary to

take into account transitions between high vibrational levels.
However, the calculations for this problem with any known method
give rise to difficulties even in the simplest case of calculations of the Franck-Condon factors.

The approach developed in [26] allows one, in a number of
cases, to avoid the difficulty of numerical integration. It is
based on the representation of molecular states as a result of a
perturbation of zeroth order approximate adiabatic states whose
vibrational-rotational structure is determined by model potentials. A perturbation theory developed on the basis of model
vibrational eigenfunctions has a constructive character. It
makes it possible to calculate transition probabilities both in
model and quantitative forms including calculations with higher-
order time dependent perturbation theory and nonadiabatic calculations. In the model approximation the matrix elements have
been calculated analytically in a general way for any process of
excitation, dissociation or ionization of the molecule as well
as for nonadiabatic transitions (e.g. autoionization or predissociation) [26]. The accurate values of the transition matrix
elements are determined using expansion over model matrix elements.

The elementary formulas obtained for model matrix elements
make it possible to calculate the compound matrix elements of the
multiphoton processes by direct summation (especially for quasi-
resonance processes) or by the use of vibrational Green functions
of model terms [25].

In any consequent approximation the accurate matrix elements
are expressed through linear combinations of model matrix elements for transitions between all possible vibrational-rotational
states of the related terms. All these matrix elements appear
in the compound matrix element of a multiphoton transition. Thus,
the accurate calculation of multiphoton transition probabilities
turns out to be reduced to the model calculations.

Multiphoton processes in molecules proceeding via highly
excited vibrational or electronic states are strongly related to
the nonadiabatic phenomena [3]. The calculation of probabilities
of nonadiabatic transitions is much more complicated than that of
probabilities of electronic-vibrational transitions. The num-
of works devoted to these questions is limited and, in general,
they deal with the simplest molecules. Transitions between high
vibrational levels or continuum states for which the matrix
elements should be integrated over a wide range of internuclear
distances [27], are of most interest. The method of calculation
of vibrational matrix elements of nonadiabatic transitions is
contained in the framework of the approach developed.

356   I. I. Tugov

The formalism developed in [26] was applied to some classical problems of molecular spectroscopy [28] as well as to transition probability calculations of a number of quasiresonant processes [29]. The accuracy of this method was compared with the results of standard methods based on numerical solutions of wave equations. For the vibrational-rotational energy levels of some diatomic molecules (the only spectroscopic property which can be determined by usual methods with sufficiently high accuracy for the simple diatomics) the approach developed provides accuracy equivalent to that of the standard methods ($\sim 10^{-7}$ a.u.) [26].

The probabilies of quasiresonant multiphoton processes have been calculated for real diatomic molecules and for quasi-diatomic models of polyatomic molecules. The quasiresonant multiphoton process of excitation of high vibrational levels of a molecule by the IR laser radiation $|V_0,J_0> + n\hbar\omega \to |V_0 + n,J_f>$ followed by irreversible disintegration of the molecule was considered and proposed [30] as a probable mechanism responsible for the "instantaneous" dissociation of polyatomic molecules observed in [21]. It was shown in [30] that this multiphoton process can provide greater isotopic selectivity than the one-photon vibrational excitation. Indeed, for one-photon excitation of the band of rotation sublevels ($J' \leqslant J_0 \leqslant J''$) in the Σ-term ($\Delta J = \pm 1$) being isotopically selective, it is necessary that $2B_e(J''-J') < \Delta_i$; (Δ_i - isotope shift). In the case of the Q-branch of multiphoton odd order process ($J_f-J_0 = 0$) the condition of selective excitation of the same rotational band is $\alpha_e(V_f-V_0)(J''-J')(J''+J'+1) < \Delta_i$.

The multiphoton odd-order quasiresonant excitation of HCl was considered and shown to be at $T = 300°K$ two times more effective selective-excitation process than one-photon process [30]. Let us note that the P, Q and R branches of multiphoton process are generally more probable than those related to the greater change of rotational energy (on the analogy of the Franck-Condon principle). The sequence of the most important rotational sublevels of intermediate resonant states may be easily established [30] for the process considered.

The numerical calculations of the quasiresonant multiphoton processes $|v_0> + n\hbar\omega \to |v_0+n>$ in diatomic molecules and for quasi-diatomic models of polyatomic molecules presented in [30,31,32] show that the intensities necessary for observation of the effects are of the order $10^7 - 10^9$ Wcm^{-2}.

REFERENCES

1. G. A. Askar'yan, JETP <u>46</u>, 403 (1964); JETP <u>48</u>, 666 (1965).

2. F. V. Bunkin, R. V. Karapetian, A. M. Prokhorov, JETP $\underline{47}$, 216 (1964).

3. G. Mainfray, C. Manus, I. I. Tugov, JETP Lett. $\underline{16}$, 19 (1972).

4. M. Lu Van, G. Mainfray, C. Manus, and I. I. Tugov, Phys. Rev. Lett. $\underline{29}$, 1134 (1972); Phys. Rev. $\underline{A7}$, 91 (1973).

5. Y. Gontier and M. Trahin, Phys. Rev. $\underline{A7}$, 1899 (1973).

6. B. A. Zon, N. L. Manakov, and L. P. Rapoport, JETP $\underline{56}$, 400 (1969); $\underline{61}$, 968 (1971).

7. P. Lambropoulos, Phys. Rev. $\underline{A9}$, 1992 (1974).

8. A. M. Prokhorov, V. D. Shigorin, G. P. Shipulo, Doklady $\underline{175}$, 793 (1967).

9. V. L. Talrose, P. P. Barashev, Zh. Vces. Khim. ob. im D.I. Mendeleeva $\underline{18}$, 15 (1973).

10. K. B. Eisenthal, W. L. Peticolas, K. E. Pieckhoff, J. Chem. Phys. $\underline{44}$, 4492 (1966).

11. D. U. Goodall, R. C. Greenhow, Chem. Phys. Lett. $\underline{9}$, 583 (1971).

12. M. C. Borde, A. Henry, M. L. Henry, Compt. Rend. Acad. Sci. $\underline{B262}$, 1389 (1966).

13. V. V. Losev et al., Khim. Vys. Energ. $\underline{3}$, 331 (1969).

14. N. V. Karlov, Yu. N. Petrov, A. M. Prokhorov and O. M. Stel'Makh, JETP Lett. $\underline{11}$, 220 (1970).

15. J. L. Lyman, R. J. Jensen, Chem. Phys. Lett. $\underline{13}$, 421 (1972).

16. N. G. Basov, E. P. Oraevski, A. V. Pankratov, Doklady $\underline{198}$, 1043 (1971).

17. N. R. Isenor and M. C. Richardson, Appl. Phys. Lett. $\underline{18}$, 224 (1971).

18. V. S. Letokhov, E. A. Ryabov, O. A. Tumanov, Opt. Comm. $\underline{5}$, 1968 (1972).

19. N. K. Berezhetskaya et al., JETP, $\underline{58}$, 753 (1970).

20. G. S. Voronov, G. A. Delone, N. B. Delone, O. V. Kudrevatova, JETP Lett. $\underline{2}$, 377 (1965).

21. R. V. Ambartzumian, N. V. Chekalin, V. S. Doljikov, V. S. Letokhov, and E. A. Ryabov, Chem. Phys. Lett. $\underline{25}$, 515 (1974).

22. N. R. Isenor, V. Merchant, R. S. Hallsworth, M. C. Richardson, Can. J. Phys. $\underline{51}$, 1281 (1973).

23. R. V. Ambartzumyan et al., JETP $\underline{69}$, 72 (1975).

24. R. V. Ambartsumyan, Yu. A. Gorokhov, V. S. Letokhov and G. N. Makarov, JETP Lett. $\underline{21}$, 375 (1975); JETP $\underline{69}$, 1956 (1975).

25. F. V. Bunkin and I. I. Tugov, Phys. Rev. $\underline{A8}$, 601 (1973); $\underline{A8}$, 620 (1973); $\underline{A11}$, 1765 (1975); I. I. Tugov, Phys. Rev. $\underline{A8}$, 612 (1973).

26. I. I. Tugov, Perturbation Theory for Molecules, I, Lebedev Institute Report, 1977.

27. R. S. Berry, and S. E. Nielsen, Phys. Rev. $\underline{A1}$, 383, 395 (1970); $\underline{A4}$, 865 (1971).

28. I. I. Tugov, Opt. and Spectr. (to be published).

29. I. I. Tugov, Perturbation Theory for Molecules, II, Lebedev Institute Report, 1977.

30. I. I. Tugov, Europhysics Study Conference on Multiphoton Processes, invited paper, Seillac, April, 1975.

31. V. E. Merchaut and N. R. Isenor, IEEE J. Quantum Electron., Vol. QE-$\underline{12}$, 603 (1976).

32. I. I. Tugov, Perturbation Theory for Molecules, III, Lebedev Institute Report, 1977.

Molecular Beam Studies on Multiphoton Dissociation of Polyatomic Molecules

E. R. GRANT,* P. A. SCHULZ,[†] Aa. S. SUDBO,[†]
M. J. COGGIOLA,[‡] Y. R. SHEN,[†] AND Y. T. LEE*
Materials and Molecular Research Division
Lawrence Berkeley Laboratory
University of California
Berkeley, California

The possibility that an isolated molecule placed in the intense field of an infrared laser can absorb many photons (enough to dissociate) in a very short span, is first suggested in 1971 by Isenor and Richardson [1] based on their experimental observation of prompt luminescence. This infrared multiphoton dissociation process was subsequently shown to be isotopically selective in the dissociation of SF_6, BCl_3 and other molecules in 1975 by Ambartzumian et al. [2] and Lyman et al.[3], in a series of experiments, and has now become one of the most extensively pursued subjects in recent years. The process has been shown to be quite efficient, and the physical principles and phenomena involved in multiphoton absorption by isolated molecules have been elucidated through many experimental and theoretical investigations [4,5]. However, there are many questions on the dynamics of excitation and decomposition which remain to be answered; especially, the extent of vibrational excitation and the degree of intramolecular relaxation before molecular decomposition, i.e., the question of whether the excitation energy is completely randomized such that a simple RRKM theory [6] can be used for the description of decomposition dynamics.

---

*Also associated with the Department of Chemistry, University of California, Berkeley, California.

[†] Also associated with the Department of Physics, University of California, Berkeley, California.

[‡] Present address; Stanford Research Institute, Menlo Park, California.

Actually, a considerable amount of interest and excitement
was generated with the suggestion that multiphoton dissociation
is a very unusual method for energizing molecules, one that of-
fered the potential for vibrational mode control of unimolecular
reactions. For the present situation, this means deciding the
dissociative outcome of a reaction by not only controlling how
much energy is supplied to each molecule, but also, loosely speaking, by directing which vibrational modes the energy is to go into
and producing dissociation products different from those of thermal
activation. Initial evidence tended to support this hypothesis.
The first reported primary product analysis for the multiphoton
dissociation of SF_6 indicated that this molecule dissociated directly to SF_4 and F_2 [7], bypassing the lower energy SF_5 and F
fragmentation channel. Results of gas cell experiments with
$CFCl_3$ were interpreted to evidence direct dissociation to CFCl
and Cl_2 [8]. In addition, the observation of C_2 and CN from
CH_3CN and CH_3NH_2 [9] and prompt luminescence for a number of molecules [10] under conditions which were supposedly collision-free
suggested explosive dissociation of very highly excited molecules.
     If indeed such phenomena are a direct consequence of mode-
specific vibrational pumping, the implications, both fundamental
and applied, would be very important. The measurements of such
unusual dynamical effects could significantly aid in the fundamental understanding of energy flow in highly excited polyatomic
molecules. Also, in a practical sense, a technique for controlling the path of a chemical reaction, bypassing, say, an undesired
low-energy channel, could prove to be of considerable utility for
synthetic applications.
     For these reasons we were attracted to a study of the multi-
photon dissociation of SF_6 and other molecules under the collisionless molecular beam conditions used routinely in our laboratory.
A molecular beam experiment on CO_2 laser-induced photodissociation
of SF_6 was performed previously by Brunner et al. [11], in which
the influence on the molecular beam of SF_6 itself by the CO_2 laser
was studied using the mass spectrometer detector. In our experiments, using a rotatable mass spectrometer, direct identification
of fragment species as well as the measurements of angular and
velocity distributions of dissociated products were performed under various laser excitation conditions. The results of our experiments, mainly derived from the dissociation pattern and the
relation between the shape as well as the magnitude of translational energy distributions, are unambiguously consistent with a
far more orderly picture of the excitation and dissociation process than that suggested by many earlier experiments and theoretical constructs. Recent work of Bloembergen, Yablonovitch, and
their coworkers [12] on the measurements of multiphoton absorption

by SF_6 and their theoretical interpretation of the experimental
observations are in close agreement with our conclusion that although the initial excitation is selective, nearly complete energy randomization must take place in the molecular vibrational
degrees of freedom before decomposition.

A schematic diagram of the apparatus used in our study is
shown in Fig. 1. It consists of a nozzle source for the propagation of a supersonic beam of SF_6 through three regions of differential pumping into a main chamber crossed by a focused CO_2
laser beam at the scattering center. The laser energy density
at this intersection was varied by both moving the lens to de-
focus the light and by using external attenuators.

The products of reaction were detected by means of a triple-
differentially-pumped quadrupole mass spectrometer contained
within the main chamber. The mass spectrometer was rotatable
about the vertical axis through the scattering center. This feature combined with internal defining slits made the detection
system angularly resolved to within 0.5°.

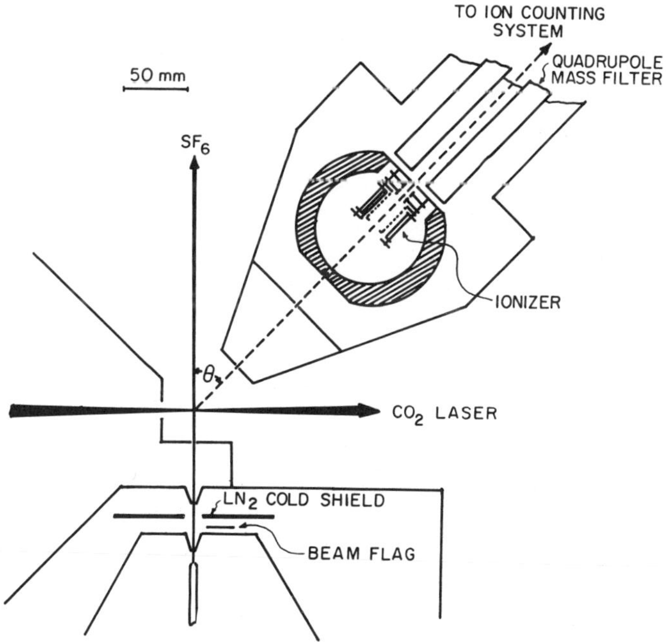

FIGURE 1. Schematic diagram of the molecular beam apparatus
used for the study of multiphoton dissociation.

The experiments were performed in two modes of operation. To simply obtain the product angular distribution, the laser trigger pulse was used to start a gate and delay timing circuit that regulated the detection electronics. An initial delay of generally 300–400 μsec between the firing of the laser and the enabling of the detector was used to allow for the flight time of the fragments traveling over a distance of 21.4 cm. Then, a gate of approximately 500 μsec was opened to insure that all fragments arriving at the detector were counted. A subsequent gate of equal duration provided a background count for the determination of net signal. Signal-to-background in this mode at the optimum angle ran as high as four to one. Time-of-flight measurements were made at several angles by using the laser pulse to define a time origin in starting a 256-channel multiscaler. Channel widths of 8 μsec generally provided adequate velocity resolution and enough of the time-of-flight spectrum to establish a baseline. The resolution of time-of-flight measurements is approximately 30 μsec, mainly due to the finite width caused by the ionizer. These angular and time-of-flight data were converted to velocity space and transformed into the center of mass to provide the recoil energy distribution for the multiphoton dissociation of SF_6.

For all the polyatomic molecules we have investigated so far, the molecules are seen to decompose through the channel that requires the least amount of energy, without exception. For example, CF_3Br and CF_3I dissociate solely into $CF_3 + Br$ and $CF_3 + I$. The high energy channels such as $CF_2Br + F$ or $CF_2 + BrF$ have not been observed. Contrary to the earlier suggestion that CCl_3F dissociates into $CClF + Cl_2$ [8], the only channel which was detected in our collisionless molecular beam experiment was $CCl_2F + Cl$. No F_2 was detected in the decomposition of SF_6. These observations are in agreement with what one would expect from a statistical theory of unimolecular decomposition. A situation exists in which the intramolecular energy transfer is faster than the chemical decomposition such that an isolated molecule serves as its own heat bath for thermal decomposition. These results also suggest that some of the observations made previously in cell experiments, even at pressures below 1 Torr, might not be entirely "collisionless."

The translational energy distributions of fragments for those systems mentioned above have one common feature; namely, the average kinetic energy released is relatively low, no more than a few kcal/mole, and the shape of distributions are strongly peaked at zero kinetic energy. This is also a feature one would expect from a statistical unimolecular decomposition. This is true because of the fact that if the excitation energy is completely randomized, the probability of a molecule decomposing into product molecules with specific translational and internal energies will depend on the phase space volume available. Hence,

for polyatomic molecules with many vibrational degrees of freedom,
(since the state density for the internal degrees of freedom is
a steeply increasing function of energy, whereas that for translation is not), the statistical theory characteristically predicts
a recoil energy distribution that is peaked near zero, i.e.,most
of the excess energy is stored as internal excitation of fragments,
and correspondingly large amounts of translational energy release
are improbable. The average recoil energy of the fragments is a
function of laser energy, the number of vibrational degrees of
freedom, and density of states associated with those vibrational
degrees of freedom. If the statistical theory is applied to
CF_3Br and SF_6, observed translational energy distributions correspond respectively to an average absorption of 1-3, and 6-10
excess CO_2 laser photons beyond dissociation energies of CF_3Br
and SF_6. Although these numbers seem to suggest a vast difference in the ability of molecules to absorb excess photons beyond
dissociation energy, they actually represent very similar degrees
of excitation in terms of the lifetime of excited molecules. For
both CF_3Br and SF_6 molecules, the range of excitations correspond
to dissociation states with a lifetime ranging from 10^{-7} to 10^{-9}
sec, which is not entirely unrelated to the typical laser pulse
width of 5×10^{-8} sec.

Although the multiphoton dissociation of SF_6 molecules has
attracted more attention than any other polyatomic molecule, its
detailed experimental study is difficult because its dissociation
mechanism is complicated, and as a consequence, our previous experimental results require further clarification. The complications that we have observed are partly due to the unusual fragmentation pattern of sulfur polyfluoride compounds in the ionization chamber of our mass spectrometer and partly due to the fact
that the SF_5 fragment can further decompose into $SF_4 + F$ in the
intense CO_2 laser field. Both SF_6 and SF_4 are known to completely
decompose in the ionization process to give SF_5^+ and SF_3^+ ions
respectively, as major mass peaks in the mass spectrometer. It
appears that the SF_5 radical as well only produces a negligible
fraction of the parent SF_5^+ in the ionization process. Our preliminary observation [13] of SF_5^+ was found to be heavily con-
taminated by SF_6 molecules that were scattered into the detector
by molecules released from the wall of the vacuum chamber by the
laser radiation. When the experimental conditions were improved
to eliminate this background scattering, the most prominent peaks
in the mass spectrometer were found to be SF_3^+ and SF_2^+. When
the laser energy was low, namely, around 10 joule/cm^2 or below,
the laboratory angular distributions of SF_3^+ and SF_2^+ were found
to be identical as shown in Fig. 2. Irrespective of the laboratory scattering angles, the ratios of SF_3^+ to SF_2^+ were found to

be approximately 4, which is between the value of 6 and 2 given
by SF_6 and SF_4 ionization [14] and is believed to be from a single
parent molecule of SF_5. When the energy of the laser pulse was
increased, the angular distributions of SF_3^+ and SF_2^+ started to
deviate from each other at wide angles as the angular distributions broadened and the ratios of SF_3^+ to SF_2^+ at wide angles
approached ~ 2 which is the value given for ionization of SF_4.
Thus, we conclude that the signals observed at the wide angles
were exclusively due to SF_4. These observations suggest that at
higher laser energies, the SF_5 radicals produced during the laser
pulse were further decomposed into $SF_4 + F$ with additional recoil
energy resulting from absorption of additional photons. This is
supported by the facts that the production of F atoms in SF_6 dissociations have been observed from the secondary chemiluminescent
$F + H_2$ reaction in other laboratories [15] and the F_2 fragments
were not detected in spite of an exhaustive search in our experiments. We therefore conclude that $SF_6 \rightarrow SF_4 + F_2$ is not a significant channel in the multiphoton decomposition of SF_6.

The observation that SF_6 molecules decompose in a time comparable to the laser pulse width is most interesting, since it
provides further verification of the statistical model. Preliminary analysis of the time-of-flight velocity spectra of the fragments, as shown in Fig. 3 as a typical example, indicates that
the level of excitation which yields the observed translational
energies according to the statistical theory is approximately
6–10 excess photons absorbed beyond dissociation energy. This
level of excitation gives the typical lifetime of excited SF_6 of
10^{-7} to 10^{-9} sec which is comparable to the laser pulse width of
5×10^{-8} sec. Extensive data have been collected on many other
aspects of multiphoton dissociation of SF_6 molecules in our molecular beam studies, including: the dependences of dissociation
on the laser frequency, the laser energy, and the rotational and
vibrational temperatures of SF_6 molecules. These experimental
results will be presented in other publications in the near
future.

As an aid to interpreting our experiments we have developed
a very simple mathematical model for the pumping and dissociation
process. Proceeding from Bloembergen and Yablonovitch's observation [12b] that absorption in the quasi-continuum occurs sequen-
tially with decreasing cross section, and our experimental find-
ings that the dissociation process can be treated statistically,
we constructed the corresponding ordinary master equation. It
describes the pumping process in completely linear terms as absorption and stimulated emission over a series of evenly spaced
levels; irreversible reactive transitions to products are included above the dissociation threshold. This model does not
consider the possibility of coherent processes that may occur in

the first few transitions. The linear approximation, however,
should be a good one for most of the powers of interest (above a
few Mw cm^{-2}) where in a real system any discrete initial bottleneck is easily overcome.

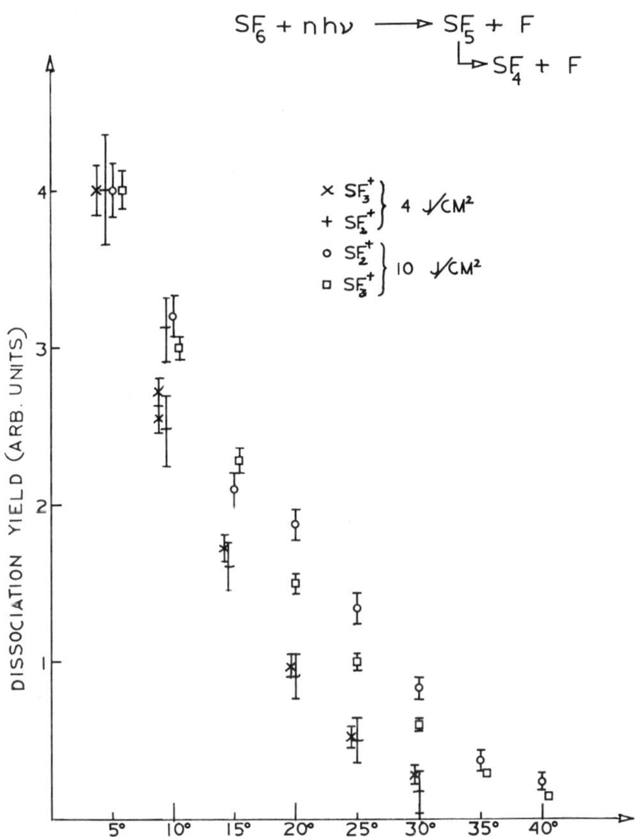

FIGURE 2. Angular distributions of SF$_3^+$ and SF$_2^+$ at higher and
lower laser intensity: lower intensity, X, SF$_3^+$; +,
SF$_2^+$; higher intensity, □, SF$_3^+$; O, SF$_2^+$.

     In our application we determined the physical parameters for
this model in a purely empirical fashion. We assumed, as the results of Bloembergen and Yablonovitch indicate, that the absorption cross section decreases exponentially with increasing molecular energization (expressed hereafter as I, the number of
photons absorbed). We thus assumed a cross section function of
the form

$$\log \sigma_I = mI + b \quad . \tag{1}$$

The parameters for Eq. (1) were determined by fitting the results of our calculation to experimental data. We found that the rate of pumping in the early stages of absorption was a sensitive function of the cross section intercept. Thus, the comparison of our calculated results for this rate to the photoacoustic data of Bloembergen and Yablonovitch served to determine a value for b (-18.361, $\sigma_1 = 4 \times 10^{-19}$cm^2). The slope of the cross section function was found by fitting our experimental measurements of the fractional yield versus incident energy influence. The result, $m = -0.03656$ gives a cross section at $I = 40$, well into the reactive manifold, of 1.5×10^{-20}cm^2. So determined, the cross section function was combined with the desired laser intensity to calculate coefficients for absorption, then weighted by the state density ratio between successive levels to produce those for stimulated emission.

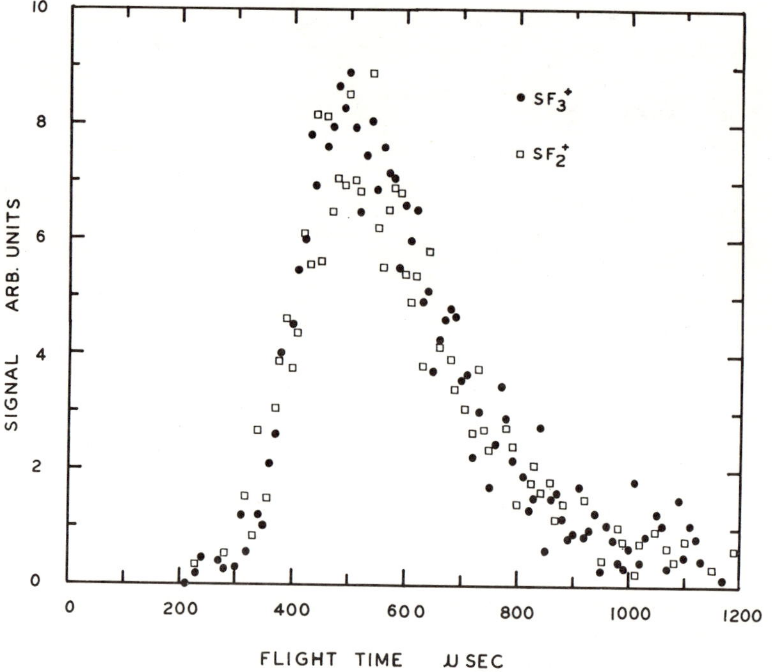

FIGURE 3.   Time of flight spectra for SF$_3^+$ and SF$_2^+$ formed in the ionizer from the multiphoton dissociative product SF$_5$: ●. SF$_3^+$; □ , SF$_2^+$.

We believe that this calibration procedure has produced a
reasonably realistic model for multiphoton absorption and dissociation. The detailed picture of this process that it presents
is quite interesting. Instantaneous level populations frozen at
several times during a laser pulse of typical power (200 MW) and
duration (100 ns) are shown in Fig. 4. We note immediately that
at this intensity multiphoton pumping to the dissociative threshold requires a time on the order of tens of nanoseconds. It is
conceivable that, at moderate levels of excitation in the dissociative manifold, unimolecular reaction could successfully compete with further pumping and the excitation process would become
lifetime limited. We see such an indication here in the high
energy tails of the distributions for 50 and 100 ns (the dissociative threshold occurs in this model at a level of 30 photons).
We observe the phenomenon experimentally as a product angular
distribution that at low powers initially broadens with increasing
power but then stabilizes to further power increases.

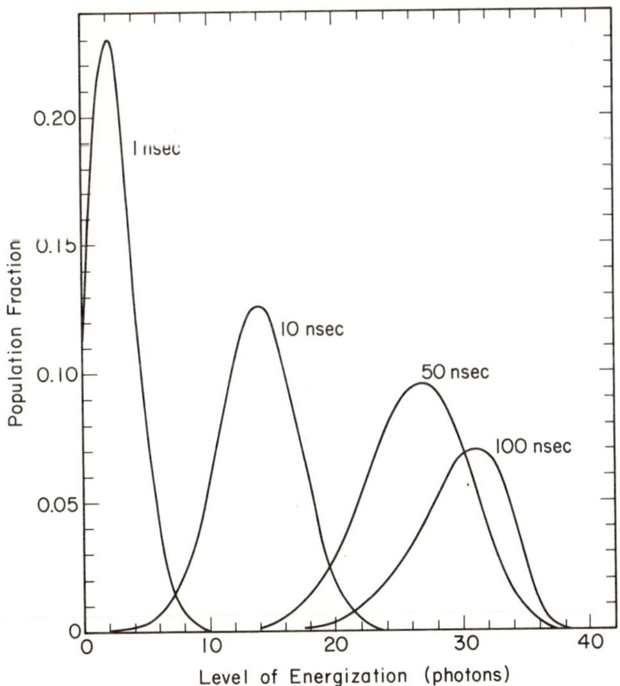

FIGURE 4.   Calculated excited state population distributions for
various times during a typical laser pulse (200 MW cm^{-2},
100 ns).

The dynamical basis for a second experimental result is also rather clearly indicated by this calculation. In the example shown in Fig. 4 the laser pulse width is 100 ns. The apparent shrinkage in the population distribution is due to a reaction that has occurred while the laser was on. Once the pulse is over, disregarding spontaneous emission and assuming a collision free environment, all the population above an energization level of 29 photons, will dissociate. The partitioning between dissociation during the laser pulse and after can be controlled in this model by appropriately choosing the laser pulse width and power. This important result is directly supported by our observation of the secondary absorption and subsequent dissociation of the newly formed SF_5. We have obtained experimental evidence that closely relates the onset of the secondary channel to the power densities predicted by the model calculation. In sum, then, all of the available experimental evidence indicates that this simple model for multiphoton dissociation is quite adequate for providing some phenomenological understanding and suggests many more detailed future experimental inquiries.

ACKNOWLEDGMENT

This work was done with support from the U. S. Energy Research and Development Administration.

REFERENCES

1.   N. R. Isenor and M. C. Richardson, Appl. Phys. Lett. $\underline{18}$, 224 (1971).

2.   R. V. Ambartzumian, Y. A. Gorokhov, V. S. Letokhov, G. N. Makarov, ZhETF Lett. $\underline{21}$, 375 (1975).

3.   J. L. Lyman, R. J. Jensen, J. P. Rink, C. P. Robinson, S. D. Rockwood, Appl. Phys. Lett. $\underline{27}$, 87 (1975).

4.   See, for example, papers in "Tunable Lasers and Applications," Proceedings of the Nordfjord Conference, edited by A. Mooradian, T. Jaeger, and P. Stokseth, (Springer-Verlag, 1976), and references therein.

5.   (a) N. Bloembergen, Optics Comm. $\underline{15}$, 416 (1975); (b) D. M. Larsen and N. Bloembergen, Optics Comm. $\underline{17}$, 254 (1976); (c) R. V. Ambartzumian, Yu. A. Gorokhov, V. S. Letokhov, G. N. Makarov, and A. A. Puretzkii, ZhETF Pis. Red. $\underline{23}$, 26 (1976); (d) S. Mukamel and J. Jortner, Chem. Phys. Lett.

$\underline{40}$, 150 (1976) and J. Chem. Phys. $\underline{65}$, 5402 (1976); (e) C. D. Cantrell and H. W. Galbraith, Optics Communications $\underline{18}$, 515 (1976); (f) V. S. Letokov and A. Makarov (to be published); (g) T. P. Cotter (to be published).

6. See, for example, P. J. Robinson, and K. A. Holbrook, "Unimolecular Reactions," John Wiley and Sons, New York, 1972.

7. K. L. Kompa in Reference 4.

8. D. F. Denver, E. J. Runwald, J. Am. Chem. Soc. $\underline{98}$, 5055 (1976).

9. K. Welge and coworkers, private communication.

10. (a) R. V. Ambartzumian, N. V. Chekalin, V. S. Letokov and E. A. Ryabov, Chem. Phys. Lett. $\underline{36}$, 301 (1975); (b) R. V. Ambartzumian, N. V. Chekalin, V. S. Doljikov, V. S. Letokov and E. A. Ryabov, Chem. Phys. Lett. $\underline{25}$, 515 (1974).

11. F. Brunner, T. P. Cotter, K. L. Kompa and D. Proch, J. Chem. Phys. (submitted 1977).

12. (a) J. G. Black. E. Yablonovitch, N. Bloembergen and S. Mukamel, paper presented in this conference; (b) J. G. Black, E. Yablonovitch, N. Bloembergen and S. Mukamel, Phys. Rev. Lett. $\underline{38}$, 1131 (1977); (c) E. Yablonovitch, paper present in Second Winter Colloquium on Laser Induced Chemistry, Park City, Utah, March 1977.

13. M. J. Coggiola, P. A. Schulz, Y. T. Lee, and Y. R. Shen, Phys. Rev. Lett. $\underline{38}$, 17 (1977).

14. E. Stenhagen, S. Abrahamssom and F. W. McLafferty, <u>Atlas of Mass Spectral Data</u> (Interscience Publishers, New York, 1969).

15. (a) C. Wittig and coworkers at University of Southern California, comments made in the Second Winter Colloquium on Laser Induced Chemistry, Park City, Utah, March 1977; (b) J. M. Preses, R. E. Weston, Jr. and G. W. Flynn, Chem. Phys. Lett. $\underline{48}$, 425 (1977).

On the Nature of the Collisionless Dissociation of the Polyatomic Molecules by the IR-Laser Field

N. V. KARLOV
P. N. Lebedev Physical Institute
Moscow, USSR

For the time being the collisionless dissociation of the polyatomic molecules by the infrared laser radiation is one of the most interesting phenomena of the resonant interaction of intense laser radiation with the matter. This phenomenon was widely discussed at every recent Quantum Electronics [1,2] and Laser Spectroscopy [3,4] Conference, Scientific Council Sessions [5] and Schools [6]. It took a good space in review papers [7,8,9].

Even without considering the practical importance of the above phenomenon for laser isotope separation (see, for example, review papers [7,8]), it is evident that its most interesting feature is the ability of the polyatomic molecules such as SF_6, SiF_4, BCl_4 and so on to gain very high energy via the collisionless process of the absorption of large numbers of low energy radiation quanta.

The polyatomic molecule absorbs many IR quanta, but this is not a multiphoton process in the common sense of the word. The process of the absorption of many quanta by a polyatomic molecule is a process of multistep, multiple quantum absorption. So it should be labeled as a multiplestep or multiplephoton process.

The main experimental features of a multiplephoton process are rather well-known [10,11]. We treat this process in the following way [12-15]. The excitation process is divided into two stages. The first stage is a frequency selective passage through a barrier, so to say, related to the anharmonicity in the sequence of the low vibrational levels of the molecule. The second stage is the effective energy gain by a previously excited molecule taking place in the quasicontinuous spectrum characteristic of the highly excited polyatomic molecules.

For the polyatomic molecules the detuning from resonance

at the excitation of the upper vibrational levels could be compensated by a great complexity of the energy levels. The overall
number of the vibrational levels, including combination ones, for
polyatomic molecules is given by binomial coefficient C^S_{s+1+I} at
the excitation of S vibrational quanta, 1 vibrational modes.
This is a big number at the high degree of excitation. Because
of the rotational structure at the excitation level corresponding
to the $(3-4)h\nu$, the averaged distance between the levels becomes
comparable with the homogeneous width of rotational sublevel. All
the levels constitute a united quasicontinuous background. The
quasicontinuum energy levels could be excited step by step with
the energy step equal to the exciting laser quantum energy $h\nu$.
Going this way a molecule could be strongly excited up to the
dissociation limit but generally speaking by forbidden transitions. The last is a weak side of the homogeneous quasicontinuum
approach.

The quasicontinuum in the excitation spectrum may cause the
collisionless ion formation by a polyatomic molecule being irradiated by an intense IR laser field [13]. The ion current could
be a very convenient means to investigate experimentally the processes to be discussed here.

For the time being the important role of the vibrational
level quasicontinuum in the process of the vibrational energy
gain by a single molecule is well established. The existence of
the quasicontinuum has been directly demonstrated by the two-
frequency technique [11]. By this technique a radiation at a frequency resonant within the linear absorption spectrum of the
molecule excites low lying vibrational levels. The intensity of
this exciting radiation should be low enough, not to cause any
measurable Stark broadening thus violating the selectivity.

At the simultaneous irradiation of the molecule by an intense enough IR laser field at some frequency ν_2, the dissociation will be observed. Taken separately the weak exciting field
ν_1 and strong dissociating field ν_2 do not dissociate the molecule
provided ν_2 is sufficiently detuned. It is evident that the non-
resonant radiation ν_2 is absorbed by the vibrational quasicontinuum.

The existence of the quasicontinuum is established with good
reliability. But we do not know how the molecules enter the
quasicontinuum (i.e. how continuous is the quasicontinuum) and
how the molecules leave the quasicontinuum for the real continuum,
i.e. how the molecules dissociate.

The simplest way to overcome the anharmonicity of the first
excited levels is to include the Stark broadening of the low lying
levels by the radiation field. But to ensure this, it is necessary
to have too high field-intensity (10^9w/cm^2). In addition, it is

difficult to expect the conservation of the isotopic selectivity
at the dissociation of the molecules having isotope shift small
as compared with the anharmonicity detuning. But in any case,
the Stark broadening technique, to pass the first few levels detuned from resonance because of the anharmonicity, is the most
general one and gives the upper estimate of the necessary field
intensity. But the IR two-frequency dissociation technique showed
that for the SF_6 molecule it is possible to overcome the anharmonicity effectively at the moderate intensity ($10^5 - 10^6 w/cm^2$).

As a simple model of the lower excited levels detuning compensation it is possible to accept the rotational compensation
model [16]. This model assumes the possibility of a transition
sequence preserving the resonance of the same laser light with
each rotation-vibration transition according to the scheme:

$$[V = 0, \quad J = J_o] \rightarrow [V = 1, \quad J = J_o - 1] \rightarrow$$

$$\rightarrow [V = 2, \quad J = J_o - 1] \rightarrow [V = 3, \quad J = J_o] \; .$$

The excitation of the third vibrational level is the result of
the sequential PQR - transitions. The resonance could be preserved if the transition frequency change due to the rotational
detuning $2B J_o$ compensated the anharmonicity detuning within the
transition linewidth.

But this model seems to be oversimplified. A polyatomic
molecule is not the result of the sum of several diatomics. The
situation is much more complicated. One should take into consideration both the intermodal anharmonicity, i.e. nonlinear
coupling between the different vibrational modes, and intramodal
anharmonicity.

We are discussing the dissociation of polatomic and sufficiently symmetric molecules. The last is very important. Polyatomic molecules exhibiting the symmetry of the 3rd and higher
order have degenerate vibrations. The degeneracy removal due to
the anharmonicity splits the levels, thus preventing anharmonicity detuning accumulation [17]. At the low vibrational levels
excitation, the choice of the initial J_o is determined by the
laser field frequency and its position in the limits of the
molecular linear absorption band. The J values for molecules interacting with the field are within the relatively narrow range
$\Delta J = Ed/2Bhc$.

So only a few molecules could participate in the sequential
excitation process.

But as we know, many recent experimental data show that the
vibrational energy absorption is very efficient. Within the
limits of the bottleneck effect [18] in the rotational distribution of the molecules, it is very difficult to explain

quantitatively the efficient energy absorption which seems to be related to the molecular interaction with the field independently of the J value. The bottleneck in the rotational distribution limits the number of the rotational states to be excited by the above discussed estimate Ed/2Bhc. For the SF_6 molecule at the laser field intensity of $10^6 w/cm^2$, it gives $\Delta J = \pm 2$.

The molecule interaction with the resonant monochromatic field, practically independently of the molecular rotational state, is a collisionless process. This is the first stage of the multiplephoton process of the selective collisionless dissociation.

Let us say that ΔJ_{int} rotational sublevels are effectively excited. The corresponding frequency range is $2B\Delta J_{int}c$. As the detuning of the rotational-vibrational transitions off the Stark broadening Ed/h becomes large the transition probability becomes quadratically small $(Ed/2B\Delta J_{int}hc)^2$.

But for the polyatomic molecule in the laser field, there is always a possibility of the fast removal of the excited molecules to the higher vibrational states and then via the quasicontinuum to the dissociation. This constant flow is very characteristic of all the processes of the excitation of polyatomic molecules. The quadratically small excitation probability of the greatly detuned rotational sublevels could be compensated by this fast removal. At the removal probability W and during the laser pulse duration τ the overall number of the rotational sublevels taken up by the radiation with the probability near one is $\Delta J_{int} = Ed\sqrt{W\tau}/2Bhc$. Strictly speaking, the probability W is not known. At the transitions in the system of the lower vibrational levels, it is possible to consider W being in order of magnitude equal to Ed/h. Then for the SF_6 at the laser radiation intensity $10^6 w/cm^2$ we have the estimate $\Delta J = \pm 30$.

This estimate shows us that the purely collisionless taking up of the great number of rotational sublevels is quite possible.

Recently [19], there has been a direct experimental observation of the vibrational excitation of large numbers of rotational states of the polyatomic molecules SF_6, SiF_4, C_2H_4 by a laser field of a moderate intensity. In this experiment is was observed that $\Delta J_{int} \gg Ed/2Bhc$ which is in good agreement with the above discussion and simple estimate.

A weak probing beam of a CW CO_2 laser was passing through a gas cell illuminated by a pulsed CO_2 laser. The probe laser intensity was below 300 MW/cm^2. The exciting pulse duration was 150 nsec, intensity $0.2 - 3$ MW/cm^2, spectral width $0.03 - 0.04$ cm^{-1}. The lasers were line tuned by gratings. The change in the attenuation of the weak probing beam has been measured with a temporal resolution of 100 nsec. Figure 1 shows the example of the

collisionless bleaching of a SF_6 cell for the probing signal at
952.9 cm^{-1} at the excitation of 944.2 cm^{-1}. The probe is resonant
to the R(109) line of the SF_6 ν_3 band, the excitation – P (63) of
the same band. The bleaching build up time is equal to the exciting pulse duration and does not depend on the SF_6 pressure in the
range 0.03 – 0.7 torr. At 0.03 torr this duration is 10 times
shorter than the SF_6 rotational relaxation time. The bleaching
became easily observable at an excitation intensity exceeding
300 kW/cm^2, its magnitude increasing approximately proportionately
to the exciting intensity.

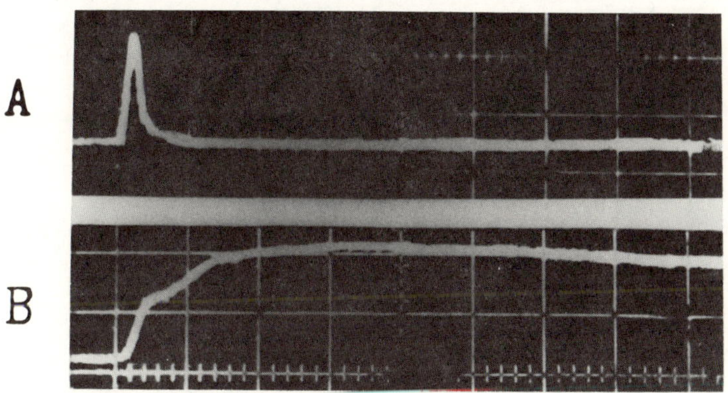

FIGURE 1.   Collisionless bleaching of the gaseous SF_6 for the
probing radiation at the frequency 952.9 cm^{-1} and
excitation at 944.2 cm^{-1}.

Figure 2 shows the bleaching magnitude dependence on the
exciting field frequency while probing at the frequency 952.9 cm^{-1}.
Below is shown the linear absorption spectrum and the spectral
dependence of the absorbed energy at a high level of excitation.
It is easy to see the correlation between absorption and bleaching
spectra.

The effect seems to be more or less general – it was observed
for the SiF_4 and C_2H_4 molecules as well. The collisionless darken-
ing observed under some spectral conditions in the experiment discussed above also gives evidence of the participation of many
rotational states.

Thus it is possible to say now that we have some degree of
understanding of the isotopically selective and efficient passing
by the low levels.

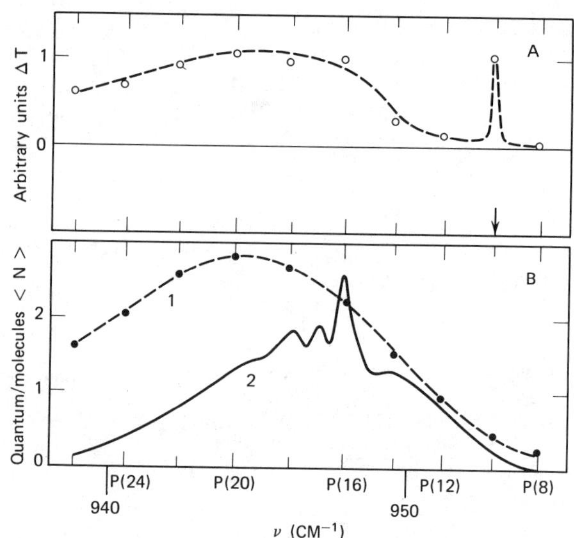

FIGURE 2.   The exciting field frequency dependence of the
bleaching.

The second stage of the process to be discussed here is the
energy absorption by a molecule in the quasicontinuum of upper
vibrational stages. The zonal structure of the quasicontinuum
[17] helps this absorption very much. It has been mentioned above
that polyatomic molecules characterized by the symmetry axis of
more than 2nd order have degenerate vibrations. The anharmonicity
removes the degeneracy.

The resulting level splitting forms the zones of the levels
which could overlap the harmonic energy positions in the quasicontinuum.

The analysis [17] has been made for the energy spectrum of
3-fold degeneracy mode ν_3 of the O_h-symmetry molecules XY_6 and
the T_d-symmetry molecules XY_4. The analysis took into account
inter- and intra- mode anharmonicity. For the T_d-case the cubic
anharmonicity was also taken into account. It turned out that
the energy zones appear in the vibrational spectrum of the ν_3
mode. If the anharmonicity constants α and β had the opposite
sign and $|\beta| > |\alpha|$, the zones are overlapping the harmonic energy
positions. For the real molecules the fulfillment of this condition seems to be realistic. The examples are $ZrCl_4$ [20] and
according to our measurements SiF_4 ($\alpha \approx -2.5$ cm^{-1}, $\beta \approx 12$ cm^{-1}).
For these molecules the zonal structure certainly overlaps
the "harmonic energy" positions.

The vibrational zones formation for the XY_6, XY_4 molecules of the O_h T_d-symmetry is the result of the nonstochastic Fermi resonance for the states of the same vibrational mode. The polyatomic molecules exhibit also Fermi resonances for the levels of the same symmetry but different vibrational modes. Because of the Fermi resonances the vibrational states are mixed up and form the vibrational levels zones. The transitions between the levels of the neighboring zones are allowed because of the anharmonicity. The zone edges are not sharp but the transition probabilities are decreasing from the zonal centre to the zonal edge.

Thus the spectrum of the three-fold degenerate vibrations of symmetry O_h and T_d molecules has zonal structure. The level density inside the zones is on the average more or less constant because the zone width and the degree of the degeneracy are both increasing proportionally to the vibrational quantum number squared. The intramodal dipole transitions are allowed and so such a structure could provide the cascade energy absorption by the molecule. The level density belonging to all polyatomic molecule modes is rapidly increasing with increasing vibrational energy. Hence the dipole moments of the intermodal transitions are increasing because of the stochastic Fermi resonances between the levels of the zones and quasicontinuum. Many levels of different modes are included into the cascade which helps to absorb the energy.

The dispersion characteristic of quasicontinuum is determined by its zonal structure. Two-frequency techniques allow the investigation of this characteristic. The remarkable increase of the dissociation of the molecules SF_6 and SiF_4 has been recently experimentally observed [21] at the large red shift of the exciting laser frequency relative to the basic vibration transition frequency.

Experimentally, two line tuned TEA-CO_2-lasers have been used. To increase the tuning range, one of the lasers employed the admixture of the $^{12}C\ ^{16}O_2$ and $^{13}C\ ^{16}O_2$ molecules in ratio 50:50. Oscillation pulse duration –100 nsec, pulse energy –1–2 J. The oscillation pulses are synchronized.

Figure 3 shows the dissociation ion current dependence for the SiF_4 molecule versus the laser frequency providing the quasicontinuum passing through. The resonant laser frequency (1029.5 cm^{-1}, P(38) of the 0.01–0.20 band) has been fixed at the maximum of the SiF_4 ν_3 linear absorption band. The tunable laser radiation power has been kept constant and equal to the 2.5 MW over the whole tuning range. The radiation has been focused into the cell by a F=13 cm lens. The resonant laser radiation was not focused, its intensity was 2 MW/cm^2. The SiF_4 pressure – 1 torr. It is seen that the tunable laser

frequency shift off the resonance towards the lower frequency by
80 cm^{-1} results in the 30-fold enhancement of the dissociation
ion current. The same is true for the visible luminescence ac-
companying the dissociation.

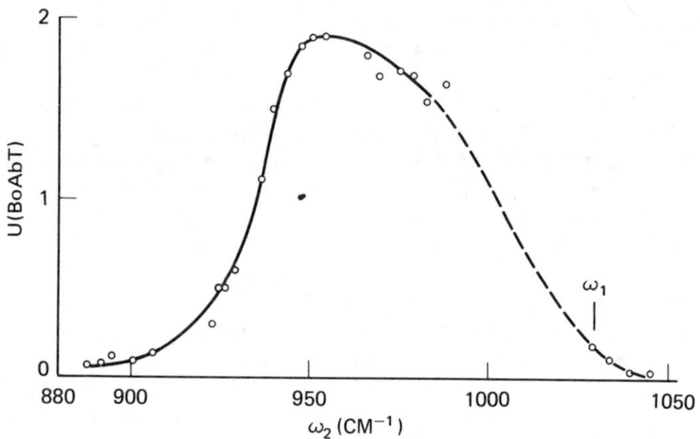

FIGURE 3. Dissociation ion current dependence upon the frequency
of the laser radiation providing the quasicontinuum
passing through for molecule SiF_4.

The dissociation ion current measurement [13] is a convenient
technique of the direct experimental study of the dissociation.
But experiments have shown that the SF_6 molecule dissociation is
not accompanied by the ion current and visible luminescence. To
investigate the red shift effect in the SF_6 dissociation the dynamic mass-spectrometer has been used. The measuring head of the
spectrometer has been inserted into the cell to be irradiated.
The isotopic content of the residual SF_6 gas has thus been analyzed. The cell (length 20 cm, diameter 1.4 cm) containing SF_6
at initial pressure 0.2 torr, has been irradiated by 100 pulses.
The resonant laser has been fixed at the frequency 944.2 cm^{-1}
(P (20)001-100 band). Its radiation has not been focused, intensity 3 MW/cm^2. The tunable laser radiation (2.5 MW) has been
focused into the cell by a F=13 cm lens. At the tuning of this
laser to the frequency 895 cm^{-1} (P(22) 001-100 band of the ^{13}C
$^{16}O_2$ molecule) the enrichment coefficient has been increased by
a factor of seven.
The red shift of the radiation frequency of the laser providing the passing through the quasicontinuum results in a
dramatic decrease of the SF_6 isotopic selective dissociation

threshold. Figure 4 shows the mass spectroscopy of the residual
SF_6 gas (fragment ion SF_5^+) after the cell irradiation by nonfocused beams of laser light with intensities $2-3$ MW/cm^2 for the
cases of the second laser tuned to the 944.2 cm^{-1} and 895 cm^{-1}.
There is not any isotopic enrichment for the first case; the
second gives an enrichment factor equal to 28 for ^{34}S (200 shots).

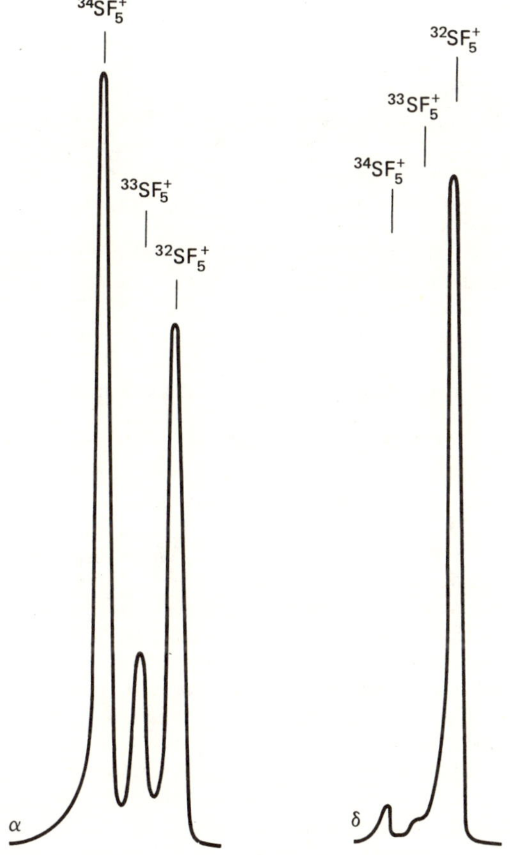

FIGURE 4.   Mass-spectra of the residual SF_6 gas after being irradiated by nonfocused laser radiation beams at the
frequencies 944 cm^{-1} and 895 cm^{-1}.

These results give evidence of the existence of the quasi-continuum zonal structure. But the results [21] are very important not only because of that. For practical applications, it is very important to have the possibility to carry out the isotopically selective dissociation by nonfocused laser beams. The right choice of the dissociating laser frequency decreases the dissociation threshold and helps the molecule to absorb energy in the zonal structure of the upper vibrational states quasicontinuum.

Thus, the quasicontinuum zonal structure approach is very important in order to understand the second stage of the polyatomic molecules collisionless dissociation - the stage of the molecular absorption of the high vibrational energy.

The frequency dependence of Fig. 3 is the dispersion curve of the dissociation process. The level density in the central area of the zones is the highest, the central position is going down the energy scale proportionally to vibrational quantum number V. Hence the observed red shift of the dispersion curve maximum could represent the zonal structure effective anharmonicity. But besides the laser radiation helping to pass through the quasicontinuum,it could also stimulate the dissociation directly [17].

The dissociation of the highly vibrationally, excited molecule is the third stage of the process under examination. This stage is far from being trivial. Obviously, the collisional dissociation of collisionless vibrationally over-excited molecules contributes significantly to the overall selective dissociation yield [14]. But during the completely collisionless phase of the process,it is possible to have only spontaneous or field induced dissociation of the over-excited molecules. This dissociation is much more probable if the molecule comes to the lower levels of the upper zones. The lower levels correspond to the less steep values of the multidimensional potential well of the molecule and thus are more closely related to the repulsion states. Hence, the red shift could increase the rate of dissociation process directly.

We are not discussing here the dissociation kinetics. It is determined by the passing through the low levels, energy absorption and dissociation kinetics itself. The common feature of all of these processes is sufficiently irreversible disappearance of the molecules including raising the molecules to the upper levels, dissociative expenditure of the molecules, chemical scavenging of the dissociation fragments and so on.

The probability of the field induced dissociation should be proportional to the field intensity multiplied by the field duration [17]. For the polyatomic heavily over-excited molecule, the complicated movement inside the multidimensional and far from

smooth potential well could be characterized by some dynamic stability. The spontaneous outlet of the molecule into the repulsion
valley may occur not necessarily just after the molecule had absorbed energy exceeding the dissociation limit. This is the reason why the field induced dissociation could be so important. The
experimental investigation of this dissociation seems to be possible by a three-frequency technique according to the three stages
of the process as a whole.

Thus in the process of the collisionless multiplephoton dissociation of polyatomic molecules by the IR intense laser field
we have the idea of the mechanisms responsible for the overcoming
of the anharmonicity and for absorbing the vibrational energy by
a molecule via stages of passing through the lower levels and
gaining the energy in the zonal structure of the upper levels
quasicontinuum. The dissociation of the heavily over-excited
molecules should be specially carefully investigated.

REFERENCES

1.   The VII-th All-union Conference on the Coherent and Non-
     linear Optics. Two volumes, Tbilisi, USSR, May, 1976.

2.   IXth International Conference on Quantum Electronics.
     Conference Digest. Optics Comm., $\underline{8}$, No I, July, 1976.

3.   Laser Spectroscopy, IInd International Conference, Megève,
     France, June, 1975. (Springer-Verlag, Berlin, Heidelbcrg,
     Ncw York, 1975).

4.   Tunable Lasers and Applications, Loen Nord Fiord Conference
     June, 1976, Norway (Springer-Verlag, Berlin, Heidelberg,
     New York, 1976).

5.   The Academy of Sciences of the USSR. Scientific Council
     Session on the Problem "Coherent and Nonlinear Optics"
     (January, 15-17, 1976). Kvantovaya Electronica, $\underline{3}$, 1841-
     1858 (1976).

6.   Physics of Quantum Electronics, Laser Photochemistry, July,
     1975. Summer School, Santa Fe, New Mexico, USA. (Addison-
     Wesley Publishing Co., Advanced Book Program, London,
     Amsterdam, Don Mills, Ontario, Sydney, Tokyo, 1976).

7.   N. V. Karlov, A. M. Prokhorov, Uspekhi Fizicheskikh Nauk,
     $\underline{118}$, 583 (1976).

8.   V. S. Letokhov, C. B. Moore, Kvantovaya Electronica, $\underline{3}$,
     248, 485 (1976).

9.   N. G. Basov, E. M. Belenov, V. A. Isakov, E. P. Markin,
      A. N. Oraevsky, V. I. Romanenko, Uspekhi Fizicheskikh Nauk,
      <u>121</u>, 427 (1977).

10.   N. Bloembergen, C. D. Cantrell, D. M. Larsen, see Ref. 4,
      p. 162.

11.   R. V. Ambartsumian, see Ref. 4, p. 150.

12.   V. M. Akulin, S. S. Alimpiev, N. V. Karlov, L. A. Shelepin,
      JETP, <u>69</u>, 836 (1975).

13.   V. M. Akulin, S. S. Alimpiev, N. V. Karlov, N. A. Karpov,
      Yu. N. Petrov, A. M. Prokhorov, V. A. Shelepin, JETP
      Letters, <u>22</u>, 100 (1975).

14.   V. M. Akulin, S. S. Alimpiev, N. V. Karlov, B. G. Sartakov,
      L. A. Shelepin, JETP , <u>71</u>, 454 (1976).

15.   N. V. Karlov, see Ref. 5, p. 1848.

16.   R. V. Ambartsumian, Yu. A. Corokhov, V. S. Letokhov, G. N.
      Makarov, JETP Letters, <u>22</u>, 96 (1975).

17.   V. M. Akulin, S. S. Alimpiev, N. V. Karlov, B. G. Sartakov,
      JETP, <u>72</u>, 88 (1977). See also Ref. 1, Vol. 2, p. 112;
      Ref. 5, p. 1848.

18.   V. S. Letokhov, D. D. Makarov, JETP, <u>63</u>, 2064 (1972).

19.   S. S. Alimpiev, V. N. Bagratashvily, N. V. Karlov, V. S.
      Letokhov, V. V. Lobko, A. A. Makarov, B. G. Sartakov,
      E. M. Khokhlov (to be published).

20.   K. T. Hecht, J. Mol. Spectr., <u>5</u>, 355 (1960).

21.   V. M. Akulin, S. S. Alimpiev, N. V. Karlov , A. M.
      Prokhorov, V. G. Sartakov, E. M. Khokhlov, JETP Letters,
      <u>25</u>, 422 (1977).

COLLISIONS IN INTENSE FIELDS

Collisions in Strong Electromagnetic Fields

KENNETH M. WATSON*
Department of Physics
and
Lawrence Berkeley Laboratory
Berkeley, California

I. INTRODUCTION

Collisions between atoms and/or molecules can be modified in several ways by an intense laser field. A rather large number of papers have appeared in the past few years analyzing theoretical aspects of such collisions and some related experiments have been done, or are in progress. We shall attempt to describe briefly here the highlights of work presently published on the subject.

The collisions which we are discussing are supposed to occur in the focal spot of a laser beam. These are presumably gaseous kinetic collisions, but may occur in crossed atomic or molecular beams.

II. RADIATIVE TRANSITION ASSOCIATED WITH COLLISIONS

During the collision of two atoms (or molecules) A and B, the composite system has additional levels, not present in A or B individually. These additional levels are available for radiative transitions. For example, the collision with B may merely enhance the absorption of one or more photons by A. On the other hand, a more complex transfer of excitation may be involved. Such processes have been studied by Gudzenko and Yakovlenko [1,2] and by Harris and Lidow and their collaborators [3,4].

We illustrate this process with the simple theoretical model of Harris and Lidow [3]. The interaction Hamiltonian is taken as

---

*Supported in part by the Energy Research Development Agency at Lawrence Berkeley Laboratory and Physical Dynamics, Inc., Berkeley, California.

$$H' = -e\, \underset{\sim}{r}_A \cdot \underset{\sim}{E}\, \cos\omega t - e\, \underset{\sim}{r}_B \cdot \underset{\sim}{E}\, \cos\omega t$$

$$+ [e^2/R^3(t)]\, [x_A x_B + y_A y_B - 2\, z_A z_B] \ . \tag{1}$$

Here $\underset{\sim}{E}$ is the electric field of the laser beam, ω its frequency, x_A ... an electron coordinate, and $R(t)$ the internuclear spacing gi-en as a function fo time. The last term represents the Van der Waals interaction for a not-too-close collision.

Time-dependent perturbation theory is used to calculate the transition probability.

An experiment to study transfer of excitation from Sr to Ca was performed by Lidow, et al. [4]. This is illustrated schematically in Fig. 1. The first step is two photon absorption ($2\hbar\omega_1$) from level 1 to level 2 of Sr. This radiatively decays to level 3 emitting a photon of frequency ω_2. During the collision with Ca a photon of frequency ω_3 is absorbed with transfer of excitation and a transition from level 4 to level 5 of Ca. This then decays to level 6 with emission of a photon of frequency ω_4. The observed process appeared consistent with calculations based on the Hamiltonian (1).

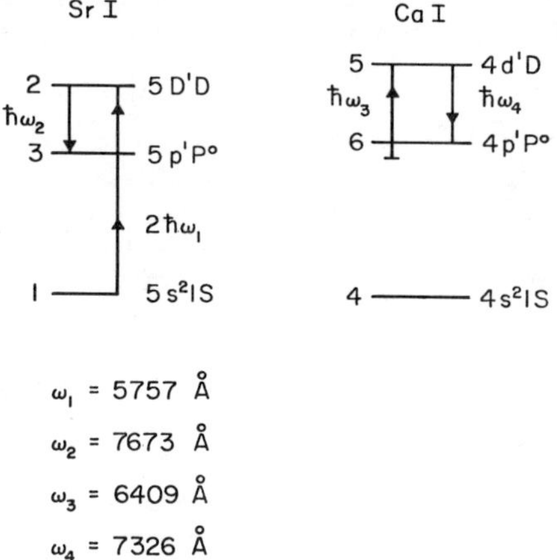

FIGURE 1.   Illustration of the experiment of Lidow, et al.
             (Ref. [4]).

For close approaches of A and B during collision, a modified process can occur. This was studied by Kroll and Watson [5] and by Lau [6] and is illustrated in Fig. 2. A pair of molecular potential curves are illustrated as a function of internuclear separation R. Resonant absorption of a photon of frequency ω can occur where

$$\hbar\omega = U_2(R) - U_1(R) \; ,$$

leading to the reaction

$$A(1) + B(1) + \hbar\omega \rightarrow A(2) + B(2) \; . \tag{2}$$

The cross sections for such processes have been calculated [5,6] to be comparable to the elastic scattering cross sections for sufficiently strong fields.

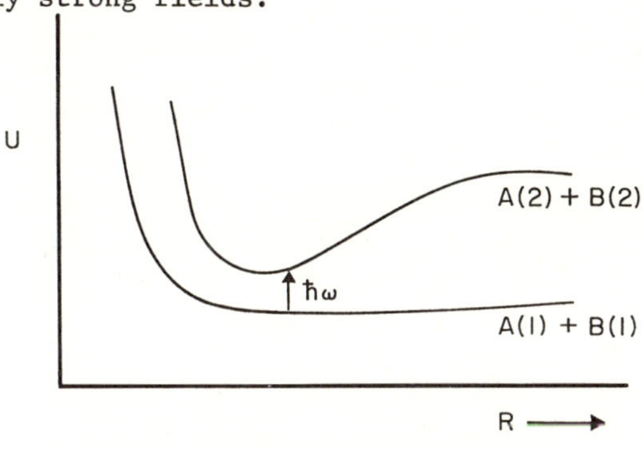

FIGURE 2.   Illustration of a radiative transition occurring during a collision.

III.   THE CASCADE BREAKDOWN LIMIT

If the experiments being considered are carried out by focusing a laser beam in a gas, it is necessary to avoid cascade breakdown. For air the laser power threshold for breakdown is [5]

$$P_B = 1.44\,[p_R^2 + 2.2 \times 10^5/\lambda^2]\;\text{MW/cm}^2 \; . \tag{3}$$

Here λ is the laser beam wavelength in μm and p_R is the pressure in atmospheres. This relation is illustrated in Fig. 3. It is seen that for $\lambda < 10$ μm the pressure term tends to be negligible near atmospheric pressure.

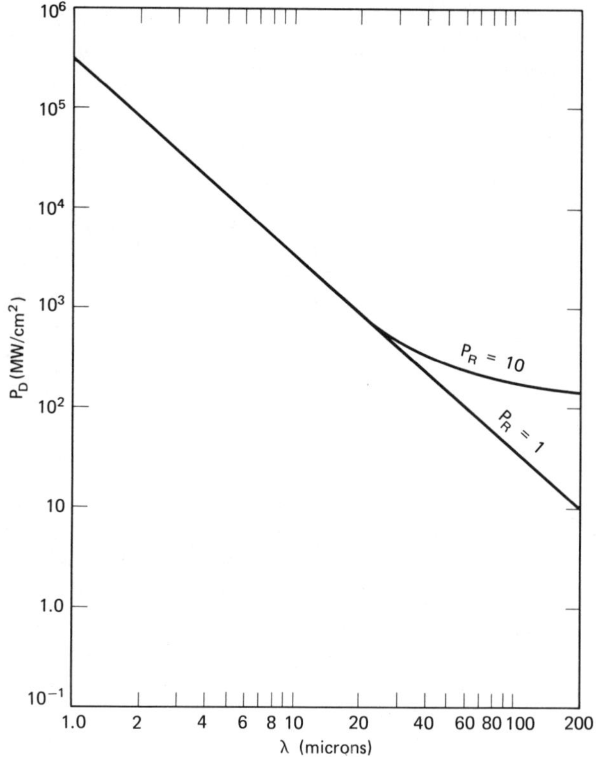

FIGURE 3.   Breakdown power in air as a function of photon
            wavelength.

The expression (3) gives the breakdown power in a large
volume of air. When a small spot is irradiated the power P_B may
be substantially increased as was studied in some detail in reference [5]. The results of reference [5] may be approximately
stated by the relation

$$P_B = \frac{0.41}{\lambda^2 [R_s P_R^2]^{5/4}} \quad MW/cm^2 \; . \tag{3'}$$

Here R_s is the radius in cm. of the illuminated spot and we have
neglected the P_R^2 - terms in Eq. (3). Equation (3') is valid for
$10^{-7} < R_s \, P_R^2 < 10^{-5}$ cm. For $R_s \, P_R^2 > 4 \times 10^{-5}$ cm, Eq. (3) should

be used.

IV. STRONG FIELD EFFECTS

When the electromagnetic field is sufficiently intense one is led to consider adiabatic eigenstates of the quasi-molecule and field [1,2,6-10]. In this case it is convenient to consider the collision of A and B as occurring in a resonant cavity in which a small number, M_c, of electromagnetic modes are excited. The Hamiltonian for the radiation field is then

$$h_\gamma = \sum_{\lambda=1}^{M_c} \hbar\omega_\lambda a_\lambda^+ a_\lambda , \tag{4}$$

where λ labels the cavity modes and a_λ^+, a_λ are respective photon creation and annihilation operators. The adiabatic molecular states ϕ_α are determined from the Schrödinger equation

$$h_\gamma \phi_\alpha = U_\alpha(R)\phi_\alpha , \quad \alpha = 1,2,\ldots S . \tag{5}$$

A finite number S of discrete molecular states are considered to contribute [11]. The interaction of the molecule with the electromagnetic field is represented by the Hamiltonian

$$h_{int} = \sum_j e_j \, \underset{\sim}{x}_j \cdot \underset{\sim}{E}(0) . \tag{6}$$

(Magnetic dipole and electric quadrupole interactions are considered in Ref. [8]).

The state function Ψ is expanded in terms of the states ϕ_α , [12]

$$\Psi = \sum_{\alpha=1}^{S} C_\alpha(t)\phi_\alpha \tag{7}$$

and the resulting Schrödinger equation is then

$$i\hbar\,\dot{C}_\alpha = \left[U_\alpha(R) + \sum_{\lambda=1}^{M_c} \hbar\omega_\lambda a_\lambda^+ a_\lambda \right] C_\alpha$$

$$+ \sum_{\beta(\neq\alpha)=1}^{S} \left\{ \sum_{\lambda=1}^{M_c} g_{\alpha\beta}^{(\lambda)} (a_\lambda + a_\lambda^+)C_\beta - i\hbar(\phi_\alpha,\dot{\phi}_\beta)C_\beta \right\} , \tag{8}$$

where $g_{\alpha\beta}^{(\lambda)}$ is the appropriate dipole matrix element. The

390   K. M. Watson

collision is defined by specifying $R(t)$ as a function of time. The adiabatic eigenstates are obtained from (8) by setting

$$i\hbar \, \dot{C}_\alpha = \left[E(R) + \sum_{\lambda=1}^{M_c} N_\lambda \hbar\omega_\lambda \right] C_\alpha \; . \tag{9}$$

Here N_λ (assumed to be $\gg 1$) represents a nominal value for the number of photons in mode λ in the cavity. On adding a state label "m" we obtain

$$\left[E_m - U_\alpha - \sum_\lambda \hbar\omega_\lambda (a_\lambda^+ a_\lambda - N_\lambda) \right] C_\alpha^{(m)}$$

$$= \sum_\beta \sum_\lambda g_{\alpha\beta}^{(\lambda)} (a_\lambda + a_\lambda^+) C^{(m)} \; . \tag{10}$$

This corresponds to the linear coupling of M_c harmonic oscillators to a Hermitian matrix $\tilde{g}^{(\lambda)}$. The eigenstates and eigenfunctions can be easily evaluated numerically for a limited number, $M_c S$, of cavity modes and molecular states [6-10].

To simplify our discussion, we suppose there to be only one cavity mode excited, so $M_c = 1$. The state m may be labeled by the corresponding state in the limit that $g_{\alpha\beta} = 0$. Then

$$m = (\alpha_o, \nu), \quad E_m = U_{\alpha_o} - \nu\hbar\omega \; , \tag{11}$$

where α_o is a given adiabatic molecular state and ν represents the number of photons *removed* from the cavity (of the N originally present).

An illustration of the adiabatic eigenstates is shown in Fig. 4, taken from Ref. [8]. In Fig. 4a "unperturbed" states corresponding to $g_{\alpha\beta} = 0$ are shown. These are labelled as "11" and "22" with and without addition of a photon. The corresponding adiabatic eigenstates, or "dressed" states are shown in Fig. 4b.

A more complex set of states is illustrated in Fig. 5 (taken from Ref. [6]). A pair of molecular states are each shifted by addition or removal of several photons, corresponding to a set of values for ν in Eq. (11). When the interaction $g_{\alpha\beta}$ is "turned on", the energy levels are as shown in Fig. 6. The "crossings" of Fig. 5 become "avoided crossings". The transition probability at the avoided crossings may be calculated by the approximation of Landau and Zener [6-10]. At field intensities well below P_B the calculated transition probabilities range from very small values to values comparable to unity, depending on oscillator strengths, number of photons involved, and gap width.

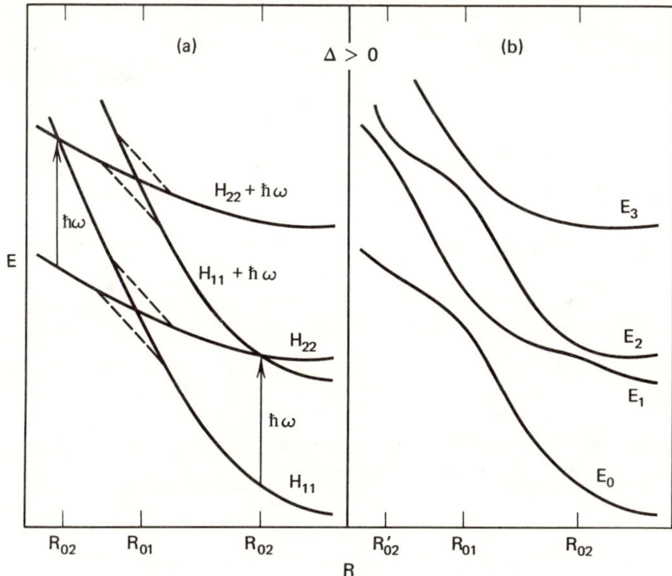

FIGURE 4.   Illustration of a pair of molecular states without
            and with field "dressing" (this figure is taken from
            Ref. [8]).

The adiabatic energies E_m determined from Eqs. (10) depend
parametrically on the laser power P_L. Thus the laser power may
be used to tune, or de-tune resonances, shift potential barriers
[7], etc. An approximate expression for the shift ΔE of the
field-free quasi-molecular energy (taken from ref. [6]) is

$$\Delta E \sim 5 \times 10^{-8} \; \frac{\lambda \; P_L}{\Delta W} \; ev. \tag{12}$$

Here λ is the wavelength in μm, P_L the laser power in MW/cm^2, and
ΔW is the distance (in ev) to the closest neighboring level. For
applications, the cascade breakdown limitation described by Eqs.
(3) and (3') must be taken into account.

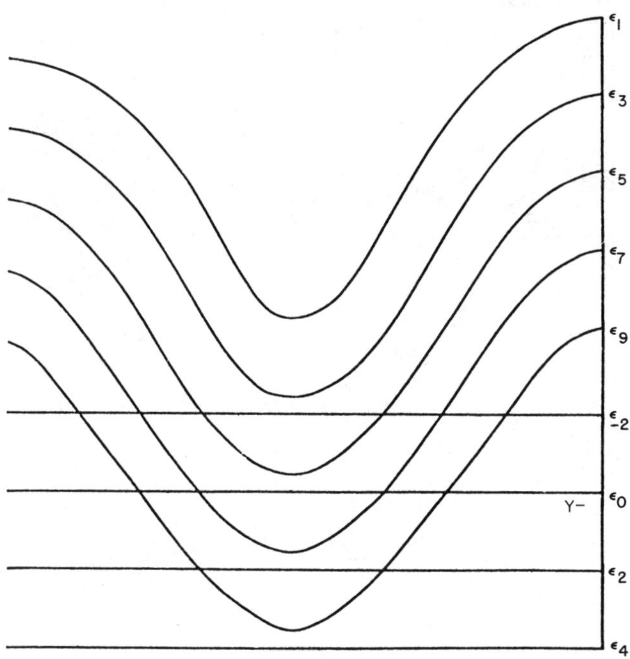

FIGURE 5.   A pair of quasi-molecular states with addition or
removal of photons in the weak field limit.

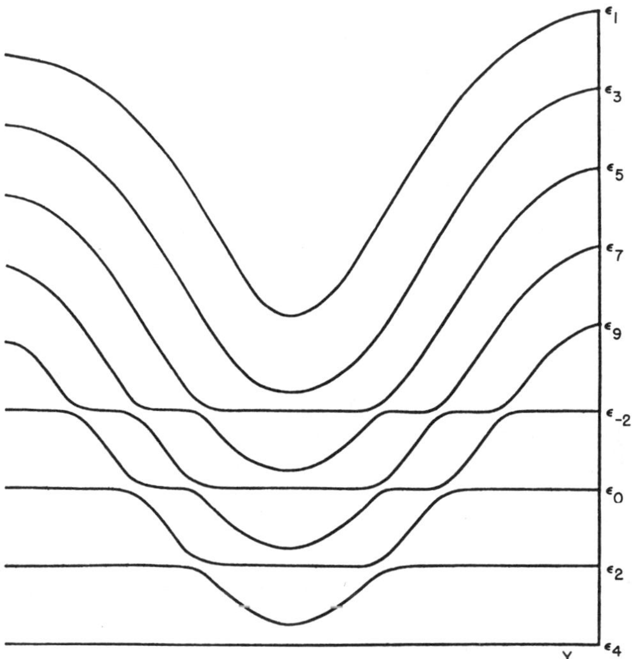

FIGURE 6.   The adiabatic states corresponding to Fig. 5 when the
electromagnetic field is "turned on".

V.   FIELD INDUCED AVOIDED CROSSINGS

Lau and Rhodes state in Ref. [10][13] that, "Our work is
motivated by the use of laser radiation to initiate and to control nonreactive and reactive molecular processes and to direct
energy flow pathways, especially those that involve electronic
transitions."

Figure 7 is taken from Ref. [10] and illustrates a crossing
for which no field-free transitions occur because $\Delta\Lambda = \pm 2$.   A
small field-induced coupling can produce inelastic transitions
into new channels.

Figure 8 is also taken from Ref. [10].   Here a field-free
transition can occur, but may be greatly enhanced by the laser
field.   Lau and Rhodes suggest that the isotopic enrichment process for Br reported by Leone and Moore [14] may be enhanced by
this mechanism.

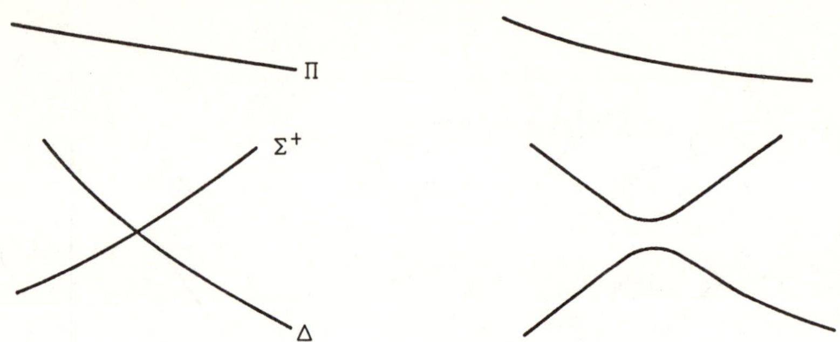

FIGURE 7.   Field induced avoided crossing for which no field-free transition occurs (from Ref. [10]).

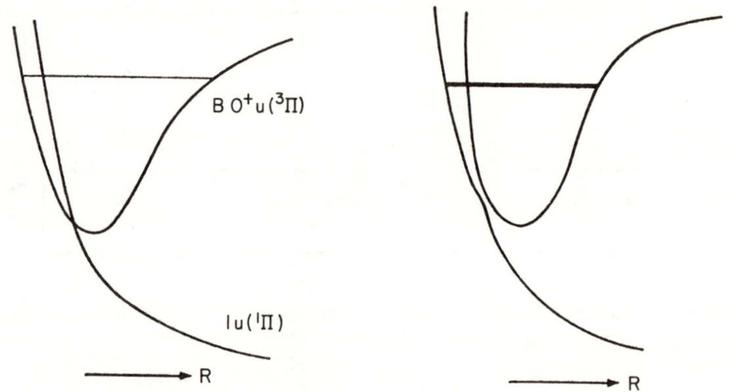

FIGURE 8.   Field induced avoided crossing for which a field-free transition occurs (from Ref. [10]).

REFERENCES

1.   L. I. Gudzenko, V. S. Lisitsa and S. I. Takovlenko, Sov. Phys. JETP $\underline{35}$, 877 (1972): $\underline{30}$, 759 (1975).

2.   S. I. Yakovlenko, Sov. Phys. JETP $\underline{37}$, 1019 (1973).

3.   S. E. Harris and D. B. Lidow, Phys. Rev. Lett. $\underline{33}$, 674 (1974); $\underline{34}$, 172(E) (1975).

4.   D. B. Lidow, R. W. Falcone, J. F. Young, and S. E. Harris, Phys. Rev. Lett. $\underline{36}$, 462 (1976).

5.   See, for example, N. M. Kroll and K. M. Watson, Phys. Rev. $\underline{A5}$, 1883 (1972).

6. N. M. Kroll and K. M. Watson, Lawrence Berkeley Laboratory Report LBL-1587, February 1973, and Phys. Rev. $\underline{A13}$, 1018 (1976).

7. A. M. Lau, Phys. Rev. $\underline{A13}$, 139 (1976); $\underline{A14}$, 279 (1976).

8. J. M. Yuan, J. R. Laing, and T. F. George, J. Chem. Phys. $\underline{66}$, 1107 (1977).

9. A. M. Lau and C. K. Rhodes, Phys. Rev. $\underline{A15}$, 1570 (1977).

10. A. M. Lau and C. K. Rhodes, submitted to Physical Review A.

11. For simplicity of presentation we exclude here excitation of continuum states.

12. We are here following the notation of Ref. (6).

13. There is considerable overlap in the content of Refs. (8) and (10).

14. S. R. Leone and C. B. Moore, Phys. Rev. Lett. $\underline{33}$, 269 (1974).

Inelastic Collisions by Intense Laser Radiation

J. F. YOUNG AND S. E. HARRIS
Edward L. Ginzton Laboratory
W. W. Hansen Laboratories of Physics
Stanford University
Stanford, California

SUMMARY OF PRESENTED PAPER

When the energy defect ΔE between the initial and final states of two colliding atoms is large with respect to kT, the cross section for inelastic collision is quite small. In this presentation we consider processes which utilize one or more photons to conserve energy, i.e., $n\hbar\omega = \Delta E$. In particular, we will describe the experimental observation of two such processes, and compare the results to the calculated magnitude and lineshape.

Such collision processes are of interest for a number of potential applications. In the absence of the applied field the initial atomic state will have a very small collision cross section, and thus a long lifetime. Thus energy could be pumped into the system and stored in such a state over a relatively long period of time. When the photon field is applied, the collision cross section will rise to a large value with a rise time equal to that of the laser, permitting rapid extraction of the stored energy. For this reason the process is often called a switched collision. In addition, the reverse process in which a photon is emitted as the two atoms collide, termed a radiative collision, may permit one to construct variable gain lasers with a center frequency corresponding to the energy difference of the levels of two different atomic species.

Collision processes of this type have been predicted by Gudzenko and Yakovlenko [1] and by Harris and Lidow [2]. Recent theoretical work has also been done by Payne and Nayfeh [3], Geltman [4], and George, et al [5]. In this presentation we consider primarily one photon, dipole-dipole induced collisions, although the process may be considerably more general. It may involve the absorption of several photons instead of one; the coupling may be dipole-quadrupole or quadrupole-quadrupole; processes involving charge exchange, spin exchange, and ionization are also possible.

397

   We have performed three different experiments using a Sr-Ca
system in an attempt to observe laser induced collisions. Figure
1 is an energy level diagram illustrating these experiments. In
all cases energy was first stored in the radiatively trapped
$5p\,{}^1P^0$ Sr level, and the transfer laser was tuned to induce an inelastic collision into a particular Ca level. The occurrence of
transfer was determined by measuring the resulting fluorescence
to the lower $4p\,{}^1P^0$ state. The first experiment [6] involving
6409 Å transfer to the Ca $4d\,{}^1D$ experienced a difficulty because
of an inadvertent coincidence (to within 0.1 Å) between the
transfer wavelength and a "bare" transition in Sr alone. The two
other experiments, however, were successful [7,8]; details of the
experimental set-up, diagnostics, and results will be discussed.
Figure 2 shows the wavelength dependence of the induced collision
cross section to the Ca $4p^2\,{}^1S$ state.

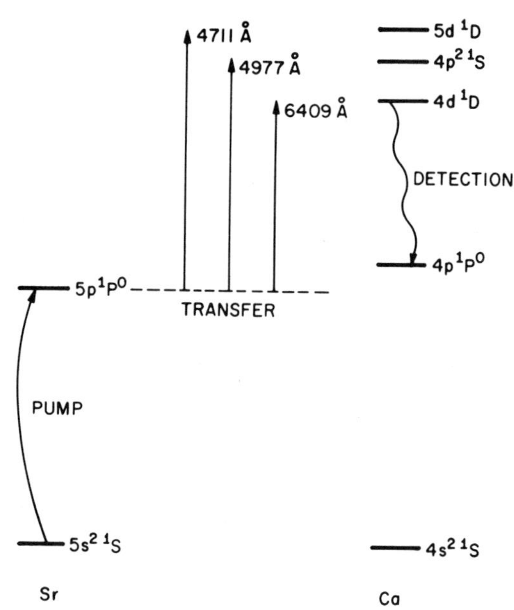

FIGURE 1.   Partial energy level diagram of the Sr-Ca system
            showing transfer wavelengths.

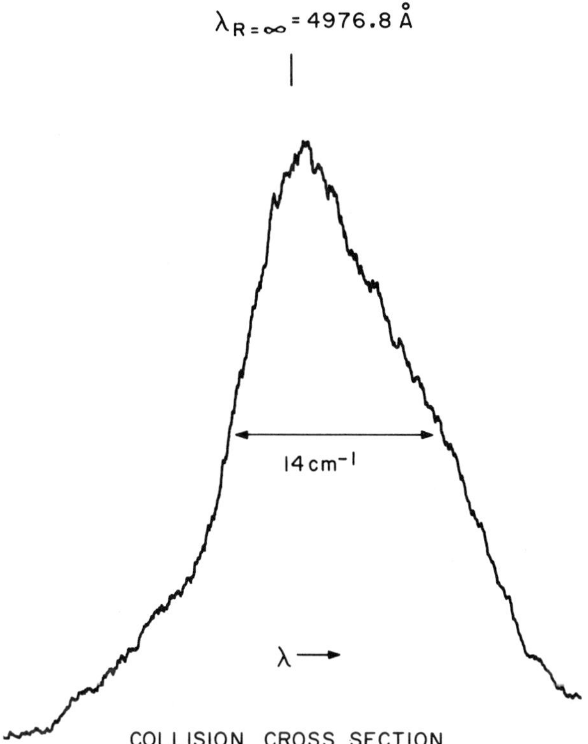

FIGURE 2. Relative induced collision cross section to the
Ca $4p^2$ ^1S state vs. wavelength.

The laser induced collision process can be usefully viewed
physically from three related viewpoints: (1) a collision into
a resonant, virtual level followed by absorption of a photon;
(2) an electromagnetic transition between the states of a quasi-
molecule formed during the time of a collision; and (3) as a
transition caused by the application to the second atom of the
near electric field of the first atom induced by the applied
field. The implications and limitations of each of these approaches will be discussed.

More rigorously, we treat the problem by expanding the wave
function in a basis set of product eigenfunctions of the separated
atoms. We assume a straight-line trajectory at velocity v and
impact parameter ρ. We calculate the classical dipole-dipole
interaction Hamiltonian, substitute into Schrödinger's equation,
and take matrix elements using a fixed atom approximation, and

random orientation of the dipole moment of the p state atom with
respect to the s state atom. By making a slowly varying approximation, the description reduces to a set of two coupled differential equations for the probability amplitude of the initial and
final product states.

We have solved these equations numerically to obtain the
wavelength dependence of the induced cross section, as shown in
Fig. 3. The peak may be understood from an impact theory point
of view, while the red wing can be derived from a Landau-Zener
curve crossing model. Calculated results for the peak magnitude
of the induced cross section, Fig. 4, show a saturation at about
10^9W/cm^2. Experiments to verify these results are underway and
will be described.

The lineshapes calculated here will also apply to other types
of laser induced processes, including multiphoton processes. Such
lineshapes should also be observable via spontaneous emission in
the reverse, radiative collision process. Several examples of
other processes will be presented.

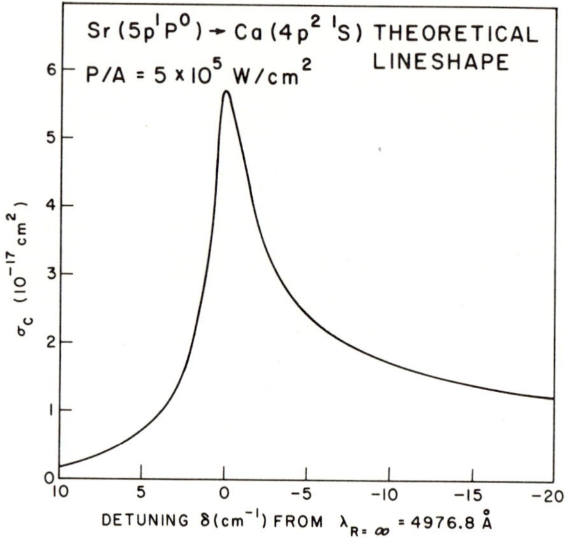

FIGURE 3. Calculated lineshape of induced collision.

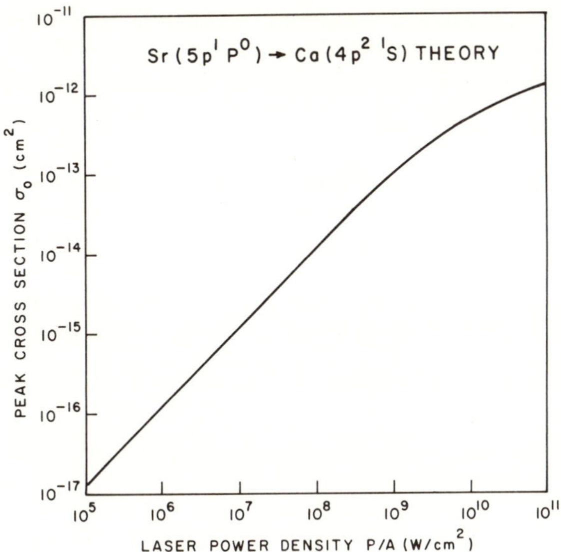

FIGURE 4. Calculated peak cross section as a function of power
density.

ACKNOWLEDGMENTS

The numerical calculations were performed by Mr. J. C. White.
The authors acknowledge important discussions with Dr. Alan
Gallagher and Dr. Sydney Geltman. This work was supported in
part by the Energy Research and Development Administration, the
Office of Naval Research, and the Air Force.

REFERENCES

1.   L. I. Gudzenko and S. I. Yakovlenko, Zh. Eksp. Teor. Fiz.
      62, 1686 (1972) [Sov. Phys. JETP 35, 877 (1972)].

2.   S. E. Harris and D. B. Lidow, Phys. Rev. Lett. 33, 674
      (1974) and 34, 172(E) (1975).

3.   M. G. Payne and M. H. Nayfeh, "Laser Enhanced Collisional
      Energy Transfer" (to be published).

4.   Sydney Geltman, "Theory of Laser-Stimulated Collisional
      Excitation Transfer" (to be published).

5. Thomas F. George, Jian-Min Yuan, I. Harold Zimmerman, and
    John R. Laing, "Radiative Transitions for Molecular Col-
    lisions in an Intense Laser Field," Disc. Faraday Soc. No.62.

6. D. B. Lidow, R. W. Falcone, J. F. Young, and S. E. Harris,
    Phys. Rev. Lett. 36, 462 (1976). [Erratum: Phys. Rev. Lett.
    37, 1590 (1976)].

7. S. E. Harris, D. B. Lidow, R. W. Falcone, and J. F. Young,
    "Laser Induced Collisions," in Tunable Lasers and Applica-
    tions, A. Mooradian, T. Jaeger, and P. Stokseth, eds.
    (Springer-Verlag, New York, 1976).

8. R. W. Falcone, W. R. Green, J. C. White, J. F. Young, and
    S. E. Harris, "Observation of Laser Induced Inelastic
    Collisions," Phys. Rev. A15, 1333 (1977).

Effect of the Strong Electromagnetic Field on the Electron Scattering Process

AN. V. VINOGRADOV
P. N. Lebedev Physical Institute
Moscow, USSR

The interest in the study of electron scattering in the presence of the intense electromagnetic field is connected with two kinds of problems. First, when the electron is scattered in the presence of the intense electromagnetic wave, the processes of induced absorption or emission, leading to the electron gas heating, are possible. If the field is strong enough, from a quantum mechanical viewpoint, these processes are of a multi-photon nature. A great amount of work has been recently devoted to a study of these processes, because they lie at the basis of such important phenomena as plasma heating by laser radiation, laser spark, optical breakdown of transparent materials.

Second, under the action of the strong electromagnetic field the conditions of the scattering process itself change. Thus the strong electromagnetic field affects various physical phenomena connected with electron scattering, namely the influence of a strong electromagnetic wave on the absorption in a plasma of a weak electromagnetic wave of another frequency, modification of the bremsstrahlung spectrum in the presence of a strong electro-magnetic wave, phonon damping in semiconductors and others. In spite of the physical differences, all these problems are mathe-matically very much similar. From a quantum mechanical viewpoint this similarity lies in the fact that the field action on the electron during scattering is taken into account by using an exact wave function of a free electron in the field,

$$\psi_p(\vec{r},t) \sim \exp\left\{\frac{i}{\hbar}\vec{p}\cdot\vec{r} - \frac{i}{\hbar}\int_{-\infty}^{t}\frac{1}{2m}\left(\vec{p} - \frac{e\vec{E}}{\omega}\sin\omega t'\right)^2 dt'\right\} \tag{1}$$

shown in Eq. (1), whereas the scattering itself is treated in the first Born approximation. At present, such an approach is used extensively. For the first time this approach was applied by Bunkin and Fedorov [1] in the study of bremsstrahlung in a strong radiation field. It leads to an appearance of the same expression in all of the problems under consideration and is shown in Eqs. (2) and (3),

$$F(\vec{K}) = \sum_{n=-\infty}^{n=+\infty} J_n^2 \left(\frac{\lambda_K}{\hbar\omega}\right) \delta(\Delta_K - n\hbar\omega) = \tag{2}$$

$$= \frac{1}{2\pi} \int_{-\infty}^{+\infty} dx J_o \left(\frac{\lambda_K}{\hbar\omega} 2\sin\frac{\hbar\omega}{2} x\right) \exp\{i\Delta_K x\}, \tag{3}$$

where J_n is the Bessel function of order n,

$$\Delta_K = \varepsilon_{\vec{p}+\vec{K}} - \varepsilon_p , \qquad \varepsilon_p = \frac{p^2}{2m} ,$$

and

$$\lambda_K = \frac{e\vec{E}\cdot\vec{K}}{m\hbar\omega^2} .$$

Let us consider two examples: Eq. (4) is the rate of the electron energy change due to collisions in the presence of the field where C_K is the matrix element of the scattering potential. Equation (5) is the cross-section for the bremsstrahlung of the photon $\hbar\omega_1$, at the nucleus in the presence of the strong electromagnetic field.

$$\frac{d\varepsilon_{\vec{p}}}{dt} = \frac{2\pi}{\hbar} \sum_K \frac{|C_K|^2}{V} \Delta_K F(\vec{K}) =$$

$$= \frac{2\pi}{\hbar} \sum_K \frac{|C_K|^2}{V} \sum_{n=-\infty}^{n=+\infty} n\hbar\omega J_n^2 \left(\frac{e\vec{E}\cdot\vec{K}}{m\hbar\omega^2}\right) \delta(\varepsilon_{\vec{p}+\vec{K}} - \varepsilon_p - n\hbar\omega) , \tag{4}$$

$$d\sigma(\vec{p}) = d\omega_1 \frac{e^6 z^2}{\hbar c^3 m\omega_1} \frac{8}{3\pi} p^{-1} \int d\Omega \int_o^\infty dK \cdot$$

$$\cdot \sum_n J_n^2 \left(\frac{e\vec{E}\cdot\vec{K}}{m\hbar\omega^2}\right) \delta(\varepsilon_{\vec{p}+\vec{K}} - \varepsilon_p + \hbar\omega_1 - n\hbar\omega). \tag{5}$$

It is possible to give some other examples. Formally the formulas of that type solve the problem of the influence of the strong electromagnetic field on the processes of electron scattering. But the result containing the infinite series of Bessel functions is not satisfactory from the physical point of view. So it is necessary to learn to analyze such expressions when the argument of Bessel function is not small and, consequently, it is impossible to restrict the infinite series in n to the first few terms. Several ways to solve this problem, except for numerical approaches, have been proposed. A general and most essential feature of all of them is that the behavior of the Bessel function for large values of the argument depends on the ratio of order to the argument. When the absolute magnitude of the order is greater than that of the argument, that is $|n| \gg |x|$, the Bessel function is exponentially small as shown in Eq. (6);

$$J_n^2(x) \simeq \frac{1}{2\pi|n|} \exp\left\{-2|n| - 2|n|\ell n2\left|\frac{n}{x}\right|\right\} . \tag{6}$$

Thus in the series in n, it is possible to omit those terms for which Eq. (6) holds. But this approximation is not yet sufficient to perform calculations and to find the field dependence of concrete physical values. Therefore subsequent approximations are needed. In the works of Pert [2] and Elutin [3], who have considered the nonlinear absorption of an intense electromagnetic wave due to the inverse bremsstrahlung, Bessel functions were replaced by the asymptotic formula of Eq. (7);

$$J_n^2(x) \approx \frac{1}{\pi|x|} , \quad |n| < |x|, \quad |x| \gg 1 . \tag{7}$$

After that, calculations can be readily made and the nonlinear absorption coefficient, α, for the case of Coulomb scattering, that is $|C_K|^2 \sim K^{-4}$ and is given by

$$\frac{e^2 E^2}{m\omega^2} \gg \max\{\hbar\omega, K_B T_e\} \tag{8}$$

$$\alpha = \frac{4\pi N_e e^2}{n c m\omega^2} \nu_{eff}^{(-4)}(E) , \tag{9}$$

$$\nu_{eff}^{(-4)}(E) = 8zNe^4 m\left(\frac{eE}{\omega}\right)^{-3} \ell n^2 \frac{eE}{\omega K_{min}} , \tag{10}$$

$$K_{min} = \begin{cases} \sqrt{2m\hbar\omega} & \text{at } K_B T_e \ll \hbar\omega , \\ m\hbar\omega/\sqrt{2mK_B T_e} & \text{at } K_B T_e \gg \hbar\omega . \end{cases} \tag{11}$$

Another method has been proposed by Seely and Harris [4]. Seely
and Harris have noticed that when the argument of the Bessel function is large, in the whole series only those terms are essential
for which Eq. (12) holds ,

$$\frac{\lambda_K}{\hbar\omega} = \frac{e\vec{E}\cdot\vec{K}}{m\hbar\omega^2} \approx n = \frac{\Delta_K}{\hbar\omega} \; , \tag{12}$$

and they approximated a typical series of Bessel functions by the
sum of two δ-functions as shown in Eq. (13);

$$F(\vec{K}) = \sum_n J_n^2(\frac{\lambda}{\hbar\omega})\delta(\Delta-n\hbar\omega) = \frac{1}{2}[\delta(\Delta+\lambda) + \delta(\Delta-\lambda)]. \tag{13}$$

Let us note that the right hand side of Eq. (13) does not depend
on the Planck constant $\hbar$. The Seely and Harris approximation is
attractive for its simplicity and clarity, and has been used in a
number of works. It means that the electron absorbs or emits
such a number of photons, that the resulting energy change is
given by the classical value $\pm e\vec{E}\cdot\vec{K}/m\omega$. But from the mathematical
viewpoint this approximation is not quite correct. Obviously,
such an approximation, which makes it possible to obtain the main
term of the asymptotic expansion of concrete physical values as
the magnitude of the electric field strength goes to infinity,
will be mathematically correct. The argument of the Bessel function will also be large as the Planck constant goes to zero. So
it's natural to suppose that finding the above mentioned asymptotic value is equal to a transition to the classical limit. The
transition to the classical limit is demonstrated by Eq. (14).

$$\lim_{h\to 0} \sum_n J_n^2(\frac{\lambda}{\hbar\omega})\delta(\Delta-n\hbar\omega) =$$

$$= \lim_{\hbar\to 0} \frac{1}{2\pi} \int_{-\infty}^{+\infty} dx\, J_0(\frac{\lambda}{\hbar\omega}\, 2\sin\frac{\hbar\omega}{2}\, x)\, \exp\{i\Delta x\} =$$

$$= \frac{1}{2\pi} \int_{-\infty}^{+\infty} dx\, J_0(\lambda x)\exp\{i\Delta x\} = \frac{1}{\pi}\, [\lambda^2 - \Delta^2]^{-1/2} \; . \tag{14}$$

This transition leads to Eqs. (15) and (16) for the rate of the
electron energy change [5].

$$\frac{d\varepsilon_{\vec{p}}}{dt} = \frac{2}{\hbar} \sum_k \frac{|C_K|^2}{V}\, (\varepsilon_{\vec{p}+\vec{K}}-\varepsilon_{\vec{p}}) \left[\left(\frac{e\vec{E}\cdot\vec{K}}{m\omega}\right)^2 - (\varepsilon_{\vec{p}+\vec{K}}-\varepsilon_{\vec{p}})^2\right]^{-1/2} \; , \tag{15}$$

$$\left(\frac{e\vec{E}\cdot\vec{K}}{m}\right)^2 \geq (\varepsilon_{\vec{p}+\vec{K}} - \varepsilon_{\vec{p}})^2 \,. \tag{16}$$

The straightforward proof that Eq. (15) really describes the strong field limit of the corresponding quantum formula was given in ref. [6]. The quantum formula for the rate of the electron energy change was transformed as shown in Eq. (17);

$$\frac{d\varepsilon_{\vec{p}}}{dt} = \frac{2\pi}{\hbar} \sum_K \frac{|C_K|^2}{V} \sum_n n\hbar\omega J_n^2\left(\frac{e\vec{E}\cdot\vec{K}}{m\hbar\omega^2}\right) \delta(\varepsilon_{\vec{p}+\vec{K}} - \varepsilon_{\vec{p}} - n\hbar\omega) =$$

$$= -\frac{4\pi}{\hbar} \sum_K \frac{|C_K|^2}{V} \sum_{\ell=0}^{\infty} \delta^{(2\ell+1)}(\varepsilon_{\vec{p}+\vec{K}} - \varepsilon_{\vec{p}}) \frac{(\hbar\omega)^{2\ell+2}}{(2\ell+1)!} T_{2\ell+2}\left(\frac{e\vec{E}\cdot\vec{K}}{m\hbar\omega^2}\right)\,. \tag{17}$$

In order to obtain Eq. (17) it is necessary to expand the δ-function in the quantum formula into the Taylor series of $n\hbar\omega$. It turns out, that the infinite series in n, given by Eq. (18),

$$T_{2\ell+2}(a) = \sum_{n=1}^{\infty} n^{2\ell+2} J_n^2(a) \tag{18}$$

which appears in Eq. (17), is equal to the polynomial of order $(2\ell+2)$, special cases of which are:

$$T_2(a) = \frac{a^2}{4} \,; \quad T_4(a) = \frac{a^2}{4} + \frac{3a^4}{16} \,; \quad T_6(a) = \frac{a^2}{4} + \frac{15a^4}{16} + \frac{5a^6}{32} \,.$$

Equation (17) may be simplified when the argument of the polynomial T in it is much greater than unity. In this case, approximation (19) – the strong field approximation – for the polynomial T may be used;

$$\left|\frac{eE\,K}{m\hbar\omega^2}\right| \gg 1, \quad T_{2\ell+2}(a) \sim a^{2\ell+2} \,. \tag{19}$$

Introducing this approximation into Eq. (17) it is easy to see that the Planck constant $\hbar$ is cancelled, and the series of ℓ in Eq. (17) is summarized into the classical formula. It is Eq. (15). The expression under the square root in Eq. (15) must be positive. That gives inequality (16). Inequality (16) has a simple physical meaning. It restricts the energy change of the electron under the field action in a single encounter. Comparing inequalities (16) and (19), it is easy to see that the classical formula

(15) describes the multiphoton processes, but with a restricted
photon number. The integration over the impulse $\vec{K}$ in Eq. (15)
may be simply performed. For instance, in the case of electron
scattering by longitudinal acoustic phonons in semiconductors,
$(|C_K|^2 \sim K)$, one obtains Eqs. (20) and (21);

$$\frac{d\varepsilon}{dt} = \left\langle \frac{d\varepsilon_{\vec{p}}}{dt} \right\rangle_{\vec{p}} = \frac{e^2 E^2}{2m\omega^2} \nu_{eff}^{(1)}(P) \psi_1 \left(\frac{eE}{\omega P}\right), \tag{20}$$

$$\nu_{eff}^{(1)}(P) \sim P \; ; \; \psi_1(x) = 1 + \frac{9}{20} x^2 . \tag{21}$$

Equation (20) is written for the isotropic part of the rate of
energy change, which does not depend on the angle between the impulse of the electron and the polarization vector of the electromagnetic wave. It is easy to see that when the argument of the
function δ in Eq. (20) is small, the rate of the energy change
has the form of the ordinary Drude formula. In the opposite case
the additional field dependence in the Drude formula appears.
This conclusion is also valid for other scattering mechanisms.
In the case of Coulomb scattering, Eq. (15) leads, under the
strong field limit, to the same formula for the nonlinear absorption coefficient, as was mentioned above. It is connected with
the fact that in the case of Coulomb scattering the integral over
$\vec{K}$ in Eq. (15) diverges at small impulse transfer. In order to
eliminate this divergence it is necessary to use the screened
Coulomb potential, or to take into account the fact that in the
process of light quantum absorption the impulse transfer has the
minimum value. So another large parameter appears: $eE/\omega K_{min} \gg 1$.
It may be shown that obtaining the main asymptotic term relative
to this parameter is equivalent to the use of the asymptotic formula of Eq. (7) from the very beginning. So the method used in
the works of Pert [2] and Elutin [3] is applied to the case of
Coulomb scattering.

Now let us discuss Eq. (15) from another point of view. As
was mentioned above, Eq. (15) describes the multiphoton processes.
It is the semiclassical one, because the scattering act in it is
treated in the first Born approximation, whereas the interaction
of an electron with the field is treated classically. However,
Eq. (15) may be written in the completely classical form

$$\frac{d\sigma}{d\Omega} = m^2 |C_K|^2 / 4\pi^2 \hbar^4 . \tag{22}$$

For that purpose, let us express the matrix element C_K through

the differential cross section as in Eq. (22). The equation for
the rate of the energy change now is as shown in Eq. (23);

$$\frac{d\varepsilon_{\vec{p}}}{dt} = \frac{1}{\pi m^2} \int d^3K \frac{d\sigma}{d\Omega}(\varepsilon_{\vec{p}+\vec{K}}-\varepsilon_{\vec{p}}) \left[\left(\frac{e\vec{E}\cdot\vec{K}}{m\omega}\right)^2 - (\varepsilon_{\vec{p}+\vec{K}}-\varepsilon_{\vec{p}})^2\right]^{-1/2} \tag{23}$$

where $\tau = \rho/v$ is the scattering duration, and ρ is the impact
parameter ($\rho \sim \hbar/k$). In the strong field; $v \sim eE/m\omega$, $K \sim eE/\omega$.
Equation (23) may, of course, be obtained in terms of the com-
pletely classical approach. For this it is necessary to use the
assumption of the instantaneous scattering, meaning that the
phase of the wave does not change during the collision. As is
shown in Eq. (24), the condition of the instantaneous scattering
does not differ from the quantum condition for Eq. (15).

$$\frac{1}{\omega} >> \frac{\rho}{v} = \frac{\hbar m\omega^2}{e^2 E^2} \qquad \text{or} \qquad \frac{e^2 E^2}{m\hbar\omega^3} << 1. \tag{24}$$

So we come to the conclusion that in order to determine the field
action on the electron scattering process, the scattering itself
should be considered as instantaneous. It should be noted that
the classical approach based on the instantaneous scattering has
been considered earlier in the works of Pert [2] and Bunkin et al
[7]. But Eq. (23) had not been obtained, and what is more essential, the connection of this approach with the quantum one had
not been realized.
   The last question I'd like to discuss is how the classical
approximation I have spoken about is connected with that of Seely
and Harris [4], which is also of a classical nature because it
does not contain the Planck constant $\hbar$. We can see how the classical approximation may be obtained in a manner analogous to the
approximation of Seely and Harris [4]. One should approximate
by the δ-function the series containing not the squares of Bessel
functions but the Bessel functions themselves.

$$J_n^2(Z) = \frac{1}{\pi} \int_0^\pi J_{2n}(2z\cos\theta)d\theta \tag{25}$$

$$F(\vec{K}) = \frac{1}{\pi} \int_0^\pi d\theta \sum_{n=-\infty}^{n=+\infty} J_{2n}(2\frac{\lambda}{\hbar\omega}\cos\theta)\delta(\Delta - n\hbar\omega) \tag{26}$$

$$\sum_n J_{2n}(2\frac{\lambda}{\hbar\omega}\cos\theta)\delta(\Delta - n\hbar\omega) \approx \delta(\lambda\cos\theta - \Delta) \tag{27}$$

$$F(\vec{K}) = \frac{1}{\pi} \int_{0}^{\pi} d\theta \; \delta(\lambda\cos\theta - \Delta) = \frac{1}{\pi} [\lambda^2 - \Delta^2]^{-1/2}. \qquad (28)$$

The result is given by Eqs. (27) and (28) which are obtained through the use of Eqs. (25) and (26). Unlike the approximation of Seely and Harris [4], in the classical approximation the fact that the phase of the wave at the moment of scattering may be of arbitrary value, is taken into account and the averaging over the phase is performed.

Now we can come to a conclusion. As I have mentioned at the beginning of the report, the general quantum method leads to the result containing the infinite series of the Bessel functions. Such an answer is not satisfactory from the physical point of view. The aim of this report was to show that this answer is not only unsatisfactory, but also superfluous in detail, because in the area of strong fields, when the multiphoton processes take place, the classical approach should be used.

REFERENCES

1. F. V. Bunkin, M. V. Fedorov, Zh. Eksp. Teor. Fiz. **49**, 1215 (1965).

2. G. J. Pert, J. Phys. **A5**, 1221 (1972).

3. P. V. Elutin, Zh. Eksp. Teor. Fiz. **65**, 2196 (1973).

4. J. F. Seely, E. G. Harris, Phys. Rev. **A7**, 1064 (1973).

5. An. V. Vinogradov, Zh. Eksp. Teor. Fiz. **68**, 1091 (1975).

6. An. V. Vinogradov, Zh. Eksp. Teor. Fiz. **70**, 999 (1976).